基于 Android 的物联网应用开发

廖忠智　王　华　高晓惠　万　杰　主　编

刘建峰　沈志华　马　祥　方　武
洪顺利　周胜利　副主编

彭坤容　魏美琴　蔡　敏　黄非娜
邹宗冰　叶　宁　彭海玲　参　编

清华大学出版社

北　京

内 容 简 介

本书编者总结了10多年的物联网专业教学和指导学生参加竞赛的经验，精心选择物联网方面典型的项目展开分析，根据项目需求设计工作任务，采用任务式结构编写，通过引导读者完成不同的物联网程序任务，对 Android 物联网应用程序开发的各方面知识进行讲解。全书共 11 个项目，包括智慧园区系统项目概述、第一个 Android 应用程序、智慧园区界面的实现、界面显示与切换、数据存储的实现、线程与消息处理、服务与广播、媒体动画的实现、室内环境采集系统和园区监控系统的实现、网络与定位技术的使用、园区环境实时监测系统，建议课时为 120 学时。

本书既可以作为高等院校本科和高职物联网工程及相关专业学生的教材，又可以作为物联网相关从业者和爱好者的参考书。

图书在版编目(CIP)数据

基于 Android 的物联网应用开发/廖忠智等主编. —北京：清华大学出版社，2021.6（2025.1 重印）
ISBN 978-7-302-58302-8

Ⅰ. ①基…　Ⅱ. ①廖…　Ⅲ. ①物联网—程序设计　Ⅳ. ①TP393.4 ②TP18

中国版本图书馆 CIP 数据核字(2021)第 097308 号

责任编辑：梁媛媛
封面设计：李　坤
责任校对：李玉茹
责任印制：杨　艳
出版发行：清华大学出版社
　网　　址：https://www.tup.com.cn, https://www.wqxuetang.com
　地　　址：北京清华大学学研大厦 A 座　　邮　　编：100084
　社 总 机：010-83470000　　邮　　购：010-62786544
　投稿与读者服务：010-62776969, c-service@tup.tsinghua.edu.cn
　质量反馈：010-62772015, zhiliang@tup.tsinghua.edu.cn
　课件下载：https://www.tup.com.cn, 010-62791865

印 装 者：三河市龙大印装有限公司
经　　销：全国新华书店
开　　本：185mm×260mm　　印　张：22　　字　数：535 千字
版　　次：2021 年 6 月第 1 版　　印　次：2025 年 1 月第 6 次印刷
定　　价：66.00 元

产品编号：093190-01

前　言

物联网被看作是继计算机、互联网与移动通信之后的又一次信息产业浪潮，将成为未来带动中国经济发展的主要生力军。2009 年，在美国总统奥巴马与工商业领袖举办的圆桌会议上，IBM 首席执行官首次提出了“智慧地球”(Smart Earth)的构想。同年，欧盟发布了物联网研究战略路线图(Internet of Things Strategic Research Roadmap)。在我国，物联网同样得到了高度重视，在 2010 年“两会”期间，物联网已经被写入政府工作报告，确立为国家五大战略新兴产业之一。在 2016 年国务院印发的《“十三五”国家信息化规划》中，特别提出要加快信息化和生态文明建设深度融合，利用新一代信息技术，促进产业链循环化。推进物联网感知设施规划布局，发展物联网开发应用；实施物联网重大应用示范工程，推进物联网应用区域试点，建立城市级物联网接入管理与数据汇聚平台，深化物联网在城市基础设施、生产经营等环节中的应用。

本书编者总结了 10 多年的物联网专业教学和指导学生参加竞赛的经验，精心选择物联网方面典型的项目展开分析，根据项目需求设计工作任务，采用任务式结构编写，通过引导读者完成不同的物联网程序任务，对 Android 物联网应用程序开发的各方面知识进行讲解。

本书的特点如下。

(1) 紧密结合物联网开发。本书以智慧园区项目为中心，将 Android 物联网应用程序开发的知识以知识解析的形式拆分到不同的物联网开发情境中，让读者能够及时将所学的知识运用到实际开发中，提升学习兴趣，培养动手能力。

(2) 项目拆解、分任务实现。将智慧园区系统拆分成 3 个项目，分别是室内环境采集系统、园区监控系统、园区环境实时监测系统，再将这 3 个项目拆分成若干个任务，一步步实现项目功能。

(3) 综合运用、实战检验。最后通过智慧园区实时监测系统的综合开发，让读者将各部分知识综合使用、融会贯通，充分掌握 Android 物联网应用程序开发的知识。

全书共 11 个项目，建议课时为 120 学时。

项目 1　智慧园区系统项目概述。本项目介绍了智慧园区项目的需求、功能模块、运行结果以及技术选型等内容(建议课时：2 学时)。

项目 2　第一个 Android 应用程序。本项目介绍了 Android 在物联网中的应用，带领读者认识 Android，搭建 Android 开发环境，创建第一个 Android 工程，并简单地介绍如何查看 Android 日志，包含 4 个任务(建议课时：4 学时)。

项目 3　智慧园区界面的实现。本项目在学习 Android UI 基本知识的基础上，完成智慧园区的部分界面开发，包含 4 个任务(建议课时：16 学时)。

项目 4　界面显示与切换。本项目介绍了 Android 中的活动与碎片，通过 ViewPager 实现引导页的功能，通过 DrawerLayout 实现侧滑菜单以及 Toorbar 标题栏的效果，包含 4 个任务(建议课时：20 学时)。

项目 5　数据存储的实现。本项目介绍了文件存储、SharedPreferences 和 SQLite 数据库等数据存储技术，让读者掌握 Android 数据存储技术的应用，包含 3 个任务(建议课时：12

学时)。

项目 6　线程与消息处理。本项目针对 Android 开发中必须掌握的线程和消息以及异步任务进行讲解，最后实现传感器数据的实时更新，包含 4 个任务(建议课时：14 学时)。

项目 7　服务与广播。本项目介绍了如何在服务中实时监测微动开关的状态以及使用广播触发三色灯状态的实时改变，包含两个任务(建议课时：10 学时)。

项目 8　媒体动画的实现。本项目讲解如何利用手机媒体动画编写出丰富多样的物联网应用程序，包含 4 个任务(建议课时：16 学时)。

项目 9　室内环境采集系统和园区监控系统的实现。本项目在前面项目的基础上，综合所学知识，实现室内环境采集系统和园区监控系统，包含两个任务(建议课时：4 学时)。

项目 10　网络与定位技术的使用。本项目讲解如何使用 HTTP 协议和物联网云平台端进行网络交互，并对服务器返回的数据进行解析。同时讲解了利用定位模块进行定位的物联网应用，包含 4 个任务(建议课时：10 学时)。

项目 11　园区环境实时监测系统。本项目实现园区环境的实时监测。监测的数据来源有两种：第一种是采集到的传感数据经 LoRa 网关上传到物联网云平台；第二种是采集到的传感数据经 NB-IoT 上传到物联网云平台(建议课时：12 学时)。

本书适用于物联网工程以及相关专业的学生使用。

本书由廖忠智、王华、高晓惠、万杰担任主编，刘建峰、沈志华、马祥、方武、洪顺利、周胜利担任副主编，彭坤容、魏美琴、蔡敏、黄非娜、邹宗冰、叶宁、彭海玲参编。本书在编写中得到了北京新大陆时代教育科技有限公司相关人员的大力帮助和支持，在此表示感谢。

由于编者水平有限，书中疏漏之处在所难免，敬请各位读者不吝赐教，以求共同进步，感激不尽。

编　者

(扫一扫，了解本书配套资源目录)

(扫一扫，试看配套的精美课件)

目　录

项目 4 界面显示与切换......83

项目 5 数据存储的实现......121

项目1 智慧园区系统项目概述

【需求描述】

扫码观看视频讲解

某园区计划进行智能改造，最终实现下列功能。

1. 室内环境采集

(1) 实时显示室内的温湿度、光照、烟雾数据。

(2) 能根据温度高低自动判断需要开还是关通风扇。

(3) 根据湿度大小自动判断需要开还是关排气扇。

(4) 根据光照的强弱自动判断需要开还是关照明灯。

(5) 根据有无烟雾自动判断需要开还是关三色灯中的橙灯。有烟的时候发出警报提示。

(6) 微动开关被触发时打开三色灯中的绿灯，关闭三色灯中的橙灯。

2. 园区监控

(1) 在园区入口设置红外对射传感器、网络摄像头和三色灯。

(2) 红外对射传感器检测到人员由外进入园区时，三色灯中的红灯亮。当开启监控的时候，打开摄像头，可以控制摄像头上、下、左、右转动，检测到有人则拍照并存储，并可查看照片。

(3) 微动开关被触发时打开三色灯中的绿灯，关闭三色灯中的橙灯。

3. 园区环境实时监测系统

(1) 实现园区环境数据的实时采集与监测。

(2) 智能井盖监测：超声波、甲烷、三轴、井盖状态传感器数据。
(3) 水质监测：pH、浊度、电导率传感器数据。
(4) 环境数据监测：垃圾高度、PM2.5、一氧化碳、可燃气传感器数据。
(5) 将采集到的一组传感数据经 LoRa 网关上传到物联网云平台。
(6) 将采集到的另一组传感数据经 NB-IoT 上传到物联网云平台。
(7) Android 客户端程序通过 OkHttp 登录云平台。
(8) Android 客户端程序利用云平台提供的 API 从云平台读取数据进行展示。

【需求分析】

根据前文的描述，可以将用户的需求大致分为三类，即室内环境采集系统、园区监控系统、园区环境采集系统(云平台实现)。使用物联网工程实训系统 2.0 的设备实现。

1. 室内环境采集系统

1) 室内环境采集系统硬件组成

室内环境采集系统采用有线传感网络和 ZigBee 传感网实现。它由四输入模块、温湿度传感器、光照传感器、ZigBee 协调器、ADAM-4150 模块、烟雾传感器、两个风扇、LED 灯、三色灯、微动开关等组成，如图 1-1 所示。

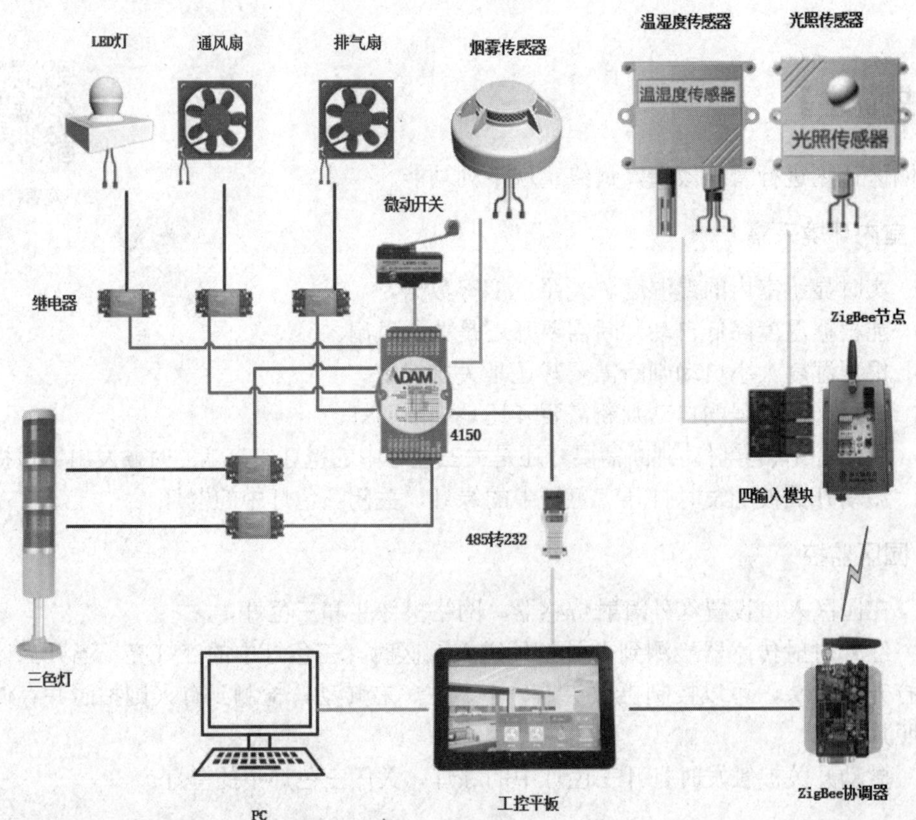

图 1-1 室内环境采集系统拓扑图

2)　室内环境采集系统软件功能

室内环境采集系统软件功能包括以下几个。

(1)　实时显示室内的温湿度、光照、烟雾数据。

(2)　单击“设置阈值”菜单可进行温湿度、光照的阈值设置。如果当前温度值不小于温度阈值，打开通风扇，否则关闭通风扇；如果当前湿度值不小于湿度阈值，打开排气扇，否则关闭排气扇；如果当前光照值不大于光照阈值，打开照明灯，否则关闭照明灯；有烟的时候三色灯中的橙灯亮，并发出警报提示；没烟的时候三色灯中的橙灯灭，并关闭警报提示。

(3)　微动开关被触发时打开三色灯中的绿灯，关闭三色灯中的橙灯。

2. 园区监控系统

1)　园区监控系统硬件组成

园区监控系统采用有线传感网和 Wi-Fi 无线网实现。它由网络摄像头、红外对射传感器和三色灯等组成，如图 1-2 所示。

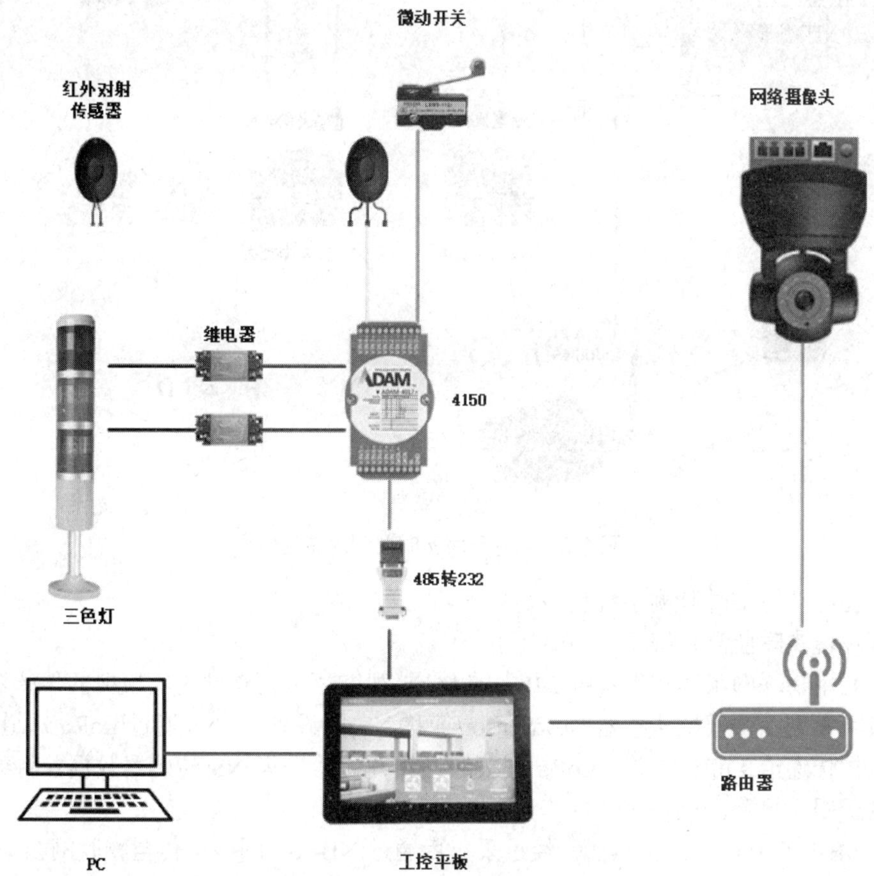

图 1-2　园区监控系统拓扑图

2)　园区监控系统软件功能

园区监控系统软件功能如下。

(1) 可进行摄像头地址设置，设置好后可开启监控，控制摄像头上、下、左、右转动。

(2) 红外对射传感器检测到人员由外进入园区时，三色灯中的红灯亮。

(3) 当开启监控的时候，打开摄像头，此时可以控制摄像头上、下转动，检测到有人，则拍照并存储。可单击菜单查看照片。

(4) 微动开关被触发时打开三色灯中的绿灯，关闭三色灯中的红灯。

3. 园区环境实时监测系统

1) 园区环境实时监测系统硬件组成

园区环境实时监测采用 LoRa 网络和 NB-IoT 网络实现，由 6 个 NS、两个 LoRa 节点、一个 LoRa 网关、两个 NB-IoT 节点组成，如图 1-3 所示。

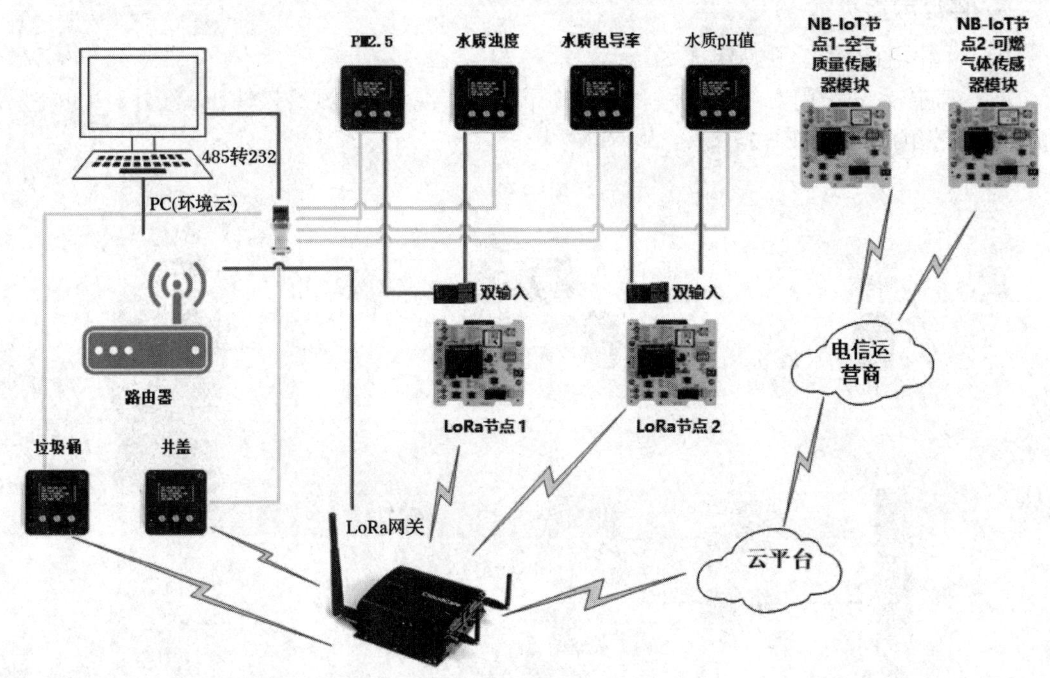

图 1-3 园区环境实时监测系统硬件

2) 园区环境实时监测系统软件功能

园区环境实时监测系统软件功能如下。

(1) 利用物联网工程实训系统 2.0 中的 PC 端智能环境云软件生成模拟的传感器数据，通过串口下发到 6 个模拟传感器 NewSensor(简称 NS)的设备上。NS 通过 LoRa 通道或 LoRa 节点把数据传递给 LoRa 网关。LoRa 网关通过轮询方式采集 NS 传感器数据并将数据上传至物联网云服务平台。

(2) NB-IoT 节点上接传感器，经过采集后通过 NB-IoT 网络上传至物联网云平台。

(3) 在物联网云平台上注册账号，添加项目和传感器，记下项目标识、设备 ID 号、传输密钥和传感器标识。

(4) 采集园区数据时，先设置云平台参数，包括云平台的域名、端口、账号、密码、要读取的设备的 ID，并保存这些参数。

(5) 利用第三方库 OkHttp 连接云平台，封装 GET 和 POST 方法。

(6) 利用云平台提供的 API 登录云平台，保存登录成功返回的访问令牌。

(7) 利用云平台提供的 API，开启线程，凭设备 ID 查询传感器数据。

(8) 查询到的传感器数据通过消息机制传到 UI 线程，解析查询到的 JSON 格式的传感器数据并显示在界面上。

【运行效果】

运行效果如图 1-4～图 1-13 所示。

图 1-4 登录界面

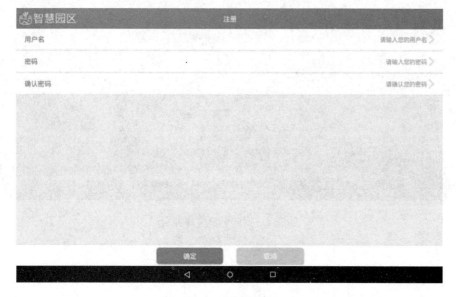

图 1-5 注册界面

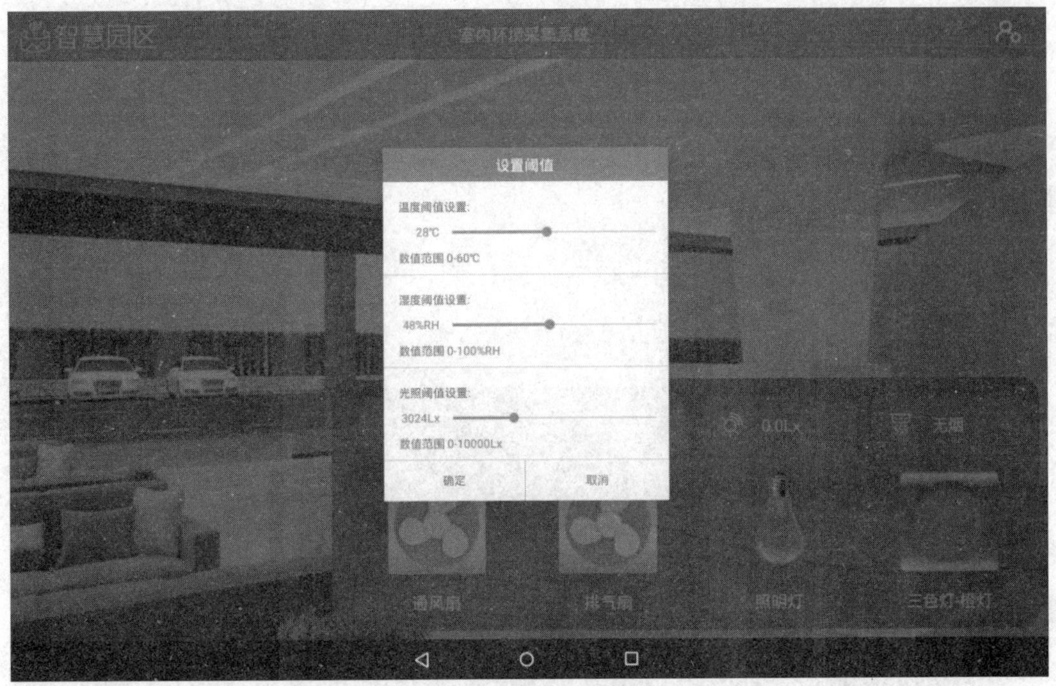

图 1-6　室内环境阈值设置界面

图 1-7　室内环境采集界面

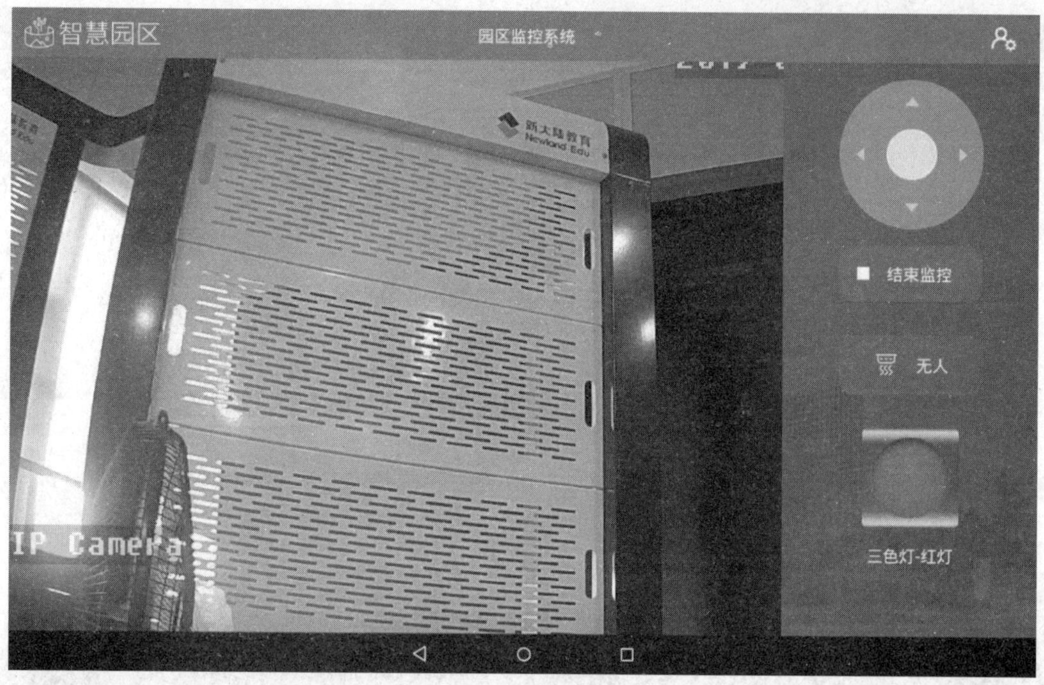

图 1-8　园区监控界面

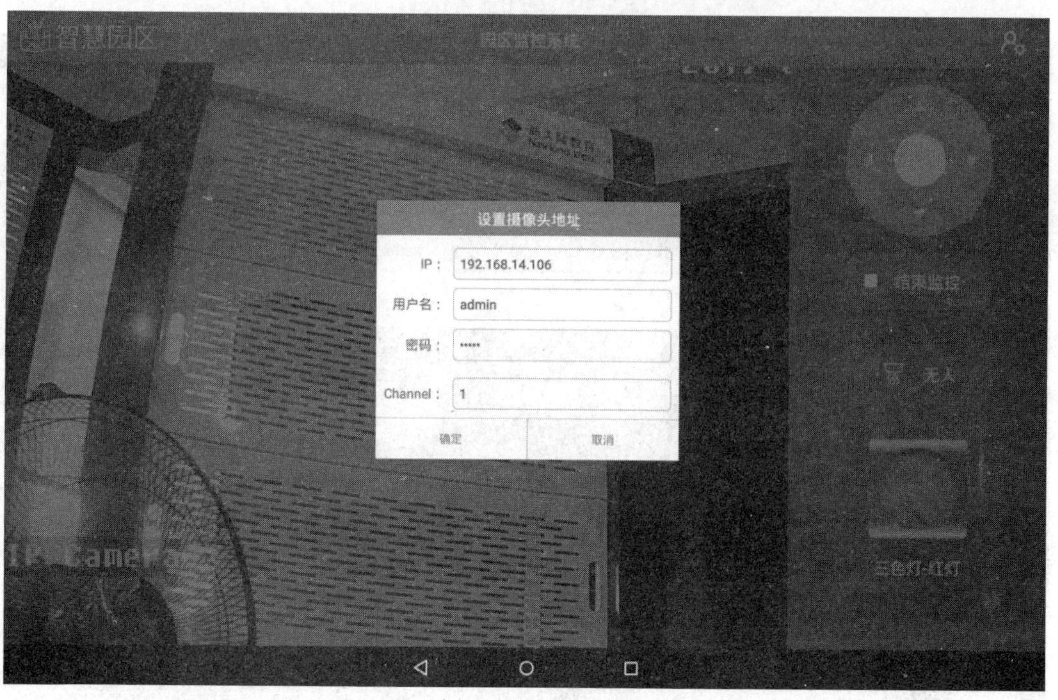

图 1-9　摄像头地址设置界面

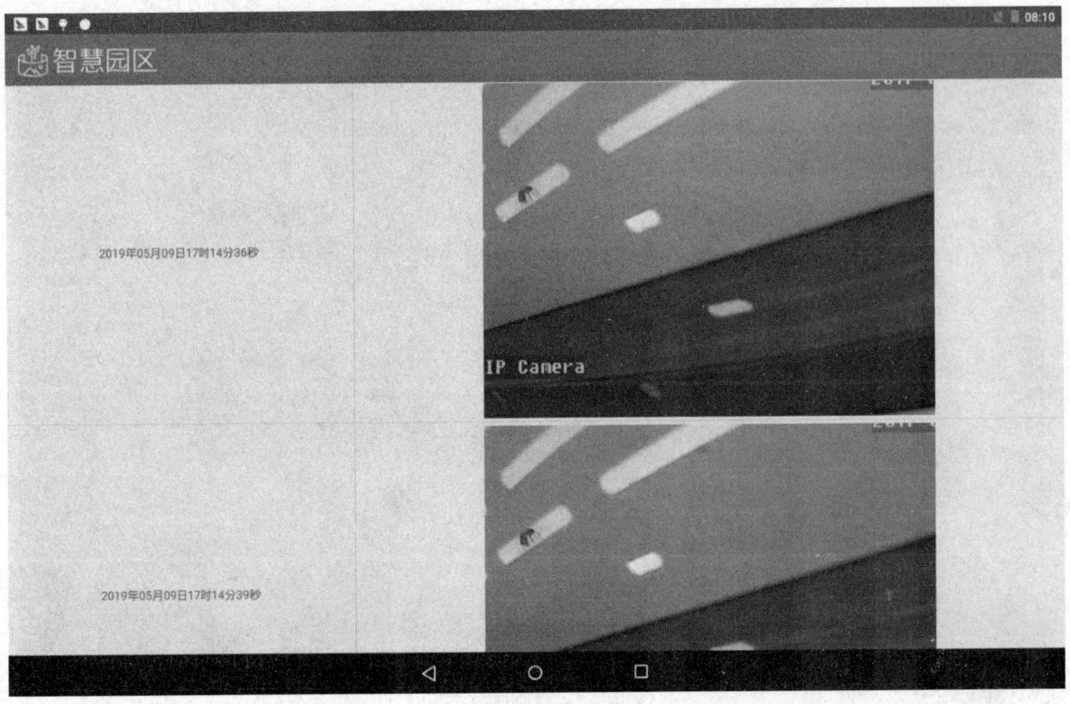

图 1-10　截图查看界面

图 1-11　通知栏烟雾警报提示

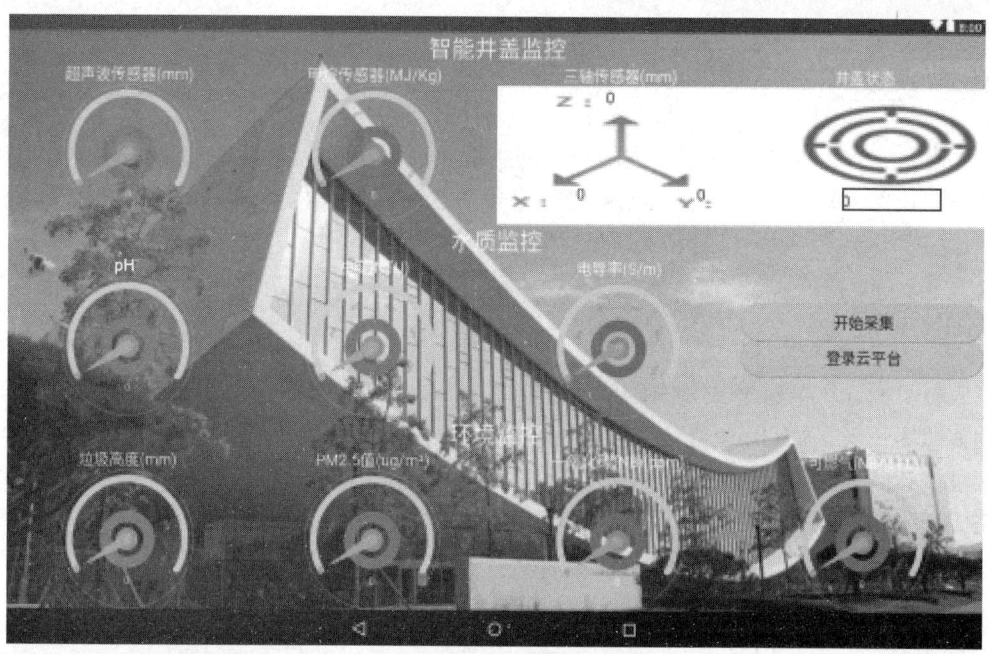

图 1-12　数据采集界面

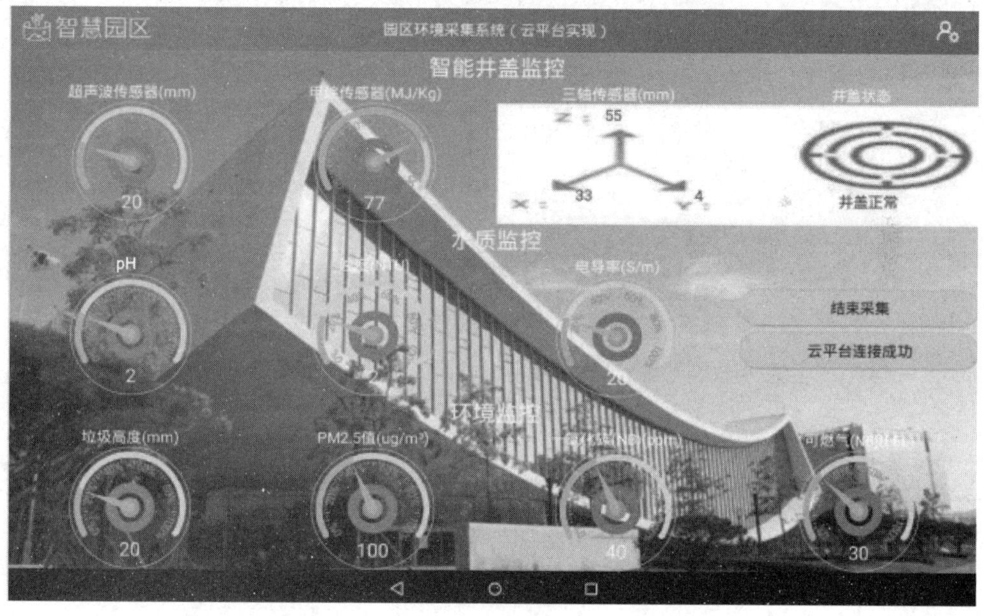

图 1-13　数据采集结果

【技术选型】

(1) 操作系统：Windows 10。
(2) JDK 版本：JDK 1.8。
(3) Android 开发环境：Android Studio 3.2.0。
(4) 运行终端：Android API26 模拟器、真实手机、物联网移动终端(工控平板)。

【设备选型】

本教材中涉及的设备选用新大陆物联网工程实训系统 2.0 套件，在任务描述中有对应的设备清单说明，读者可以按清单说明进行设备的选择，按工程实训中设备的连接要求进行接线。

【知识前提】

本教材的知识前提要求是有物联网工程实训基础和 JavaSE 基础，所以在本书中不对用到的传感器设备、ModBus 协议、ZigBee 协议、ZigBee 烧写和配置、设备连接和 JavaSE 语法做过多的阐述。本书任务代码所涉及的串口与 DI 口、DO 口以编写时的数据为准，读者在完成任务时需根据实际开发环境自行修改。

项目 2 第一个 Android 应用程序

【项目描述】

本项目介绍了 Android 在物联网中的应用，以 4 个任务带领读者认识 Android、搭建 Android 开发环境，创建第一个 Android 工程，并简单地介绍如何查看 Android 的日志，为后续的 Android 开发做好准备。具体任务列表如图 2-1 所示。

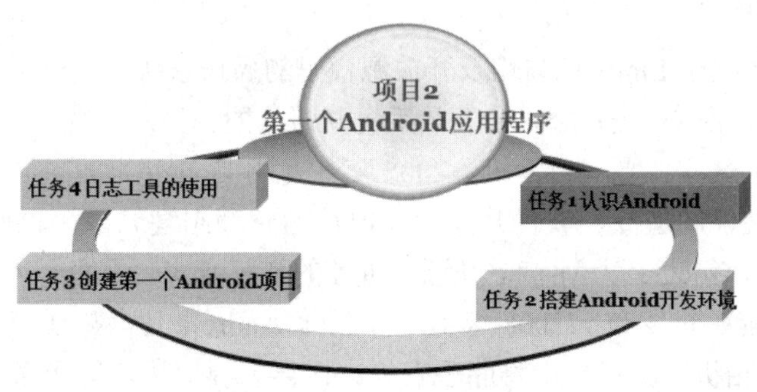

图 2-1　项目 2 任务列表

【学习目标】

知识目标：

- 了解 Android 的历史背景。
- 知道 Android 的系统架构。

- 知道 Android 与物联网的关系。
- 会进行 Android Studio 的安装与使用。
- 可熟练创建 Android 工程。
- 知道 Android 工程的目录结构。
- 掌握 Android 日志的查看与使用。

技能目标：

- 能搭建 Android Studio 开发环境。
- 能创建 Android 项目并运行。
- 能使用日志工具进行排错。

任务 1　认识 Android

【任务描述】

在进行 Android 开发前，需要知道 Android 的历史背景与系统架构，以及它在物联网项目中扮演的角色，本任务带领读者更好地认识 Android。任务清单如表 2-1 所示。

表 2-1　任务清单

任务课时	0.5 课时	任务组员数量	建议 1 人
任务组采用设备	PC 一台 模拟器或真机或工控平板		

【知识解析】

1. Android 概述

Android 是一种基于 Linux 的自由及开放源代码的操作系统，主要用于移动设备，如智能手机和平板电脑，由 Google 公司和开放手机联盟共同领导及开发。Android 操作系统最初由 Andy Rubin 开发，主要支持手机，2005 年 8 月由 Google 收购注资。2007 年 11 月，Google 与 84 家硬件制造商、软件开发商及电信运营商组建开放手机联盟共同研发改良 Android 系统。随后 Google 以 Apache 开源许可证的授权方式，发布了 Android 的源代码。第一部 Android 智能手机发布于 2008 年 10 月，后来 Android 逐渐扩展到平板电脑及其他领域，如电视、数码相机、游戏机、智能手表等。根据中国信息通信研究院公布的数据，近年来随着华为、OPPO、vivo、小米等众多国产手机的崛起，国产品牌规模优势明显。截至 2020 年 1～12 月中国国产品牌手机出货量累计达到 2.7 亿部，同比下降 23.5%，占同期手机出货量的 87.5%，较 2014 年的 78%提升近 10 个百分点，国外品牌出货量占比则从 22%下降至 12.5%，如图 2-2 所示。

每个 Android 的版本都有各自对应的 API 等级，对应着某个 Android 发布版本的名称，如表 2-2 所示。

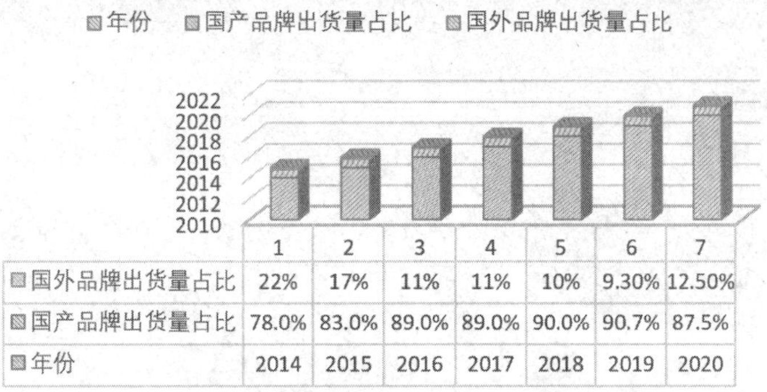

年份 国产品牌出货量占比 国外品牌出货量占比

2022
2020
2018
2016
2014
2012
2010

	1	2	3	4	5	6	7
国外品牌出货量占比	22%	17%	11%	11%	10%	9.30%	12.50%
国产品牌出货量占比	78.0%	83.0%	89.0%	89.0%	90.0%	90.7%	87.5%
年份	2014	2015	2016	2017	2018	2019	2020

图 2-2 2014—2020 年中国国产/国外品牌出货量占比

表 2-2 常用 Android 版本与对应的 API 等级

Android 版本名称	版　本	版本发布时间	对应 API
⋮	⋮	⋮	…低于 API15 的….
Ice Cream Sandwich	4.0.3～4.0.4	2012 年 2 月 6 日	API level 15，NDK 8
Jelly Bean	4.1	2012 年 6 月 28 日	API level 16
Jelly Bean	4.1.1	2012 年 6 月 28 日	API level 16
Jelly Bean	4.2～4.2.2	2012 年 11 月	API level 17
Jelly Bean	4.3	2013 年 7 月	API level 18
KitKat	4.4	2013 年 7 月 24 日	API level 19
KitKat Watch	4.4W	2014 年 6 月	API level 20
Lollipop(Android L)	5.0/5.1	2014 年 6 月 25 日	API level 21/API level 22
Marshmallow(Android M)	6.0	2015 年 5 月 28 日	API level 23
Nougat(Android N)	7.0	2016 年 5 月 18 日	API level 24
Nougat(Android N)	7.1	2016 年 12 月	API level 25
Oreo(Android O)	8.0	2017 年 8 月 22 日	API level 26
Oreo(Android O)	8.1	2017 年 12 月 5 日	API level 27
Pie (Android P)	9.0	2018 年 8 月 7 日	API level 28

本书大部分案例使用的是 API level 28，最低支持 API level 15，工控平板支持 API level 22。

2. Android 系统架构

为了能够更好地了解 Android 系统是如何工作的，首先来看一下它的系统架构。Android 大致可分为四层架构，即 Linux 内核层、系统运行层、应用框架层和应用程序层，如图 2-3 所示。

1) Linux 内核层

Android 系统是基于 Linux 内核的，这一层为 Android 设备的各种硬件提供了底层驱动，如显示驱动、音频驱动、照相驱动、蓝牙驱动、电源管理驱动等。

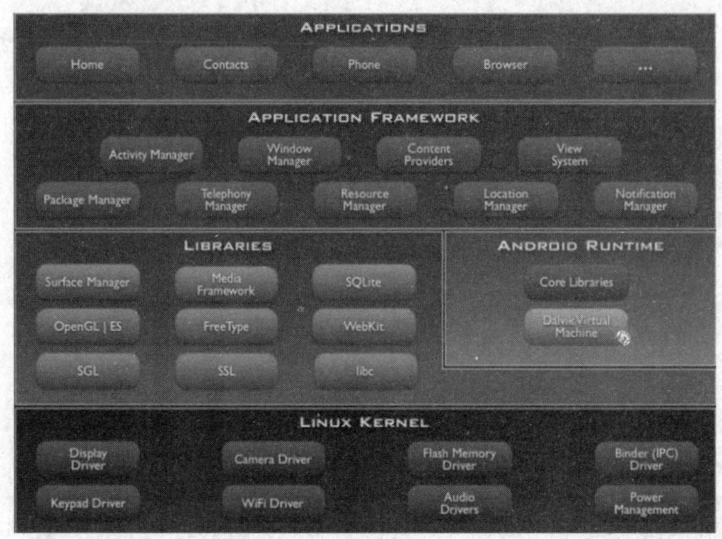

图 2-3　Android 系统架构

2)　系统运行层

这一层包括 Android 运行环境和原生态的 C/C++库。通过 C 或者 C++库为 Android 系统提供主要的特性支持等。

Android 运行时，其中包括 ART 虚拟机(Android 5.0 之前是 Dalvik 虚拟机)，每个 Java 程序都运行在 ART 虚拟机上，该虚拟机专门针对移动设备进行定制，每个应用都有其自己的 Android Runtime(ART)实例。此外，Android 运行时还包含一套核心运行时库，可提供 Java API 框架使用的 Java 编程语言的大部分功能。

3)　应用框架层

这一层主要提供构建应用程序时可能用到的各种 API，开发者通过这一层的 API 构建自己的 App，这一层也是 App 开发人员必须掌握的内容。

4)　应用程序层

所有安装在手机上的系统应用都属于这一层，用户自己开发的应用也属于这一层，如系统自带的联系人程序、短信程序或者从应用商城下载的应用程序。

3. Android 与物联网

物联网的英文名称是 Internet of things(IoT)。顾名思义，物联网就是物物相连的互联网。物联网的核心和基础还是互联网，是在互联网基础上延伸和拓展的网络。其用户端延伸和扩展到任何物品与物品之间，进行信息交换和通信，也就是物物相息。物联网通过智能感知、识别技术与普适计算等通信感知技术，广泛应用于网络的融合中，也因此被称为继计算机、互联网之后世界信息产业发展的第三次浪潮。物联网是互联网的应用和拓展，与其说物联网是网络，不如说物联网是业务和应用。

物联网的体系结构主要由 4 个层次组成，即感知层、网络层、平台层、应用层。感知层主要负责数据的采集，关键技术有 RFID 技术、传感器技术、二维码技术等。网络层主要负责数据的传输，关键技术有计算机网络技术、移动通信网技术、无线网络技术等。平台层在整个物联网体系架构中起着承上启下的作用，它实现了底层终端设备的“管、控、营”

一体化，为上层应用开发提供统一接口，构建设备和业务的端到端通道。应用层主要负责数据的处理与控制，关键技术有云计算技术、各种应用程序开发、Web 服务技术等。物联网的体系架构如图 2-4 所示。

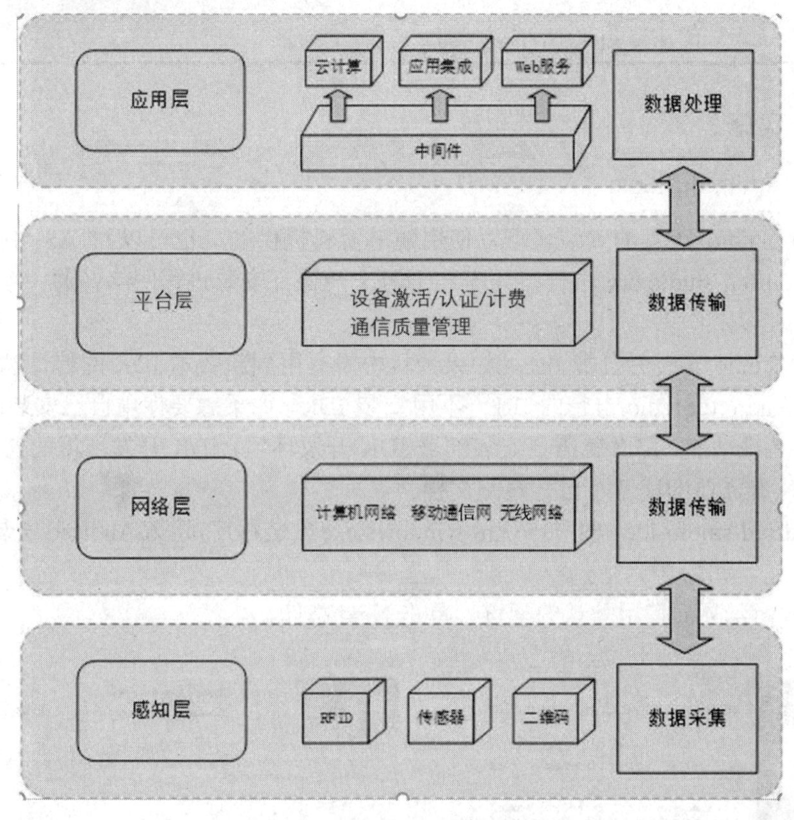

图 2-4 物联网体系架构

那么“Android 程序开发”是在物联网体系架构中的哪一层呢？答案很明显，“Android 程序开发”正是在物联网体系架构中的应用层，Android 主要负责应用层移动客户端的应用开发。例如手机、平板的应用程序，硬件平台上的 Android 嵌入式开发等。本书主要针对的是手机、平板上的 Android 应用程序开发，结合物联网设备实现 Android 程序采集物联网中各种设备的数据与控制物联网中的各种设备。例如，采集到的传感器数据传至串口或云平台后，在 Android 客户端上可以获取这些传感器的数据，同时 Android 客户端程序也可以控制各种物联网设备，如智能空调、智能电视、智能手表等。

任务 2 搭建 Android 开发环境

【任务描述】

一款好的开发工具是完成高质量、高效率的 Android 开发必不可少的。Android Studio 作为 Google 推荐的 Android 开发第一利器，为绝大多数 Android 开发人员所钟爱。本任务要求搭建 Android 开发环境，为后续的 Android 开发做准备。任务清单如表 2-3 所示。

表 2-3　任务清单

任务课时	1 课时	任务组员数量	建议 1 人
任务组采用设备	PC 一台 模拟器或真机或工控平板		

【任务实施】

1. 安装 Android Studio

Android Studio 工具的安装包可以使用随书资料提供的，也可以到 AS 中文社区官网 http://www.android-studio.org/进行在线安装(注意：应保证安装过程网络畅通，并建议计算机的内存在 8GB 以上)。

Android Studio 安装后自带 JRE，如果项目中没有用到新版本 JDK 的特性功能，那么可以直接使用 Android Studio 自带的 JRE，不用安装 JDK。如果想要使用自己安装的 JRE 和新版本 JDK 的特性功能，那么就需要安装部署 JDK 开发环境，JDK 开发环境的安装和配置此处不再展开，有需要的读者可自行查阅文档。

双击 android-studio-ide-181.5056338-windows.exe 安装程序，进入 Android 安装向导界面，单击 Next 按钮，如图 2-5 所示。

在图 2-6 所示界面使用默认的配置，单击 Next 按钮。

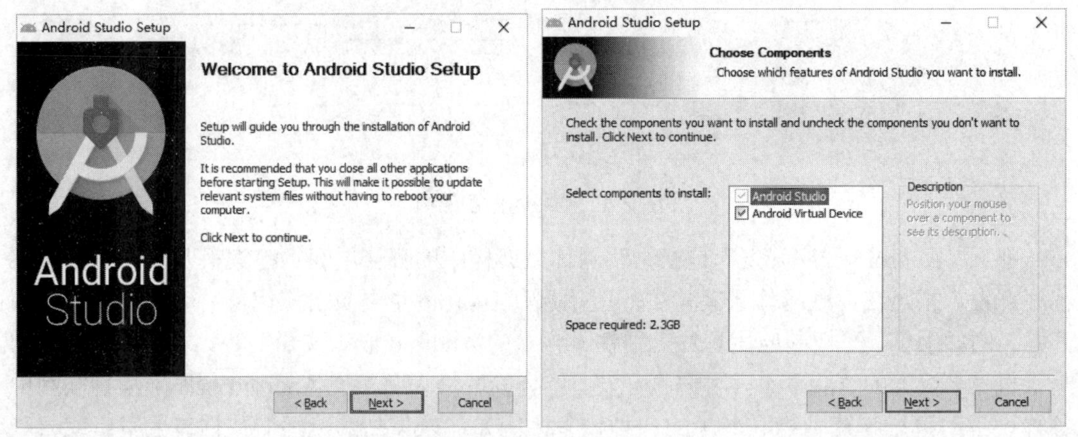

图 2-5　安装向导界面　　　　图 2-6　组件选择界面

在如图 2-7 所示界面可以设置 Android Studio 的安装路径。这里使用默认路径，单击 Next 按钮。

继续单击 Next 按钮，直到进入 Android Studio 的安装界面，安装完成后单击 Next 按钮，如图 2-8 所示。

在图 2-9 所示界面默认勾选 Start Android Studio 复选框，单击 Finish 按钮。

至此，Android Studio 的安装已经全部完成。

2. 配置 Android Studio

安装完成后运行 Android Studio，会进入导入 Android Studio 配置文件的界面，由于是第一次安装，这里选中 Do not import settings 单选按钮，单击 OK 按钮，如图 2-10 所示。

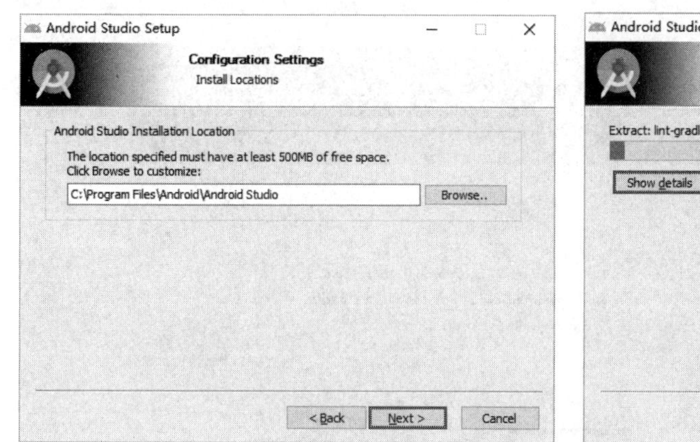

图 2-7　安装路径设置界面

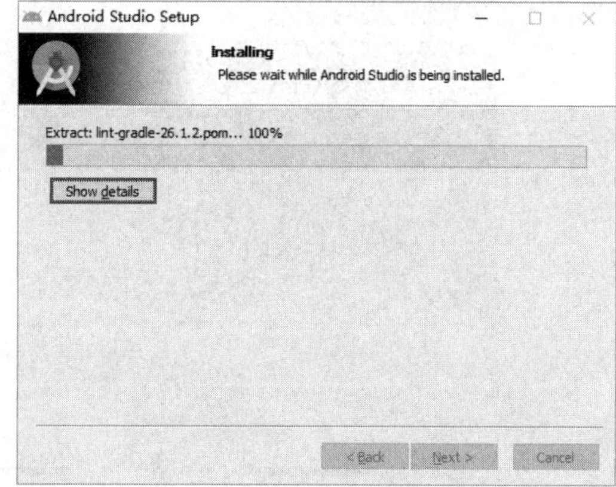

图 2-8　组件下载界面

图 2-9　完成安装界面

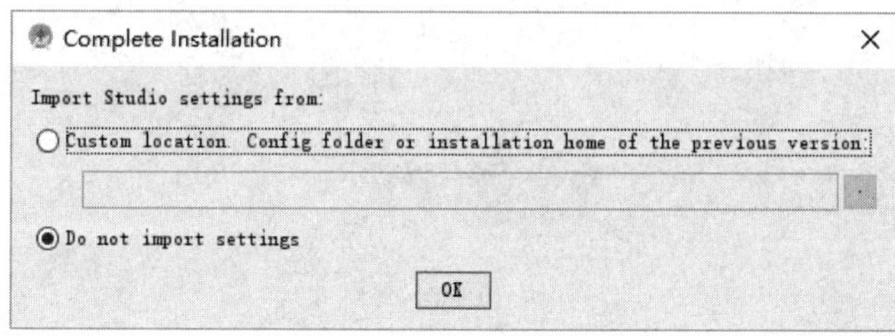

图 2-10　导入配置界面

第一次安装 Android Studio 会出现无法访问 SDK 的信息提示框，单击 Cancel 按钮取消即可，如图 2-11 所示。

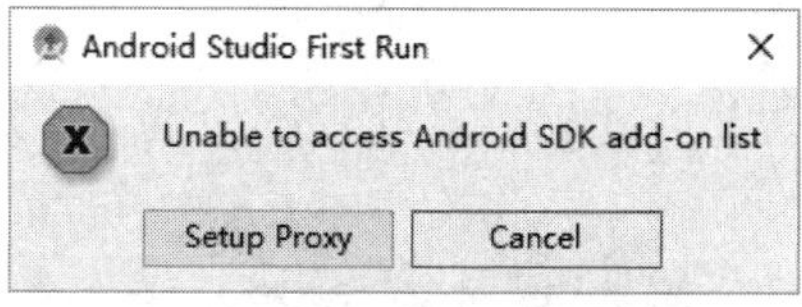

图 2-11　无法访问 SDK 信息提示框

在欢迎界面，单击 Next 按钮，如图 2-12 所示。

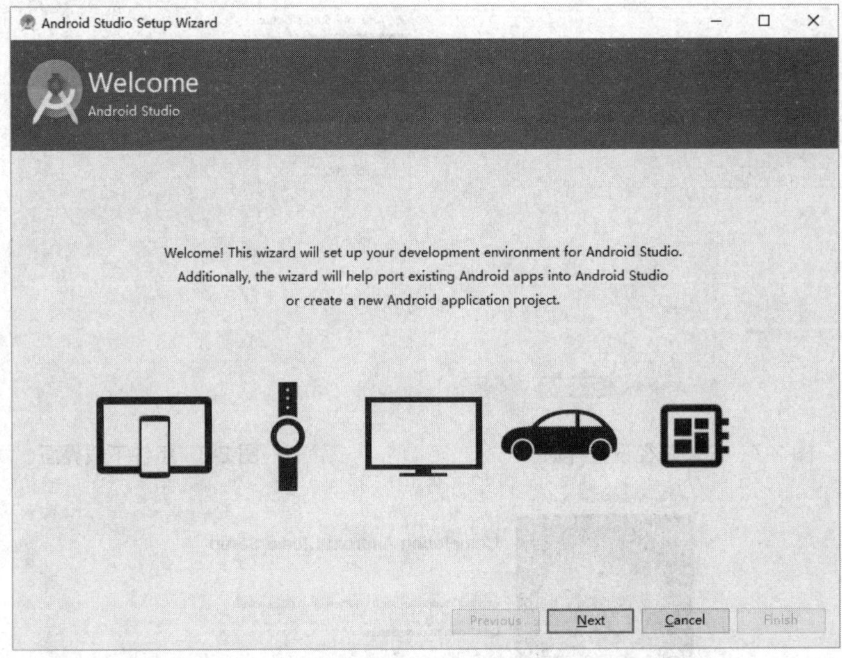

图 2-12　欢迎界面

在打开的界面中可以设置安装类型，选择 Standard(标准)模式，单击 Next 按钮，如图 2-13 所示。

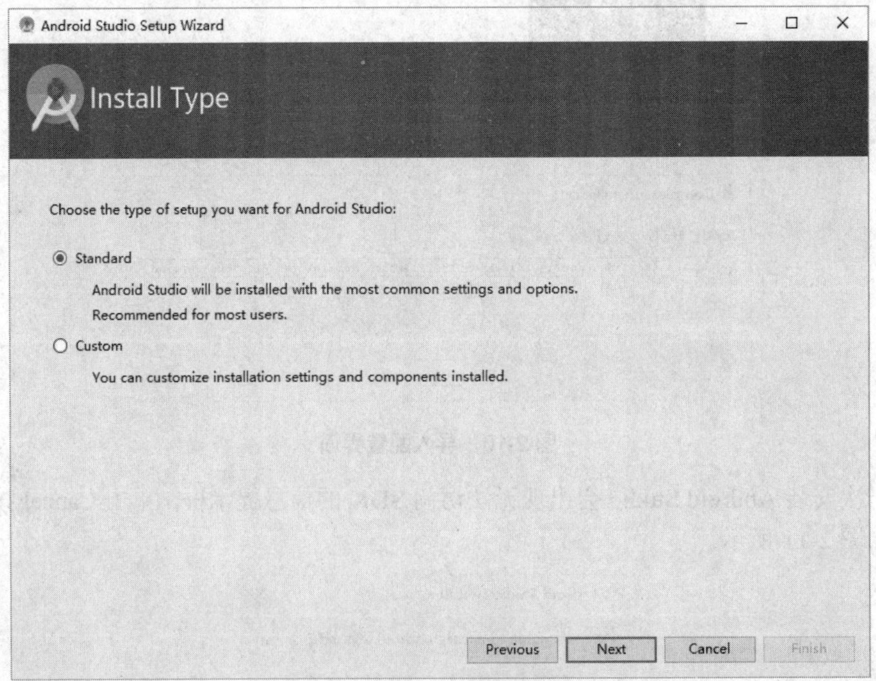

图 2-13　安装类型选择界面

在确认配置界面中单击 Finish 按钮，如图 2-14 所示。

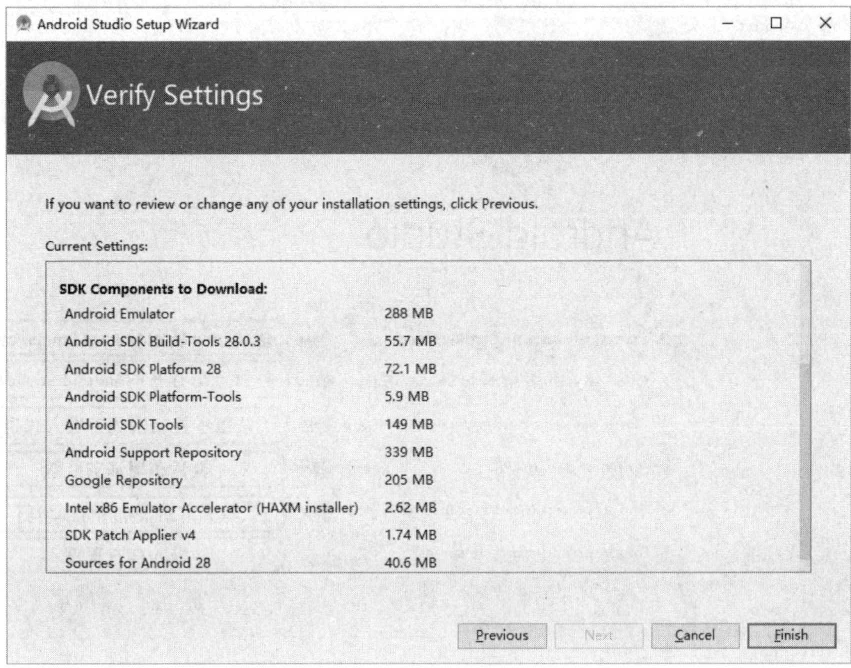

图 2-14　确认配置界面

下载组件的过程可能会比较长，需耐心等待，如图 2-15 所示。完成后，单击 Finish 按钮。

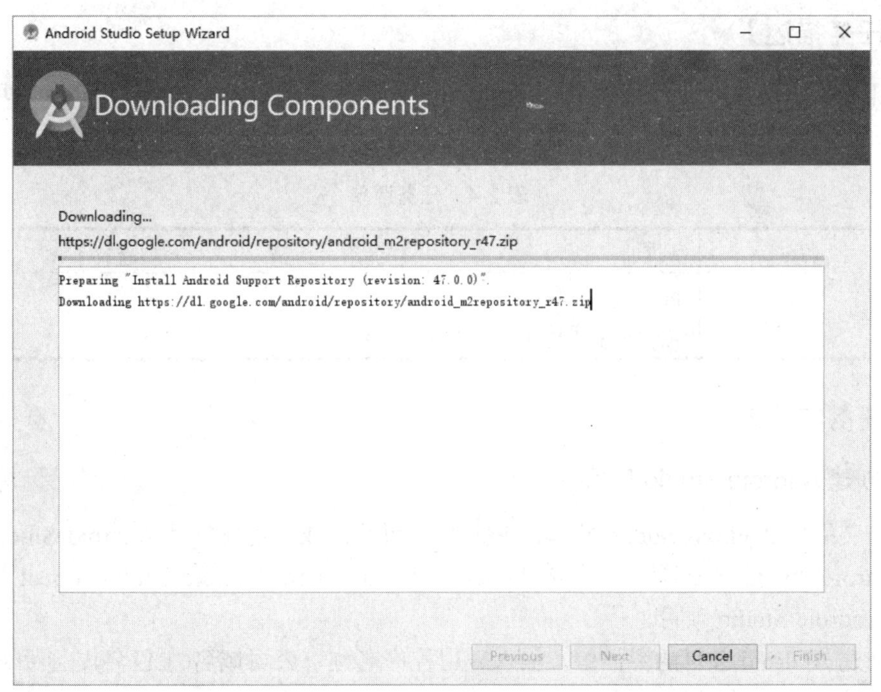

图 2-15　组件下载界面

出现 Android Studio 导航界面，如图 2-16 所示。至此，Android Studio 安装配置完毕。

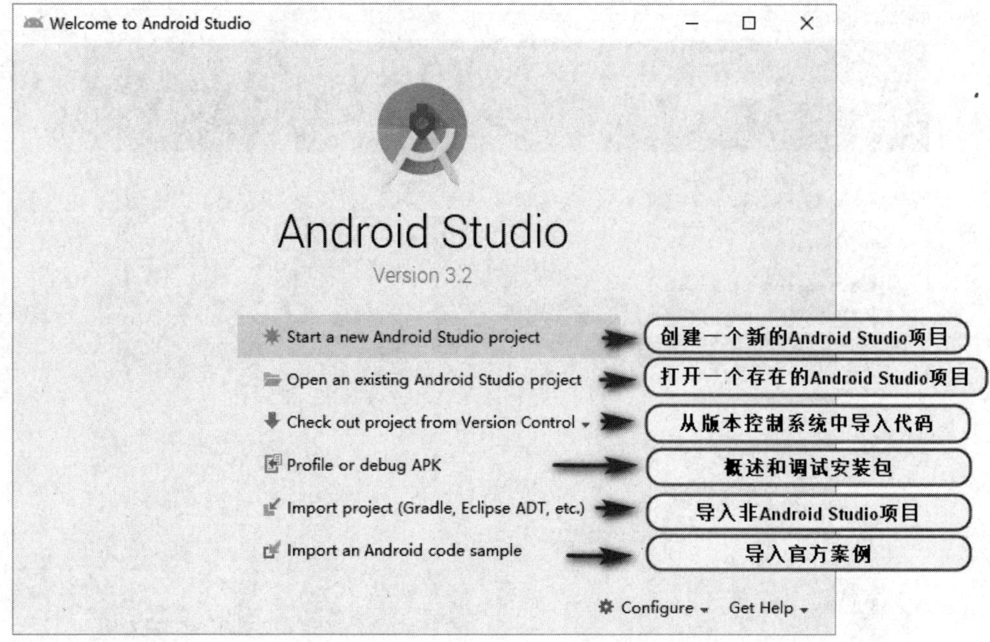

图 2-16　Android Studio 导航界面

任务 3　创建第一个 Android 项目

【任务描述】

本任务要求创建并运行第一个 Android Studio 项目，最后对 Android Studio 的项目目录结构做解析。任务清单如表 2-4 所示。

表 2-4　任务清单

任务课时	2 课时	任务组员数量	建议 1 人
任务组采用设备	PC 一台 模拟器或真机或工控平板		

【任务实施】

1. 创建 Android Studio 项目

在安装完成 Android Studio 并单击 Finish 按钮后，或者重新打开 Android Studio，均会出现 Android Studio 导航界面。单击列表第一项 Start a new Android Studio project 将创建一个新的 Android Studio 项目。

在创建 Android 项目的界面中，输入应用程序名称、公司域名(主包名)、项目路径，再单击 Next 按钮，如图 2-17 所示。

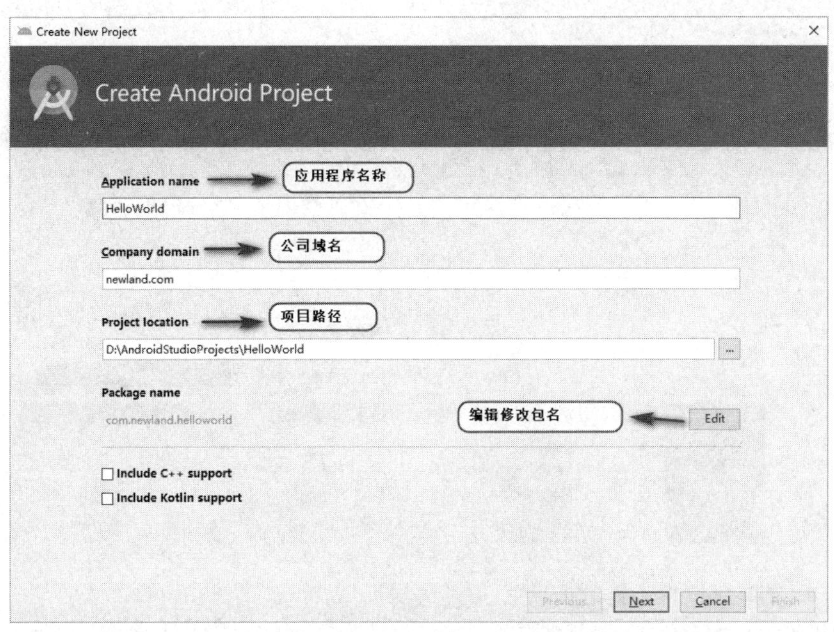

图 2-17　创建新的 Android 项目界面

在 Android 设备设置界面，选择 Phone and Tablet(针对手机和平板设置)复选框，在下拉列表中选择适配的最小版本，这里选择 API 15，单击 Next 按钮，如图 2-18 所示。

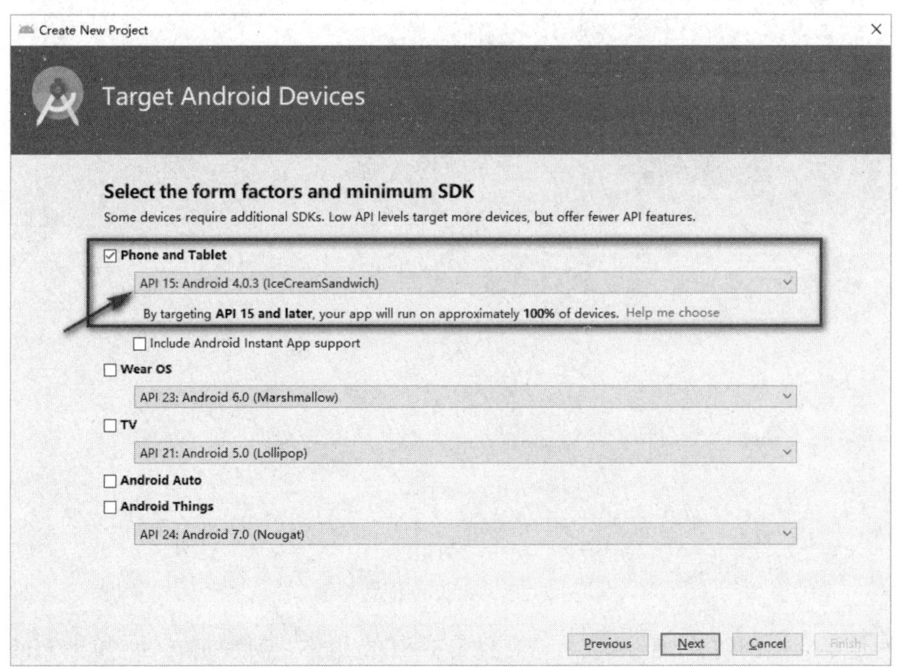

图 2-18　选择最小 SDK 版本界面

这里使用默认的 Empty Activity 模式，单击 Next 按钮，如图 2-19 所示。

在打开的界面中可以设置 Activity Name 与 Layout Name，这里使用默认设置，单击 Next 按钮，如图 2-20 所示。

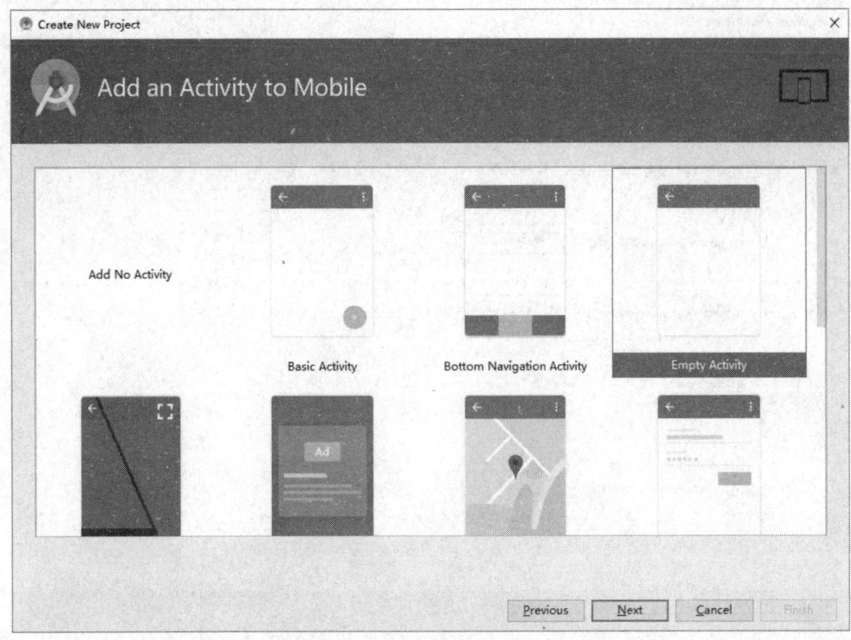

图 2-19　选择 Empty Activity 模式

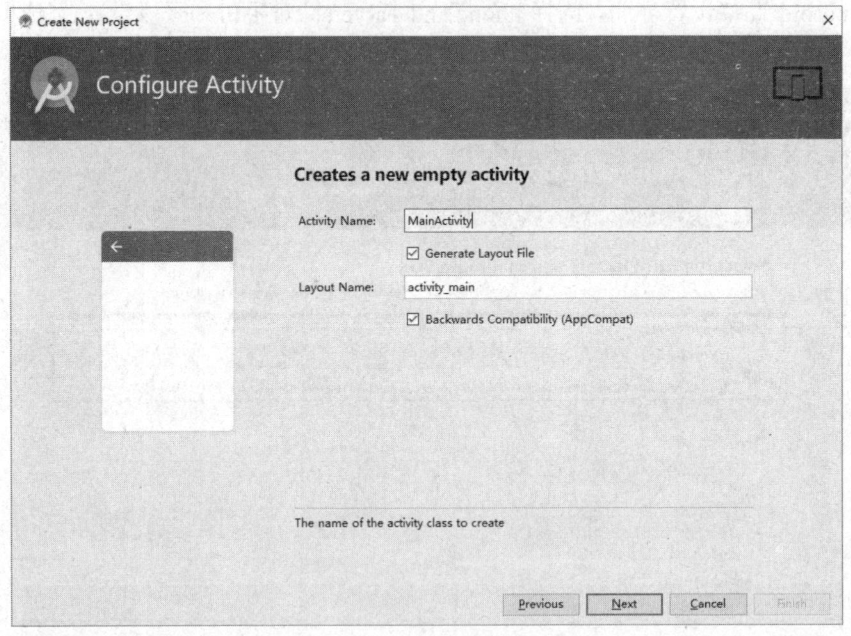

图 2-20　Activity 配置界面

因为是第一次创建 Android 项目，需要安装 SDK 组件，安装完成后，单击 Finish 按钮，如图 2-21 所示。

创建完 Android 项目后，会进入 Android Studio 工作窗口，在此窗口可能会下载项目所需的 Gradle 构建工具，只需要等待下载并构建项目即可，如图 2-22 所示。

如果出现 Failed to find Build Tools revision 28.0.2，则按要求单击蓝色的链接下载 Build Tools revision 28.0.2 即可，如图 2-23 所示。

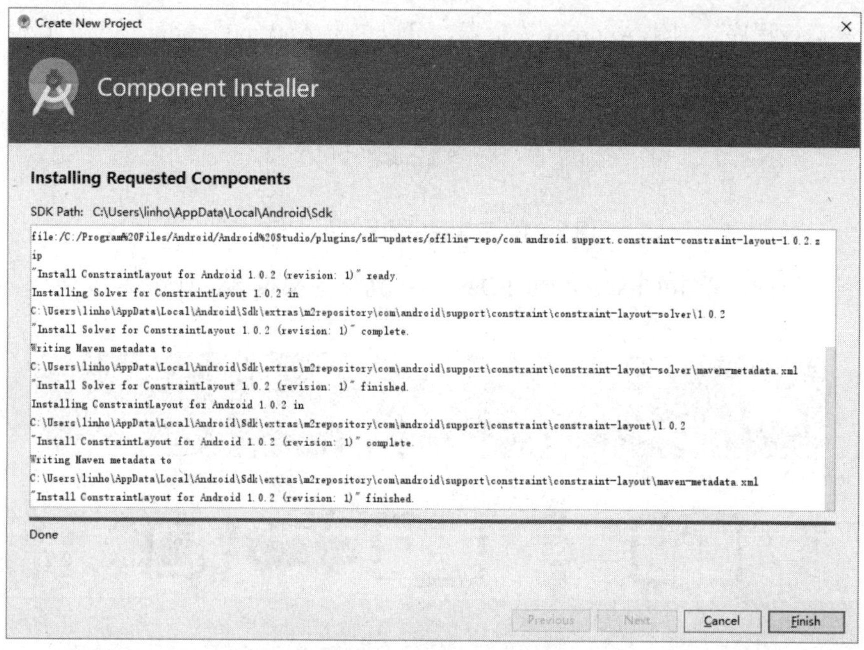

图 2-21 组件安装窗口

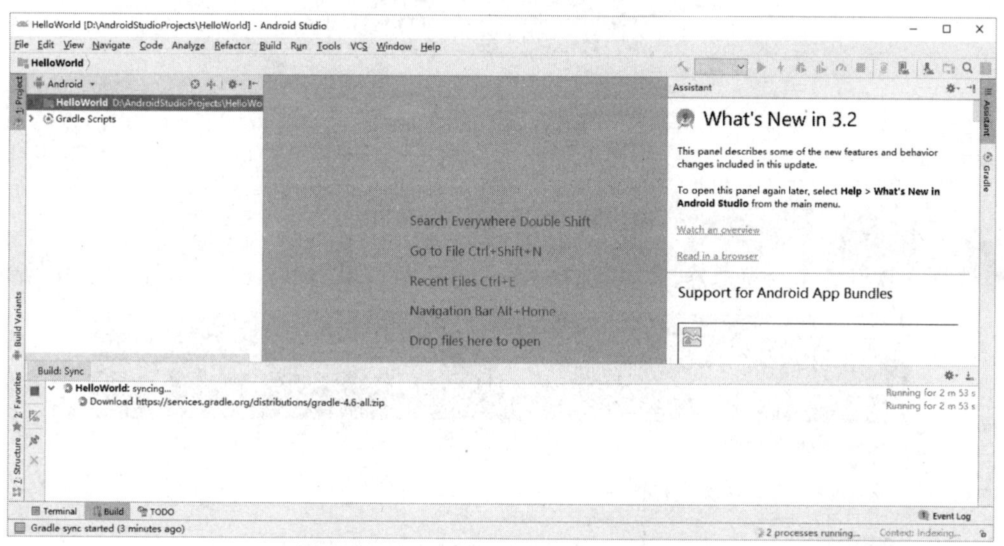

图 2-22 Android Studio 工作窗口

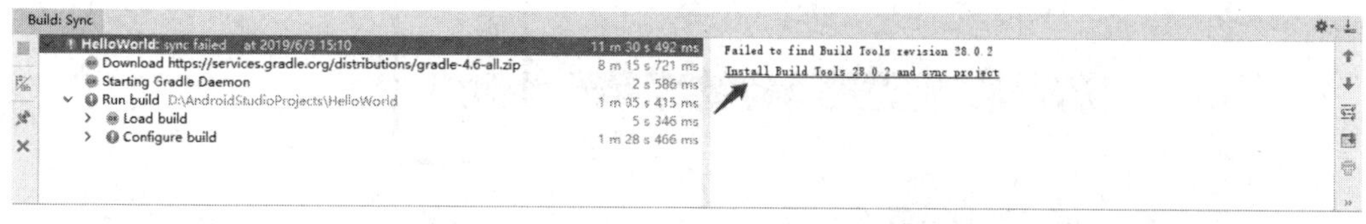

图 2-23 Failed to find Build Tools 提示

2. 在模拟器上运行 Android

Android 项目创建好后，还需要一部 Android 设备，也可以是 Android 模拟器。这里使用模拟器来运行程序。创建 Android 模拟器，可以单击 Android Studio 顶部工具栏中的相应图标进入，如图 2-24 所示。

图 2-24　工具栏中的模拟器图标

在打开的页面中单击 Create Virtual Device 按钮创建模拟器，如图 2-25 所示。

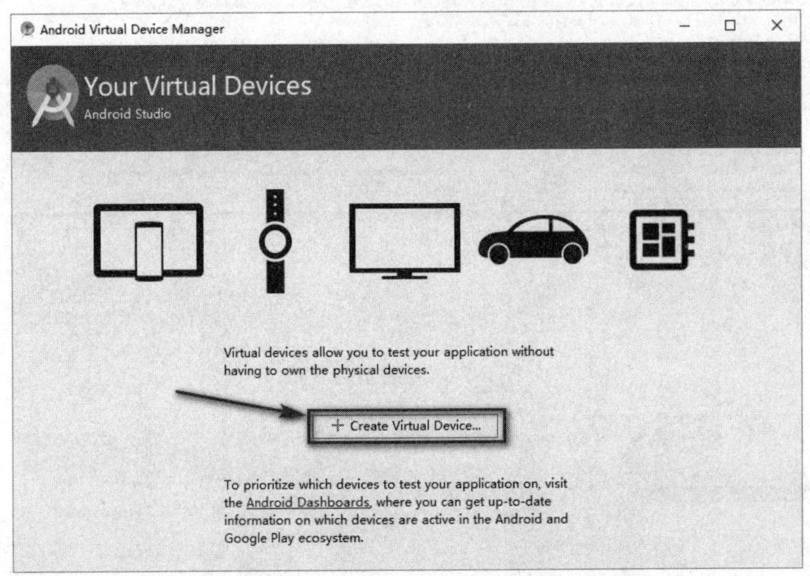

图 2-25　创建 Android 模拟器

选择一款设备作为模拟器，这里选择 Nexus 5 手机，单击 Next 按钮，如图 2-26 所示。

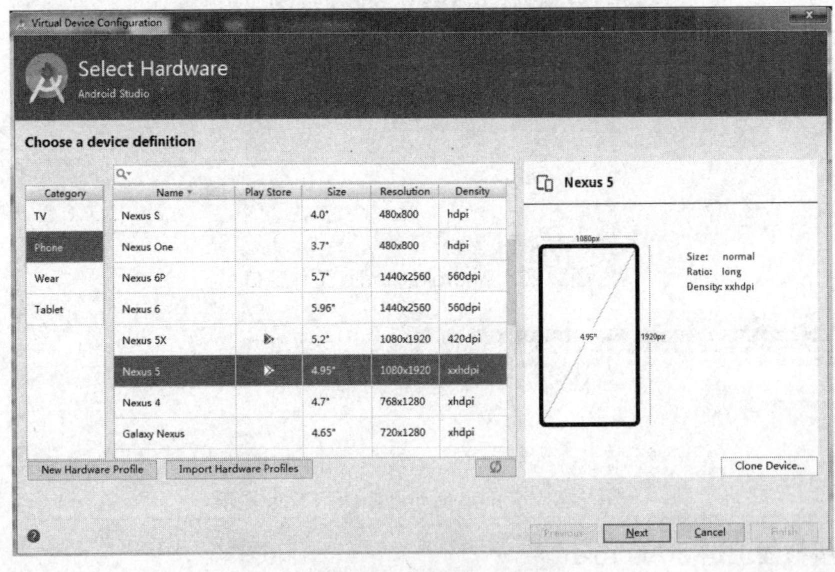

图 2-26　选择模拟器机型

选择 Android API 作为 Android 的运行环境，这里选择 Android API 28(如果没有下载需先下载)，下载完毕后单击 Next 按钮，如图 2-27 和图 2-28 所示。

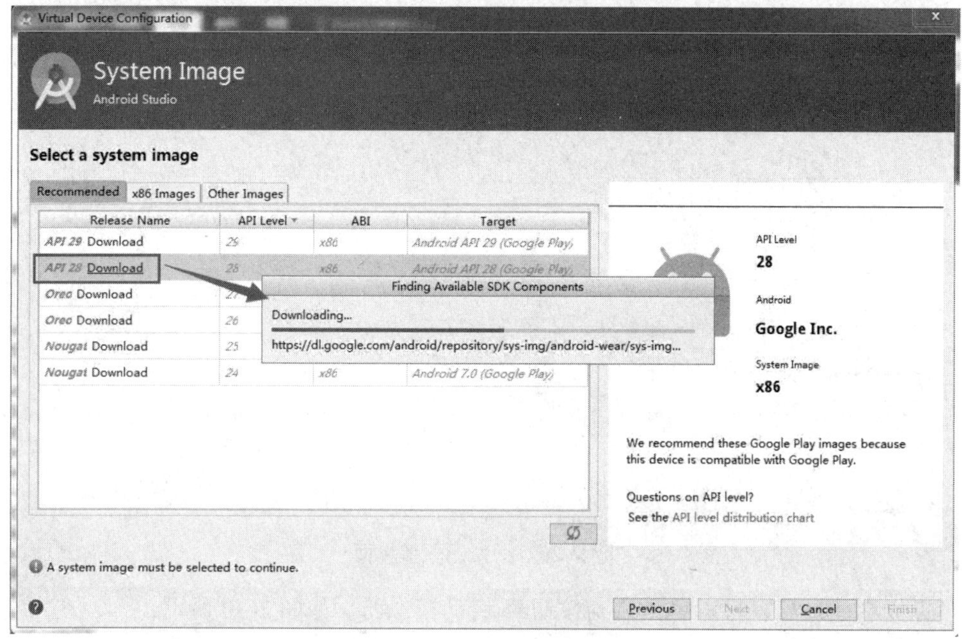

图 2-27　下载对应的 API

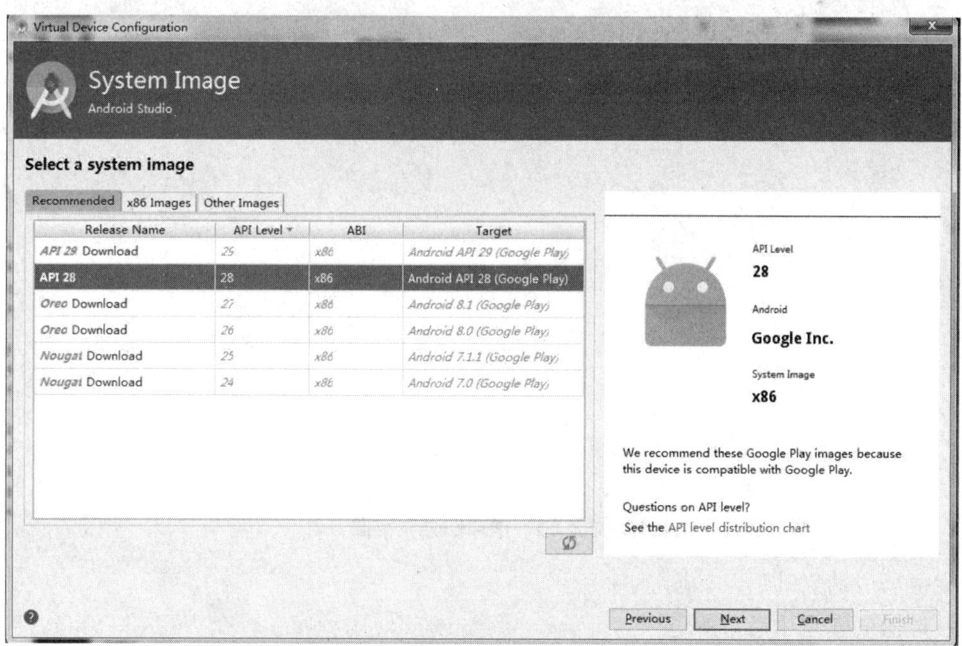

图 2-28　选择模拟器 API 版本

在这里可以设置模拟器名称以及修改模拟器参数，保持默认设置即可，单击 Finish 按钮，如图 2-29 所示。

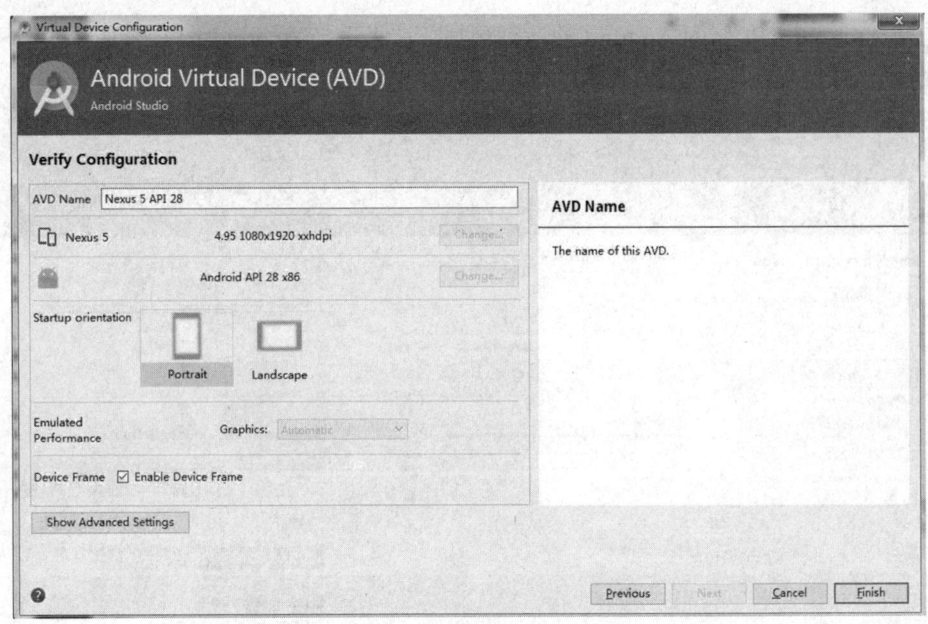

图 2-29　模拟器配置界面

模拟器创建好后，可以单击启动按钮启动模拟器，如图 2-30 所示。

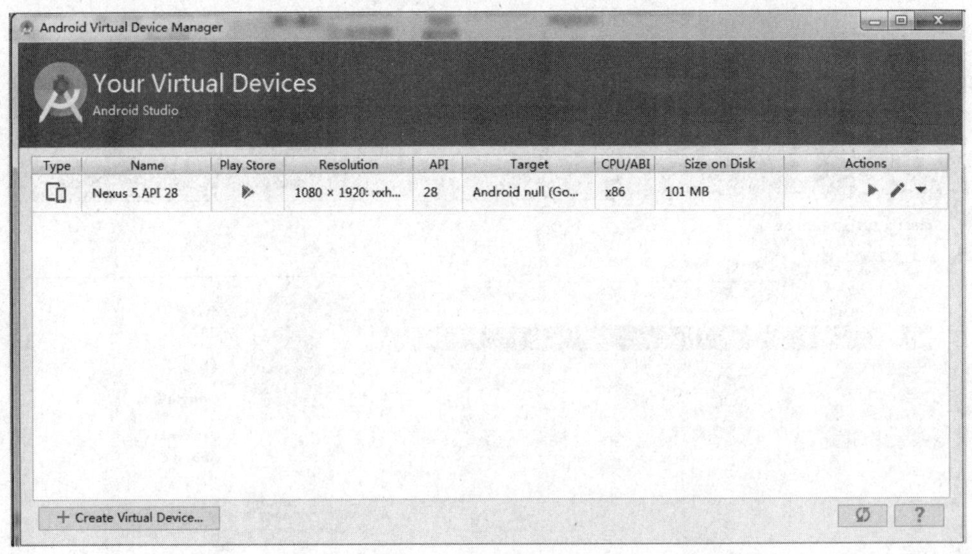

图 2-30　模拟器列表界面

模拟器成功启动后，如图 2-31 所示。

模拟器创建完毕后，就可以运行第一个 Android 应用程序了。在 Android Studio 顶部工具栏找到启动按钮，单击即可运行应用程序，如图 2-32 所示。

单击运行后会弹出选择连接的设备窗口，选中我们创建的模拟器设备，单击 OK 按钮，运行效果如图 2-33 所示。(提示：如果每次运行 Android 程序都用该设备，可以勾选左下角的 Use same selection for future launches 复选框。)

图 2-31 模拟器启动成功

图 2-32 运行 Android 应用程序

图 2-33 运行效果

3. 在真实设备上运行 Android

前面介绍的是连接模拟器运行 Android 程序，接下来介绍连接真实设备运行 Android 程序。Android 的设备有许多，厂商也可能不同，所以打开真实设备的步骤可能会有细微的区别，不过大致如下。

(1) 先将 Android 设备通过 USB 连接计算机，在设备管理器中确保驱动安装正确。

(2) 设备上弹出窗口，单击允许使用 USB 调试，并打开 Android 设备的设置界面，设置开发者模式。

(3) 找到开发者选项，并勾选 USB 调试和 USB 安装。

正常情况下，单击运行程序，在设备选择窗口中能够看到连接的真实设备，选中运行即可。

4. 分析 Android Studio 项目 app 目录结构

在 Android Studio 工作台的左侧栏中可以看到 Android 项目 HelloWorld 的项目结构，如图 2-34 所示。

Android Studio 默认使用这种 Android 模式的项目结构，但是这不是项目真实的目录结构，而是被 Android Studio 转换过的。这种项目结构简洁，但不适合初学者理解。可以单击 Android 的下拉框切换项目结构，如图 2-35 所示。

可以选择 Project，把项目目录结构切换成 Project 模式，这就是项目真实的目录结构了，如图 2-36 所示。

Android Studio 的工程目录看上去很复杂，在开发过程中我们的代码与资源基本上都是放置在 app 目录下的，接下来先对 app 目录结构进行详细讲解，如图 2-37 所示。

(1) libs。

如果项目中用到了第三方 jar 包，就需要把这些 jar 包都放在 libs 目录下。放在这个目录下的 jar 包都会被自动添加到构建路径中。

(2) java。

java 目录是放置 java 代码的地方，展开该目录可看到刚才创建的 MainActivity 文件。

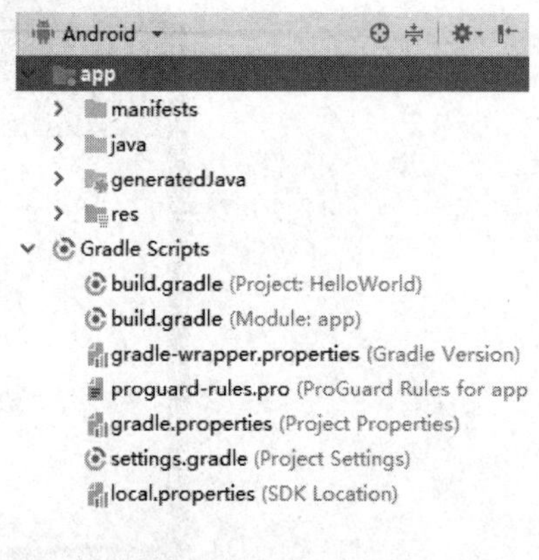

图 2-34　项目结构

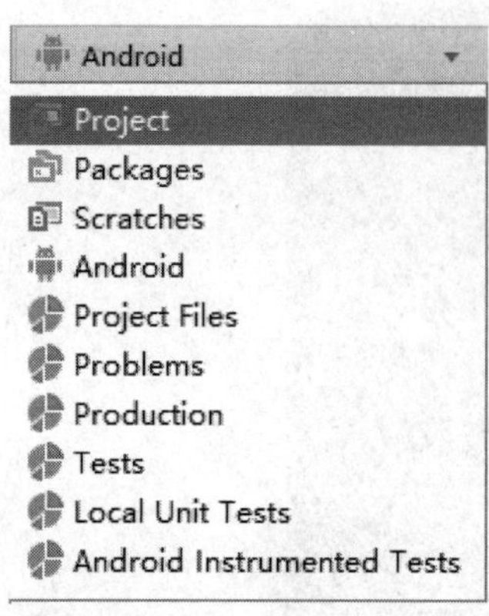

图 2-35　切换项目结构

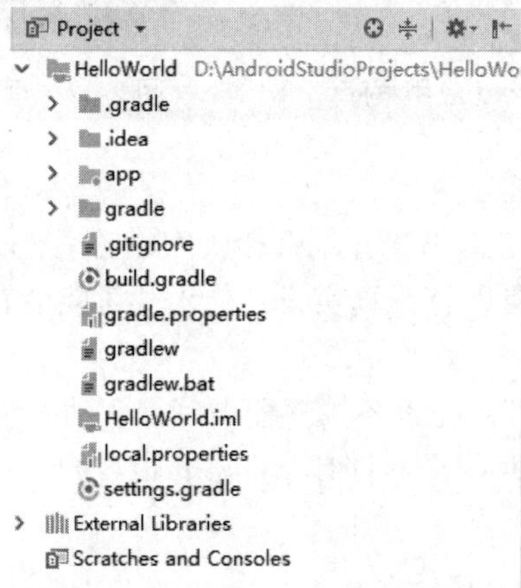

图 2-36　Project 模式的项目结构

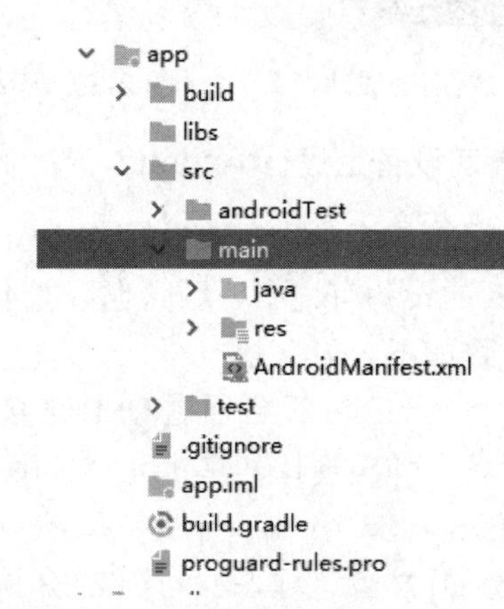

图 2-37　app 目录

(3) res。

这个目录用来放置项目中用到的所有图片、布局及字符串等资源。当然这个目录下还有很多子目录，图片放在 drawable 子目录下，布局放在 layout 子目录下，字符串放在 values 子目录下。

(4) AndroidManifest.xml。

这是整个 Android 项目的配置文件，在程序中定义的所有大组件都需要在这个文件里注册。另外，还可以在这个文件中给应用程序添加权限声明。

至此项目的目录结构就简单介绍完毕，接下来分析第一个 Android 项目 HelloWorld 是如何运行的。首先打开 AndroidManifest.xml 文件，可以找到以下代码：

```
<activity android:name=".MainActivity">
    <intent-filter>
        <action android:name="android.intent.action.MAIN" />
        <category android:name="android.intent.category.LAUNCHER" />
    </intent-filter>
</activity>
```

这段代码的意思是在 AndroidManifest.xml 中注册了一个活动。其中，intent-filter 里的 <action android:name="android.intent.action.MAIN"/>表示 MainActivity 是这个项目的主活动，

<category android:name="android.intent.category.LAUNCHER"/>表示在主界面上单击应用图标，即可启动这个活动。关于活动可以简单地理解成手机的一个屏幕，在后面会详细介绍。

接着看一下 MainActivity 中的代码，具体如下：

```
public class MainActivity extends AppCompatActivity {

    @Override
    protected void onCreate(Bundle savedInstanceState) {
        super.onCreate(savedInstanceState);
        setContentView(R.layout.activity_main);
    }
}
```

在 MainActivity 类中可以看到，MainActivity 继承自 AppCompatActivity(AppCompatActivity 是 Android 中定义好的一个活动类)。在 MainActivity 类中有一个 onCreate()方法，这个方法在一个活动被创建时会执行，onCreate()方法内有一条语句 setContentView (R.layout.activity_main)，表示绑定名称为 activity_main 的布局文件。我们找到 res 目录下 layout 目录中的 activity_main.xml 文件，打开并切换到 Text 视图模式下，就会看到在 TextView 下有一句代码 android:text="Hello World!"。这句代码正是我们之前运行的 Android 应用中屏幕上显示的文字。关于布局会在后面详细讲解。

5. 详解 Gradle

读者可能注意到，在 Android 工程项目结构中，不管是 Android 模式还是 Project 模式，可以看到很多地方都有 Gradle，如图 2-38 所示。

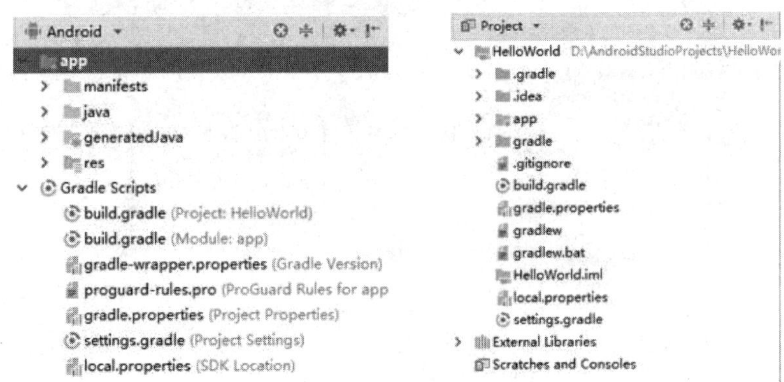

图 2-38　Android 项目中的 Gradle

那么，Gradle 到底是什么呢？Gradle 实际上是一个通用的构建工具，它不仅限于构建 Android 应用程序。Gradle 着重于构建自动化和支持多语言开发。当在任何平台上构建、测试、发布和部署软件时，Gradle 将提供一个灵活的模型，可以支持从编译和打包代码到发布的整个开发生命周期。

实际上 Gradle 本身并不能做太多，所有有用的功能都来自丰富的插件生态系统。把添加到 Android 应用程序中的所有第三方库视为插件，使用这些插件来扩展应用程序的功能，就像 Gradle 使用插件来扩展自己的功能一样。

所有带有 Gradle 字样的文件都用于为 Android 项目配置 Gradle，里面存在多个文件，并且它们都有不同的用途。下面以图形的方式讲解经常涉及的 gradle 文件的用途，如图 2-39 至图 2-43 所示。

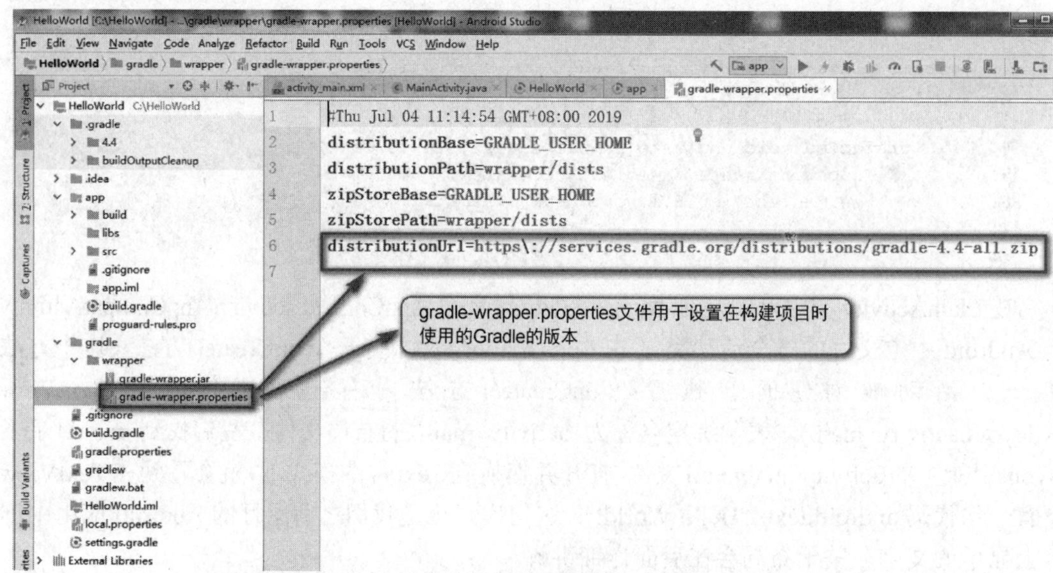

图 2-39　工程的 gradle/wrapper/gradle-wrapper.properties 文件

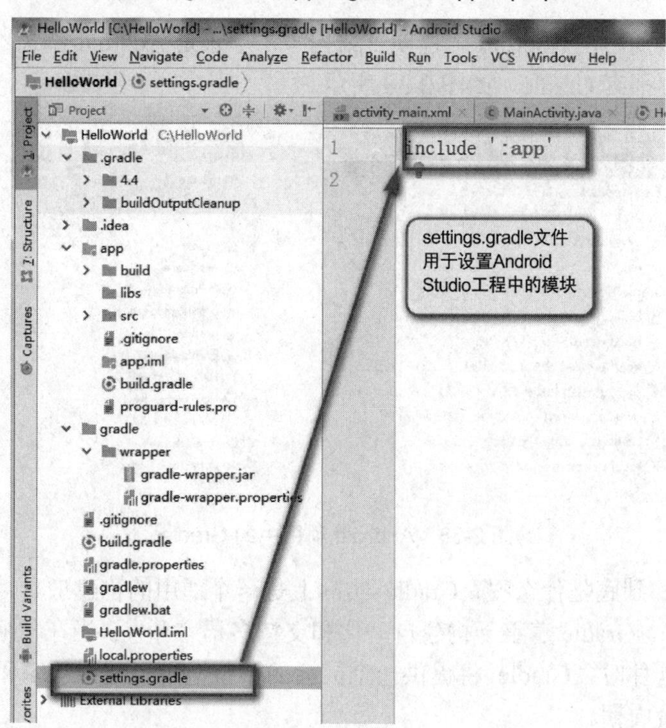

图 2-40　工程的 settings.gradle 文件

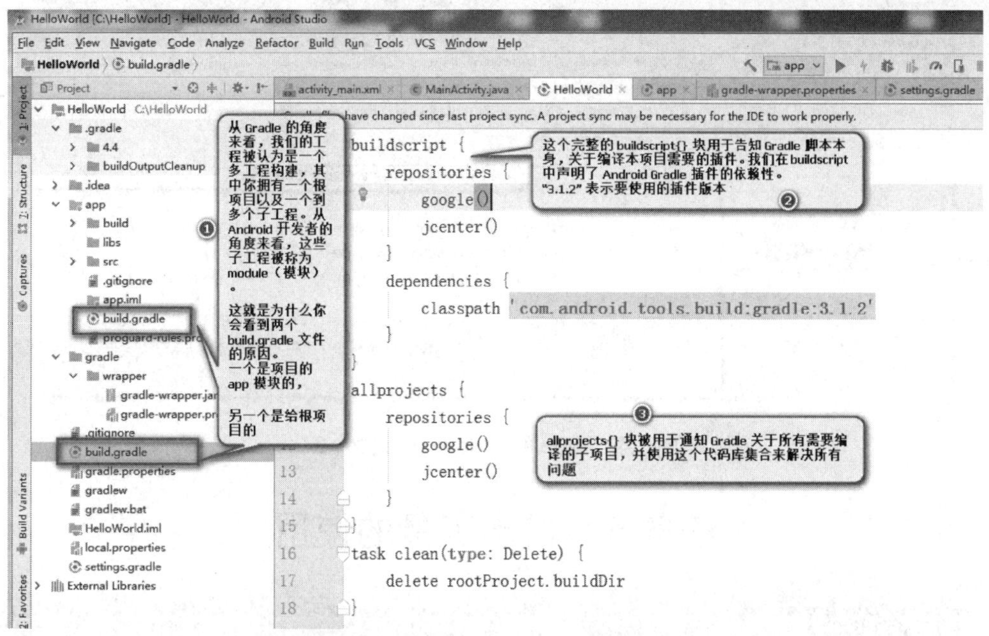

图 2-41 工程的 build.gradle 文件

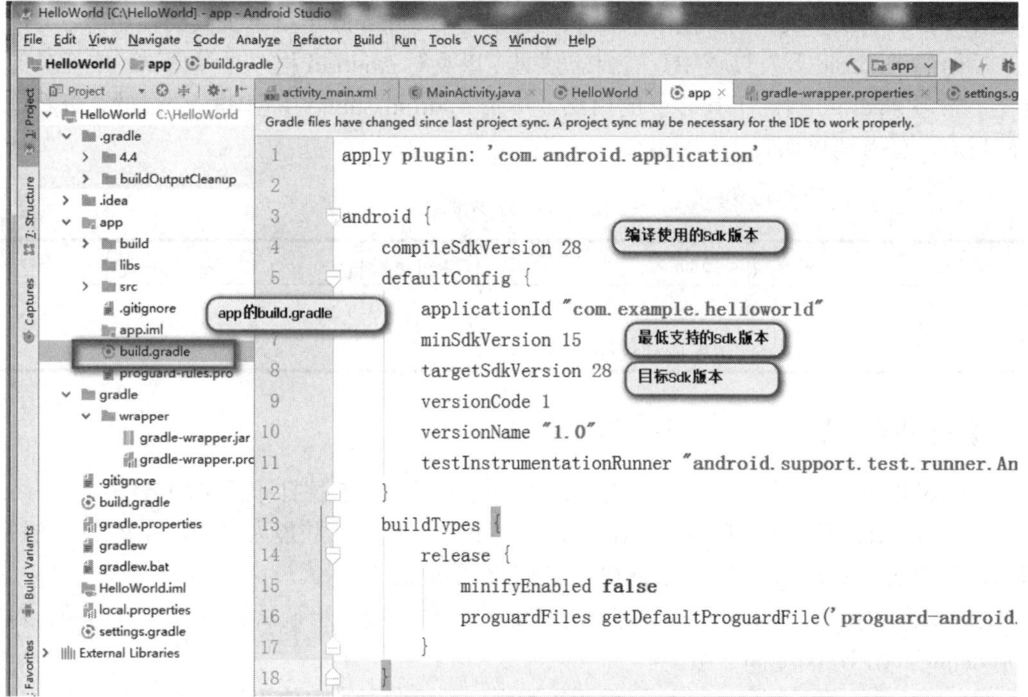

图 2-42 app 的 build.gradle 文件 1

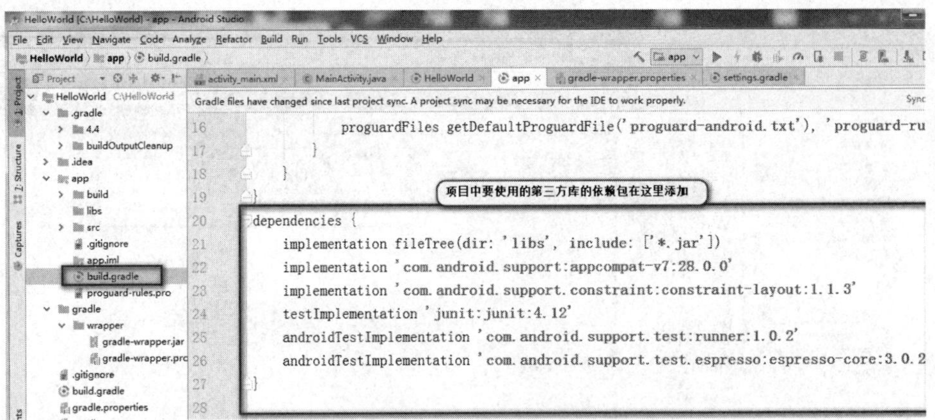

图 2-43　app 的 build.gradle 文件 2

任务 4　日志工具的使用

【任务描述】

在上一个任务中已经成功创建了第一个 Android 程序，并且对 Android 项目的目录结构和运行流程有了初步的了解。接下来在本任务中学习 Android 中日志工具的使用方法，以方便在后面的 Android 开发过程中查看日志信息，以及对 Android 程序进行调试。本任务的关键在于学习 Android 日志工具 Log 的用法，并在 Android Studio 中创建一个过滤，查看过滤后的日志。任务清单如表 2-5 所示。

表 2-5　任务清单

任务课时	0.5 课时	任务组员数量	建议 1 人
任务组采用设备	PC 一台 模拟器或真机或工控平板		

【知识解析】

Android 中提供一个日志工具类 Log，该类提供了以下 5 种打印日志的方法。

(1)　Log.v()：用于打印所有的 Android 日志信息，包括那些最为琐碎、意义较小的日志信息。对应级别 verbose，是 Android 日志里面级别最低的一种。

(2)　Log.d()：用于打印一些调试信息，这些信息一般用于调试程序和分析问题。对应级别 debug，比 verbose 高一级。

(3)　Log.i()：用于打印一些比较重要的数据，一般在程序中使用该级别日志分析用户的行为数据。对应级别 info，比 debug 高一级。

(4)　Log.w()：用于打印一些警告信息，一般在程序中使用该信息来提示程序可能存在的潜在危险。对应级别 warn，比 info 高一级。

(5)　Log.e()：用于打印程序中的错误信息。该信息用于提示程序出现错误的地方，必须尽快修复。对应级别 error，比 warn 高一级。

【任务实施】

1. 添加打印日志语句

创建了一个 Android 项目 LogDemo，在 MainActivity.java 中的 onCreate()方法中添加日志语句，具体代码如下：

```
public class MainActivity extends AppCompatActivity {
    private static final String TAG = "MainActivity";
    @Override
    protected void onCreate(Bundle savedInstanceState) {
        super.onCreate(savedInstanceState);
        setContentView(R.layout.activity_main);
        Log.v(TAG, "这是一条 verbose 日志");
        Log.d(TAG, "这是一条 debug 日志 ");
        Log.i(TAG, "这是一个 info 日志 ");
        Log.w(TAG, "这是一条 warn 日志");
        Log.e(TAG, "这是一条 error 日志");
    }
}
```

代码中使用 Log 工具编写了 5 种级别日志，这几个方法都是传入两个参数，第一个参数是 TAG，即日志的标记，后续可以通过 TAG 标记过滤日志。第二个参数是 msg，即具体打印的信息。接着运行程序，可以在 Android Studio 的 Logcat 窗口中查看日志，如图 2-44 所示。

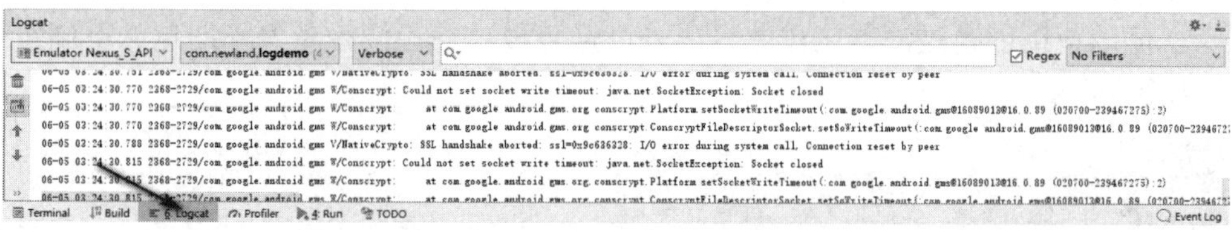

图 2-44 Logcat 日志窗口

从 Logcat 中可以看到许多日志，这些日志包括时间、包名、TAG 名、日志名等内容。但是很多日志并不是想关注的，如果只想关注 TAG 标记为 MainActivity 的日志该怎么办呢？这时就需要设置日志过滤器了。

2. 使用过滤器查看日志信息

在 Logcat 视图的右上角有一个下拉列表框，单击打开可以看到有 4 个选项：Show only selected application 表示只显示当前选中程序的日志；Firebase 是谷歌提供的一个分析工具；No Filters 相当于没有过滤器；Edit Filter Configuration 表示编辑过滤器配置。如果需要新增过滤器，需选择 Edit Filter Configuration 选项，如图 2-45 所示。

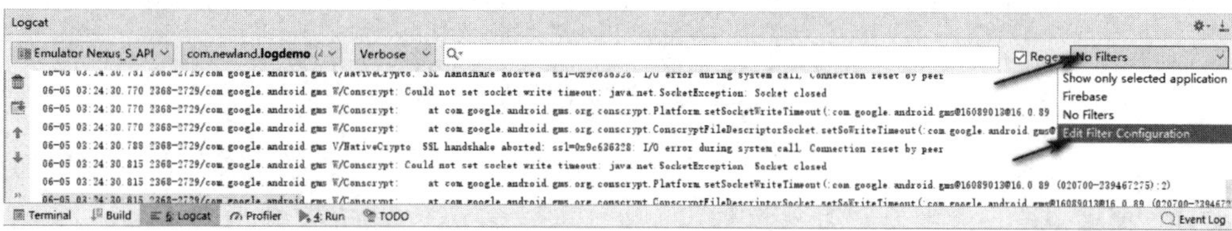

图 2-45 创建过滤器

选择 Edit Filter Configuration 选项会弹出过滤器配置界面，从中可以设置过滤器的名称及过滤信息，这里设置过滤器的名称，并将 Log Tag 设置为 MainActivity，单击 OK 按钮，如图 2-46 所示。

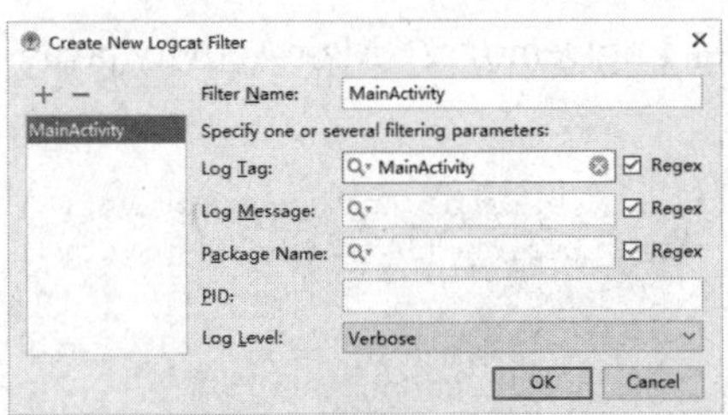

图 2-46　过滤器设置

创建过滤器后，会在 Logcat 视图看到在 MainActivity 的 onCreate()方法中打印的信息，这样就会更清晰明了，如图 2-47 所示。

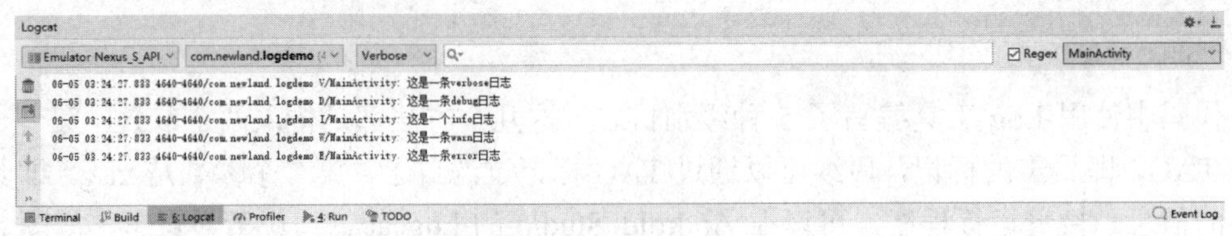

图 2-47　过滤后的日志

除了可以使用创建过滤器来过滤日志外，还可以过滤不同设备、不同包、不同级别的日志，如图 2-48 所示。

图 2-48　其他过滤说明

项 目 总 结

通过本项目的学习，你已经成为一个 Android 的开发者了，已对 Android 系统与物联网的关系有了初步的认识，并且成功搭建了 Android 的开发环境，接着还创建了第一个 Android 项目，对 Android 的项目目录结构与执行过程有了一定的认识，在最后还学习了 Android 日志工具的使用，方便后续开发过程中遇到问题进行调试。

思考与练习

2-1　简述 Android 项目的目录结构。

2-2　简述 Android 日志工具的使用。

项目3 智慧园区界面的实现

【项目描述】

在项目2中，读者已经熟悉Android Studio的项目结构，并且能够添加一些简单的代码查看打印的日志信息。本项目以4个任务的形式带领读者在学习Android UI的基本知识的基础上，完成智慧园区的部分界面开发。具体任务列表如图3-1所示。

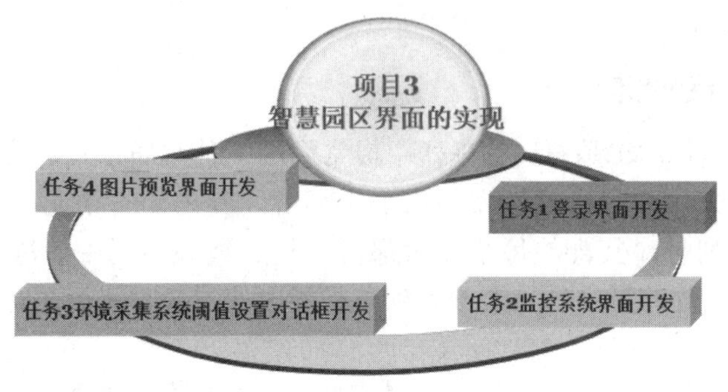

图3-1　项目3任务列表

【学习目标】

知识目标：

- 会使用Android常见的布局，包括线性布局、相对布局、帧布局。

- 能掌握约束布局的使用。
- 会使用 Android 的常见控件，包括 TextView、EditText、Button 等。
- 会使用按钮注册点击的监听器。
- 会使用常见对话框与自定义对话框。
- 会使用 ProgressBar 与 SeekBar。
- 会使用 ListView 与 RecyclerView。

技能目标：

- 能用线性布局以及常用控件完成登录界面的开发。
- 能用相对布局或约束布局完成园区监控系统界面的开发。
- 能用自定义对话框与 SeekBar 完成阈值设置对话框的开发。
- 能使用 RecyclerView 完成图片预览界面的开发。

任务 1　登录界面开发

【任务描述】

扫码观看视频讲解

本任务要求使用线性布局以及 TextView、EditText、Button 等常见控件完成智慧园区登录界面的开发。任务清单如表 3-1 所示。

表 3-1　任务清单

任务课时	4 课时	任务组员数量	建议 1 人
任务组采用设备	PC 一台 模拟器或真机或工控平板		

【知识解析】

1. LinearLayout 的使用

LinearLayout(线性布局)是布局管理器中最常用的一种布局方式，放入其中的元素可按照垂直或水平方向来排列。当线性布局水平排列时，其内的每个元素都占一列；同理，当线性布局垂直排列时，其内的每个元素都占一行。线性布局中的每个元素都位于前一个元素之后，但要注意的是，当元素排列到布局界面的边缘时，后面的元素将不会被显示出来。

线性布局中有一个非常重要的属性 android:orientation，用于控制控件的排列方向，该属性有 vertical(垂直显示)和 horizontal(水平显示，为默认)两个值。接下来通过一个例子来体会 LinearLayout 排列方向的使用，具体代码如下：

```
<LinearLayout xmlns:android="http://schemas.android.com/apk/res/android"
    android:layout_width="match_parent"
    android:layout_height="match_parent"
    android:orientation="vertical">

    <Button
```

```
            android:layout_width="wrap_content"
            android:layout_height="wrap_content"
            android:text="BUTTON1" />

        <Button
            android:layout_width="wrap_content"
            android:layout_height="wrap_content"
            android:text="BUTTON2" />

        <Button
            android:layout_width="wrap_content"
            android:layout_height="wrap_content"
            android:text="BUTTON3" />

    </LinearLayout>
```

在上述代码中，将 android:orientation 属性值设置为 horizontal，控件会水平排列显示。若将该值设置为 vertical，则控件会垂直排列显示。预览效果如图 3-2 所示。

图 3-2　控件水平排列和垂直排列

此外，上述代码中的属性 android:layout_width 用来设置布局或控件的宽度，属性 android:layout_height 用来设置布局或控件的高度，宽度和高度可以设置具体数值，单位使用 dp，也可以设置为 wrap_content(大小由包裹的内容大小自动伸缩)，或者设置为 match_parent(大小填充父容器)。在该例子中，当控件水平排列时，控件的宽度不能设置为 match_parent；否则其余的控件会被挤出屏幕右侧不显示。同理，如果控件垂直排列时，控件的高度不能设置为 match_parent；否则会出现其余控件无法显示的情况。Button 控件还有一个属性 android:text，用来设置按钮上的文本内容。至于 LinearLayout 中的属性 xmlns:android 是一个命名空间，这里不做详释。

从图 3-2 中可以看出，当控件水平排列时，3 个按钮未占满一行，右侧留有空白区域。如果想让这 3 个按钮占满一行，可以使用 android:layout_weight 属性，该属性的值为权重，通过权重比例调整布局中控件的大小，在进行屏幕适配时起到关键的作用。接下来，通过一个例子来学习 android:layout_weight 属性的使用，具体代码如下：

```
<LinearLayout xmlns:android="http://schemas.android.com/apk/res/android"
    android:layout_width="match_parent"
    android:layout_height="match_parent"
    android:orientation="horizontal">

    <Button
        android:layout_width="0dp"
        android:layout_height="wrap_content"
        android:layout_weight="1"
        android:text="BUTTON1" />

    <Button
        android:layout_width="0dp"
        android:layout_height="wrap_content"
        android:layout_weight="2"
        android:text="BUTTON2" />

    <Button
        android:layout_width="0dp"
        android:layout_height="wrap_content"
        android:layout_weight="1"
        android:text="BUTTON3" />
</LinearLayout>
```

上面的代码中，我们为 LinearLayout 内的 Button 控件加上了属性 android:layout_weight，按钮 1 的比重值为 1，按钮 2 的比重值为 2，按钮 3 的比重值为 1。需要注意的是，LinearLayout 设置的排列方向是 horizontal，所以只能设置宽度的比重。还有一个要注意的地方，既然宽度使用比重显示大小，那么控件的宽度不再由 android:layout_width 来决定，所以指定为 0dp 不会影响效果，这样写也是一种规范。预览效果如图 3-3 所示。

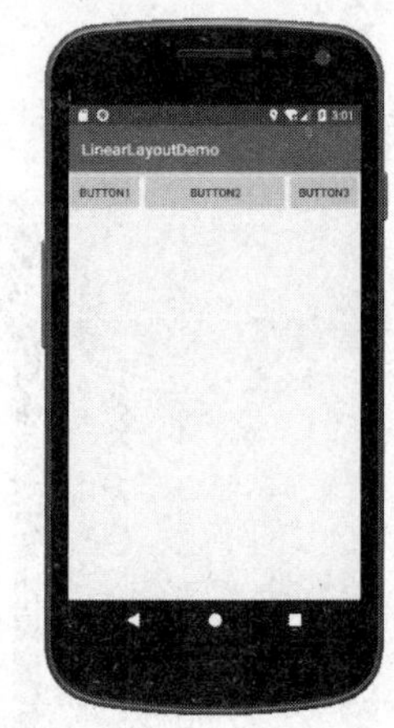

图 3-3　设置 layout_weight 属性效果

多学一招：px、dp、sp 的区别

众所周知，Android 的厂商非常多，不同尺寸的 Android 设备也层出不穷，导致 Android 的碎片化现象越来越严重。Google 公司为了解决分辨率过多的问题，在 Android 的开发文档中定义了 px、dp、sp，方便开发者适配不同分辨率的 Android 设备。对于初学者来说，只需理解这些单位的基本概念即可。

px 指的是像素单位，如通常所说的手机分辨列表 800×400 都是以 px 为单位。dp 是虚拟像素，在不同像素密度的设备上会自动适配。在设置 Android 的布局宽高、控件大小与边距时，建议使用 dp。sp 同 dp 相似，会根据用户的字体大小偏好来缩放，建议用来作为文字大小的单位。

2. TextView 的使用

TextView(文本框)可以说是 Android 中最简单的一个控件了，它主要用于在界面上显示一段文本信息。开发者可以在代码中设置 TextView 控件属性，如字体大小、文字样式等。TextView 控件的属性较多，这里列举一些常用属性，具体如表 3-2 所示。

表 3-2　TextView 的常用属性

控件属性	功能描述
android:text	设置显示文本
android:textColor	设置文本的颜色
android:textSize	设置文字大小，推荐单位为 sp
android:textStyle	设置文字样式，如 bold(粗体)、italic(斜体)、bolditalic(粗斜体)
android:height	设置文本区域的高度
android:width	设置文本区域的宽度
android:maxlength	设置文本长度，超出不显示
android:gravity	设置文本位置
android:layout_height	设置 TextView 控件的高度
android:layout_width	设置 TextView 控件的宽度

关于 TextView 更多的属性，这里就不一一列举了。接下来通过一个例子来学习如何为 TextView 控件设置宽、高、文本颜色、字体大小、文本位置属性，具体代码如下：

```
<LinearLayout xmlns:android="http://schemas.android.com/apk/res/android"
    android:layout_width="match_parent"
    android:layout_height="match_parent"
    android:orientation="vertical">

    <TextView
        android:layout_width="match_parent"
        android:layout_height="wrap_content"
        android:gravity="center"
        android:text="Hello Android!"
        android:textColor="#0000ff"
        android:textSize="25sp" />
</LinearLayout>
```

在上述代码中，通过 android:layout_width 和 android:layout_height 属性设置 TextView 的宽和高，android:gravity 属性设置 TextView 文本的位置，android:text 属性设置文本的内容，android:textColor 属性设置文本的颜色，android:textSize 属性设置字体的大小。预览效果如图 3-4 所示。

图 3-4　TextView 控件

多学一招：Android 颜色代码

Android 颜色代码用#ARGB 表示，A 为透明度，R 为 Red，G 为 Green，B 为 Blue，每个字母都用一个十六进制来表示，其中透明度 A 从 0 到 F 表示完全透明到完全不透明。例如，#FFFF 表示完全不透明的白色，也可以省略 A，使用#RGB 表示完全不透明的颜色。还可以使用#AARRGGBB 表示更高精度的颜色值(其中 A 也可省略)。

3. EditText(输入框)的使用

EditText 控件也称为输入框，它允许用户在控件里输入和编辑内容。EditText 继承自

TextView，所以 EditText 可以使用 TextView 定义的一些属性，此外，EditText 还定义了自己特有的属性。下面列举一些 EditText 的常用属性，具体如表 3-3 所示。

表 3-3　EditText 的常用属性

控件属性	功能描述
android:hint	设置 EditText 没有输入内容时显示的提示文本
android:inputType	设置输入文本的类型，如 textPassword 表示输入的文本为密码类型(文本将以“.”显示)，phone 表示电话号码类型，date 表示日期类型等

接下来，通过一个例子学习 EditText 控件的使用，具体代码如下：

```
<LinearLayout xmlns:android="http://schemas.android.com/apk/res/android"
    android:layout_width="match_parent"
    android:layout_height="match_parent"
    android:orientation="vertical">

    ...

    <EditText
        android:layout_width="match_parent"
        android:layout_height="wrap_content"
        android:hint="请输入账号"/>
    <EditText
        android:layout_width="match_parent"
        android:layout_height="wrap_content"
        android:inputType="textPassword"
        android:hint="请输入密码"/>
</LinearLayout>
```

在上述代码中，EditText 控件的属性 android:hint 用于设置没有输入内容时的提示文本。属性 android:inputType 用于设置输入内容类型，值 textPassword 表示输入的内容为文本密码，将以“.”显示。运行结果如图 3-5 所示。

图 3-5　EditText 控件

4. Button 的使用

Button(按钮)控件，也就是常说的按钮，是程序与用户交互非常重要的一个控件，其作用是响应用户的一系列点击事件。为按钮注册点击的监听器常用的方式有 3 种，分别是在布局中指定 onClick 属性、使用匿名内部类、在当前 Activity 中实现 OnClickListener 接口。接下来，通过一个例子学习按钮点击方式的使用。

1)　使用 android:onClick 属性的方式

在布局文件中加入 4 个按钮，分别给 4 个按钮设置 id，用于区分不同的按钮，并给按钮 1 加入 android:onClick 属性，值为 btn1Click。具体代码如下：

```
<LinearLayout xmlns:android="http://schemas.android.com/apk/res/android"
    android:layout_width="match_parent"
    android:layout_height="match_parent"
    android:orientation="vertical">

```

```
    ...

    <Button
        android:id="@+id/btn1"
        android:layout_width="match_parent"
        android:layout_height="wrap_content"
        android:onClick="btn1Click"
        android:text="按钮 1"/>
    <Button
        android:id="@+id/btn2"
        android:layout_width="match_parent"
        android:layout_height="wrap_content"
        android:text="按钮 2"/>
    <Button
        android:id="@+id/btn3"
        android:layout_width="match_parent"
        android:layout_height="wrap_content"
        android:text="按钮 3"/>
    <Button
        android:id="@+id/btn4"
        android:layout_width="match_parent"
        android:layout_height="wrap_content"
        android:text="按钮 4"/>
</LinearLayout>
```

将光标移至 android:onClick 的属性值最后位置，在出现的灯泡图标位置可以选择在 MainActivity 中创建一个点击的回调方法 btn1Click()，如图 3-6 所示。

图 3-6 创建点击的回调方法

在 MainActivity 中将出现添加的 btn1Click()回调方法(该方法名称与 onClick 属性的值要一致)，接下来在 MainActivity 中实现逻辑代码，具体代码如下：

```
public class MainActivity extends AppCompatActivity {

    @Override
    protected void onCreate(Bundle savedInstanceState) {
        super.onCreate(savedInstanceState);
        setContentView(R.layout.activity_main);
    }

    //通过 android:onClick 的方式注册监听器添加的回调方法
    public void btn1Click(View view) {
        //添加按钮 1 点击时的逻辑代码
        Toast.makeText(this, "按钮 1 被点击了！", Toast.LENGTH_SHORT).show();
    }
}
```

在上述代码中，执行到 btn1Click()回调方法中的 Toast.makeText(this, "按钮 1 被点击了！",

Toast.LENGTH_SHORT).show();会弹出一个“吐司弹窗”，该窗口会短暂地停留在界面上，显示一段时间后会自动消失。Toast 的 makeText()方法需要传入 3 个参数，第一个参数为 Context 上下文环境，这里可以使用 this 表示当前的 MainActivity 环境；第二个参数为提示的文本；第三个参数为“吐司弹窗”停留的时间，建议使用系统提供的 Toast.LENGTH_SHORT 或 Toast.LENGTH_LONG，后者显示的时间比前者长一些。最后，在 makeText()方法后要调用 show()方法显示弹窗。

2) 使用匿名内部类的方式

接下来，在 MainActivity 中演示使用匿名内部类的方式注册按钮的监听器，具体代码如下：

```
public class MainActivity extends AppCompatActivity {

    private Button btn2;

    @Override
    protected void onCreate(Bundle savedInstanceState) {
        super.onCreate(savedInstanceState);
        setContentView(R.layout.activity_main);
        //通过 findViewById()初始化控件
        btn2 = findViewById(R.id.btn2);
        //通过匿名内部类的方式注册按钮 2 的点击监听器
        btn2.setOnClickListener(new View.OnClickListener() {
            @Override
            public void onClick(View v) {
                Toast.makeText(MainActivity.this, "按钮 2 被点击了！",
Toast.LENGTH_SHORT).show();
            }
        });
    }

    //通过 android:onClick 的方式注册监听器添加的回调方法
    ...
}
```

在上述代码中，通过 findViewById()方法初始化控件，然后使用初始化好的控件调用 setOnClickListener()方法对按钮 2 注册点击的监听器，在监听器内实现 onClick()方法，在此方法中编写逻辑代码。

3) 在当前 Activity 中实现 OnClickListener 接口

当按钮点击的事件较多时，使用上述两种方式都会比较麻烦，这时可以使用第三种方法，即在当前 Activity 中实现 OnClickListener 接口的方式。接下来，通过修改 MainActivity 中的代码来演示第三种注册监听器的方式，具体代码如下：

```
public class MainActivity extends AppCompatActivity implements View.OnClickListener {

    private Button btn2;
    private Button btn3;
    private Button btn4;

    @Override
    protected void onCreate(Bundle savedInstanceState) {
        super.onCreate(savedInstanceState);
        setContentView(R.layout.activity_main);
        //通过 findViewById()初始化控件
```

```
        btn2 = findViewById(R.id.btn2);
        btn3 = findViewById(R.id.btn3);
        btn4 = findViewById(R.id.btn4);
        //通过匿名内部类的方式注册按钮 2 的点击监听器
        ...

        //通过当前 Activity 实现 OnClickListener()接口注册监听器
        btn3.setOnClickListener(this);
        btn4.setOnClickListener(this);
    }

    @Override
    public void onClick(View v) {
        //通过参数 v 调用 getId()方法可以获取触发点击事件的控件 id，用于区分不同的按钮
        switch (v.getId()){
            case R.id.btn3:
                Toast.makeText(MainActivity.this, "按钮 3 被点击了！",
Toast.LENGTH_SHORT).show();
                break;
            case R.id.btn4:
                Toast.makeText(MainActivity.this, "按钮 4 被点击了！",
Toast.LENGTH_SHORT).show();
                break;
        }
    }

    //通过 android:onClick 的方式注册监听器添加的回调方法
    ...
}
```

在上述代码中，Activity 实现了 OnClickListener 接口并重写 onClick()方法，然后使用参数 v 调用 getId 方法可以获得触发点击事件的按钮，结合 switch 语句区分不同的按钮。需要注意的是，在注册监听器时 setOnClickListener()方法中传入 this 参数，这个 this 表示当前 Activity 的引用，由于当前 Activity 已经实现 OnClickListenr 接口，所以这里的 this 可以当作 OnClickListener 的引用。

5. ImageView 的使用

ImageView(视图控件)是用于在界面上显示图片的一个控件。在学习这个控件之前，需要准备一些图片资源放在 res/drawable 目录下。当然也可以根据不同的分辨率在 res 目录下创建以 drawable 开头的目录，如 drawable-hdpi 目录、drawable-xhdpi 目录。这里我们就不考虑分辨率的问题，把事先准备好的图片资源 bg.png 复制到 drawable 目录下(注意：图片的名称要由小写字母、数字、下划线组成，且只能以小写字母开头)。

ImageView 控件可以使用 android:src 属性把图片当作内容显示，也可以用 android:background 把图片当作背景显示。接下来，通过一个例子学习 ImageView 的使用，具体代码如下：

```
<LinearLayout xmlns:android="http://schemas.android.com/apk/res/android"
    android:layout_width="match_parent"
    android:layout_height="match_parent"
    android:orientation="vertical">

    <ImageView
        android:layout_width="wrap_content"
```

```
        android:layout_height="0dp"
        android:layout_weight="1"
        android:src="@drawable/bg"/>

    <ImageView
        android:layout_width="wrap_content"
        android:layout_height="0dp"
        android:layout_weight="1"
        android:background="@drawable/bg"/>

</LinearLayout>
```

在上述代码中，src/drawable 目录图片的引用使用@drawable 来找到文件夹，再以“/”加上图片的名称引用具体的图片。第一个 ImageView 标签使用 android:src 属性把图片当内容显示，第二个 ImageView 标签使用 android:background 属性把图片当背景显示。区别在于使用 android:src 属性显示的图片以原图大小显示，而使用 android:background 属性显示的图片会根据控件的大小进行伸缩。运行效果如图 3-7 所示。

图 3-7　ImageView 控件

当 ImageView 使用 android:src 属性显示图片时，可以配合使用 android:scaleType 属性设置显示的图片如何缩放。具体使用方式可以参考 Android 的官方文档，这里就不做详解了。

6. RadioButton 的使用

RadioButton(单选按钮)需要与 RadioGroup(单选组合框)配合使用，才能实现多个单选按钮间的互斥效果。接下来通过一个例子学习 RadioButton 的使用，具体代码如下：

```
<LinearLayout xmlns:android="http://schemas.android.com/apk/res/android"
    android:layout_width="match_parent"
    android:layout_height="match_parent"
    android:orientation="vertical">

    <TextView
        android:layout_width="match_parent"
        android:layout_height="wrap_content"
        android:text="请选择性别：" />
    <!-- RadioButton 结合 RadioGroup 实现单选效果-->
    <RadioGroup
        android:layout_width="match_parent"
        android:layout_height="wrap_content"
        android:orientation="horizontal">

        <RadioButton
            android:id="@+id/male"
            android:layout_width="wrap_content"
            android:layout_height="wrap_content"
            android:text="男" />

        <RadioButton
            android:id="@+id/female"
            android:layout_width="wrap_content"
            android:layout_height="wrap_content"
```

```
            android:checked="true"
            android:text="女" />
    </RadioGroup>
</LinearLayout>
```

图 3-8　RadioButton 控件

在上述代码中，单选组合框 RadioGroup 中加入了两个单选按钮(RadioButton，分别为：男、女)。单选按钮控件需要设置 id 属性；否则有可能不会出现互斥效果。单选按钮中 android:checked 属性用来设置是否选中，默认为 false。在第二个单选按钮中，设置了 android:checked 属性的值为 true。运行结果如图 3-8 所示。

7. CheckBox 的使用

CheckBox(复选框)又称为多选按钮，它允许用户一次选择多个选项。接下来通过一个例子学习 CheckBox 的使用方法，具体代码如下：

```
<LinearLayout xmlns:android="http://schemas.android.com/apk/res/android"
    android:layout_width="match_parent"
    android:layout_height="match_parent"
    android:orientation="vertical">

    ...

    <TextView
        android:layout_width="match_parent"
        android:layout_height="wrap_content"
        android:text="请选择爱好：" />

    <CheckBox
        android:id="@+id/cbBasketBall"
        android:layout_width="wrap_content"
        android:layout_height="wrap_content"
        android:text="篮球" />

    <CheckBox
        android:id="@+id/cbSing"
        android:layout_width="wrap_content"
        android:layout_height="wrap_content"
        android:checked="true"
        android:text="唱歌" />

    <CheckBox
        android:id="@+id/cbGame"
        android:layout_width="wrap_content"
        android:layout_height="wrap_content"
        android:text="游戏" />

    <CheckBox
        android:id="@+id/cbTour"
        android:layout_width="wrap_content"
        android:layout_height="wrap_content"
        android:checked="true"
        android:text="旅游" />
</LinearLayout>
```

在上述代码中，加入了 4 个 CheckBox，并使用 android: checked 属性将唱歌和旅游设置为选中效果，运行结果如图 3-9 所示。

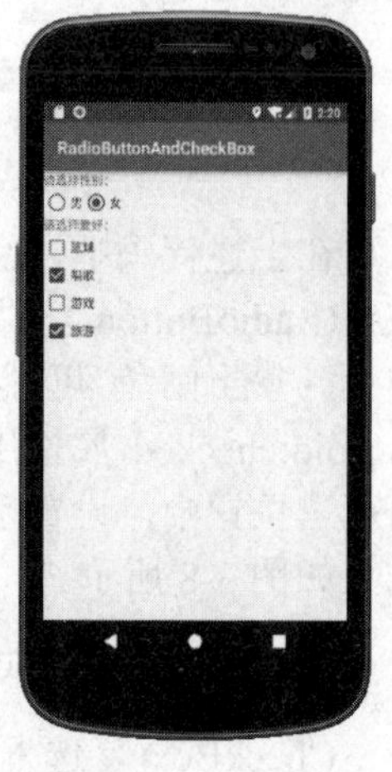

图 3-9　CheckBox 控件

8. 边距的使用

在 Android 的布局中，经常需要给控件和容器设置一些边距，如控件与控件之间的边距称为外边距，控件与其内容的边距称为内边距。可以使用属性 android: layout_margin 设置 4 个方向相同的外边距(margin)，也可以使用属性 android:layout_marginLeft、android:layout_marginTop、android:layout_marginRight、android:layout_marginBottom 分别设置左、上、右、下 4 个方向的外边距。同理，可以使用属性 android:padding 设置 4 个方向相同的内边距(padding)，也可以加上方向分别设置某个方向的内边距，如图 3-10 所示。

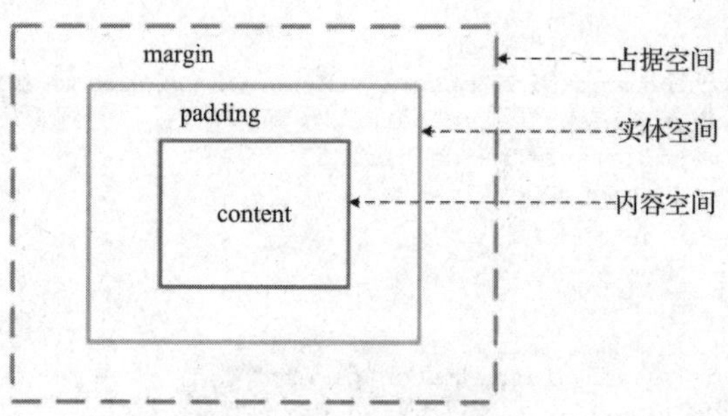

图 3-10　外边距与内边距图示

接下来通过一个例子学习外边距(margin)与内边距(padding)的使用方法，具体代码如下：

```
<LinearLayout xmlns:android="http://schemas.android.com/apk/res/android"
    android:layout_width="match_parent"
    android:layout_height="match_parent"
    android:orientation="vertical">

    <TextView
        android:layout_width="wrap_content"
        android:layout_height="wrap_content"
        android:background="#c0c0c0"
        android:gravity="center"
        android:padding="50dp"
        android:text="文本内容" />

    <TextView
        android:layout_width="wrap_content"
        android:layout_height="wrap_content"
        android:layout_margin="50dp"
        android:background="#c0c0c0"
        android:gravity="center"
        android:text="文本内容" />
</LinearLayout>
```

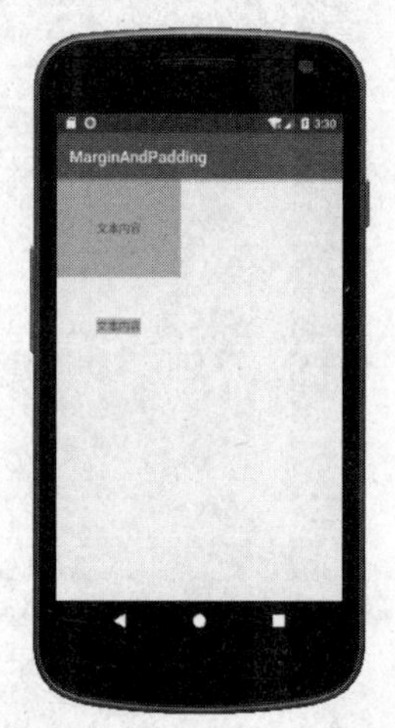

图 3-11　外边距与内边距

运行程序结果如图 3-11 所示。

从运行结果可以看出，第一个文本框我们设置了内边距为

50dp，控件的背景将包含内边距的范围。第二个文本框我们设置了外边距为 50dp，控件的背景不包含外边距的范围。

【任务实施】

1. 任务分析

本任务要求完成智慧园区登录界面的开发，界面的设计分析如下。

(1) 外层布局使用线性布局，设置背景为图片 bg_out.png，布局的内容居中显示。

(2) 内层布局使用线性布局，垂直显示控件，按效果图要求加入控件，并调整相应属性的值，完成效果如图 3-12 所示。

图 3-12 登录界面效果图

2. 任务实现

(1) 新建工程 Project3_1，将创建的 Activity 名称改为 LoginActivity，创建的布局名称改为 activity_login，如图 3-13 所示。

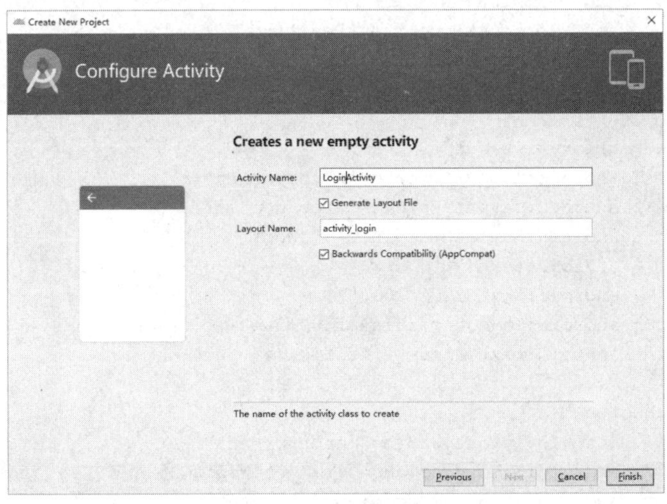

图 3-13 设置默认的 Activity 名称与布局名称

(2) 修改 activity_login.xml 布局文件的代码，具体代码如下。

```
<LinearLayout xmlns:android="http://schemas.android.com/apk/res/android"
    android:layout_width="match_parent"
    android:layout_height="match_parent"
    android:background="@drawable/bg_out"
    android:gravity="center"
    android:orientation="vertical">

    <LinearLayout
        android:layout_width="380dp"
        android:layout_height="560dp"
        android:background="@drawable/bg_login"
        android:gravity="center"
        android:orientation="vertical"
        android:padding="20dp">

        <ImageView
            android:layout_width="wrap_content"
            android:layout_height="wrap_content"
            android:src="@drawable/icon_smart_park" />

        <EditText
            android:layout_width="match_parent"
            android:layout_height="wrap_content"
            android:layout_marginTop="15dp"
            android:background="@drawable/input_normal"
            android:hint="请输入账号"
            android:paddingLeft="15dp"
            android:textSize="25sp" />

        <EditText
            android:layout_width="match_parent"
            android:layout_height="wrap_content"
            android:layout_marginTop="15dp"
            android:background="@drawable/input_normal"
            android:hint="请输入密码"
            android:inputType="textPassword"
            android:paddingLeft="15dp"
            android:textSize="25sp" />

        <CheckBox
            android:layout_width="match_parent"
            android:layout_height="wrap_content"
            android:layout_marginTop="15dp"
            android:text="记住密码"
            android:textSize="20sp"
            android:layout_marginLeft="15dp"
            android:textColor="#ffffff" />

        <Button
            android:layout_width="match_parent"
            android:layout_height="wrap_content"
            android:layout_marginTop="15dp"
            android:background="@drawable/btn_login_normal" />
```

```

        <Button
            android:layout_width="match_parent"
            android:layout_height="wrap_content"
            android:layout_marginTop="15dp"
            android:background="@drawable/btn_register_normal" />
    </LinearLayout>
</LinearLayout>
```

3. 运行结果

将应用程序在工控平板上运行，运行结果如图 3-14 所示。

图 3-14　登录界面

任务 2　监控系统界面开发

【任务描述】

本任务要求使用相对布局或约束布局以及相应控件完成智慧园区监控系统界面的开发。任务清单如表 3-4 所示。

扫码观看视频讲解

表 3-4　任务清单

任务课时	4 课时	任务组员数量	建议 1 人
任务组采用设备	PC 一台 模拟器或真机或工控平板		

【知识解析】

1. RelativeLayout(相对布局)的使用

相对布局是通过相对定位的方式指定控件的位置，即以其他控件或父容器为参照物，摆放控件位置。在设计相对布局时要遵循控件之间的依赖关系，后放入控件的位置依赖于先放入的控件。

相对布局的属性较多，但都是有规律的，下面介绍相对布局控件摆放位置的属性。

相对布局相对父容器定位的属性如表 3-5 所示。

表 3-5　设置控件相对父容器的位置

控件属性	功能描述
android:layout_centerHorizontal	设置当前控件位于父容器的水平居中位置
android:layout_centerVertical	设置当前控件位于父容器的垂直居中位置
android:layout_centerInParent	设置当前控件位于父容器的中央位置
android:layout_alignParentTop	设置当前控件与父容器顶部对齐
android:layout_alignParentLeft	设置当前控件与父容器左对齐
android:layout_alignParentRight	设置当前控件与父容器右对齐
android:layout_alignParentBottom	设置当前控件与父容器底部对齐

相对某个控件定位的属性如表 3-6 所示。

表 3-6　设置控件相对某个控件的位置

控件属性	功能描述
android:layout_above	设置当前控件位于某个控件的上方
android:layout_below	设置当前控件位于某个控件的下方
android:layout_toLeftOf	设置当前控件位于某个控件的左侧
android:layout_toRightOf	设置当前控件位于某个控件的右侧
android:layout_alignTop	设置当前控件与某个控件顶部对齐
android:layout_alignBottom	设置当前控件与某个控件底部对齐
android:layout_alignLeft	设置当前控件与某个控件左对齐
android:layout_alignRight	设置当前控件与某个控件右对齐

接下来通过一个例子来学习相对布局的使用，具体代码如下：

```
<RelativeLayout xmlns:android="http://schemas.android.com/apk/res/android"
    android:layout_width="match_parent"
    android:layout_height="match_parent">

    <Button
        android:id="@+id/button1"
        android:layout_width="wrap_content"
        android:layout_height="wrap_content"
        android:text="按钮 1"/>

    <Button
        android:id="@+id/button2"
        android:layout_width="wrap_content"
```

```
            android:layout_height="wrap_content"
            android:layout_alignParentRight="true"
            android:text="按钮 2"/>

        <Button
            android:id="@+id/button3"
            android:layout_width="wrap_content"
            android:layout_height="wrap_content"
            android:layout_alignParentBottom="true"
            android:text="按钮 3"/>

        <Button
            android:id="@+id/button4"
            android:layout_width="wrap_content"
            android:layout_height="wrap_content"
            android:layout_alignParentRight="true"
            android:layout_alignParentBottom="true"
            android:text="按钮 4"/>

        <Button
            android:id="@+id/button5"
            android:layout_width="wrap_content"
            android:layout_height="wrap_content"
            android:layout_centerInParent="true"
            android:text="按钮 5"/>

        <Button
            android:id="@+id/button6"
            android:layout_width="wrap_content"
            android:layout_height="wrap_content"
            android:layout_toLeftOf="@id/button5"
            android:layout_alignBottom="@id/button5"
            android:text="按钮 6"/>

        <Button
            android:id="@+id/button7"
            android:layout_width="wrap_content"
            android:layout_height="wrap_content"
            android:layout_toRightOf="@id/button5"
            android:layout_alignBottom="@id/button5"
            android:text="按钮 7"/>

        <Button
            android:id="@+id/button8"
            android:layout_width="wrap_content"
            android:layout_height="wrap_content"
            android:layout_above="@id/button5"
            android:layout_alignLeft="@id/button5"
            android:text="按钮 8"/>

        <Button
            android:id="@+id/button9"
            android:layout_width="wrap_content"
            android:layout_height="wrap_content"
            android:layout_below="@id/button5"
            android:layout_alignLeft="@id/button5"
            android:text="按钮 9"/>
    </RelativeLayout>
```

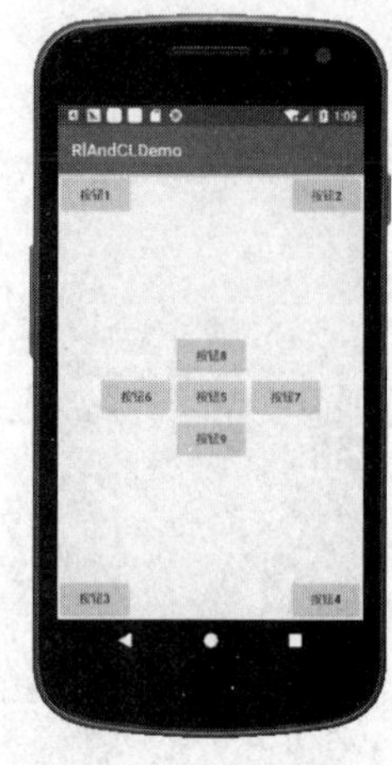

图 3-15　相对布局

在上述代码中，相对布局中默认加入的控件从左上角显示，故按钮 1 显示在布局的左上角。按钮 2 至按钮 5 分别设置了相对父容器的位置。按钮 6 至按钮 9 分别设置了相对控件按钮 5 的位置。预览效果如图 3-15 所示。

2. FrameLayout(帧布局)的使用

帧布局是 Android 中最为简单的一种布局，该布局为每个增加的组件创建一个空白的区域(称为一帧)。在帧布局中，所有控件都默认显示在屏幕左上角，并按照放入的先后顺序重叠摆放，先放入的控件显示在最底层，后放入的控件显示在最顶层。接下来通过一个例子学习帧布局的使用，具体代码如下：

```
<FrameLayout xmlns:android="http://schemas.android.com/apk/res/android"
    android:layout_width="match_parent"
    android:layout_height="match_parent">

    <View
        android:layout_width="200dp"
        android:layout_height="200dp"
        android:background="#ff0000"/>

    <View
        android:layout_width="150dp"
        android:layout_height="150dp"
        android:background="#00ff00"/>

    <View
        android:layout_width="100dp"
        android:layout_height="100dp"
        android:background="#0000ff"/>
</FrameLayout>
```

帧布局预览效果如图 3-16 所示。

从效果图中可以看出，帧布局加入的控件按先后顺序显示，该例子中后加入的控件比先加入的控件范围小，故可以看到层叠效果。若后加入的控件比先加入的控件范围大，则先加入的控件将被覆盖。

图 3-16　帧布局预览效果

3. ConstraintLayout(约束布局)的使用

ConstraintLayout 是 Google 在 2016 年的 Google I/O 大会上提出的一个可以灵活控制子控件位置和大小的新布局。它已是目前 Android Studio 中项目的默认布局。

在布局设计中，布局不宜嵌套太多层，否则会严重影响软件的性能。而约束布局提供了强大的功能，可以尽量避免开发者使用嵌套布局完成复杂界面的设计。

使用约束布局需要在 app module 的 build.gradle 中添加依赖，早在 Android Studio 3.2.0 中创建项目时已经自动添加了，如图 3-17 所示。

```
dependencies {
    implementation fileTree(dir: 'libs', include: ['*.jar'])
    implementation 'com.android.support:appcompat-v7:28.0.0'
    implementation 'com.android.support.constraint:constraint-layout:1.1.3'
    testImplementation 'junit:junit:4.12'
    androidTestImplementation 'com.android.support.test:runner:1.0.2'
    androidTestImplementation 'com.android.support.test.espresso:espresso-core:3.0.2'
}
```

图 3-17　约束布局配置

接下来介绍约束布局的几种常见用法。

1)　Relative positioning(相对定位)

约束布局中的相对定位与相对布局中的相对定位有些相似，控件是相对另一个位置的约束，如图 3-18 所示。

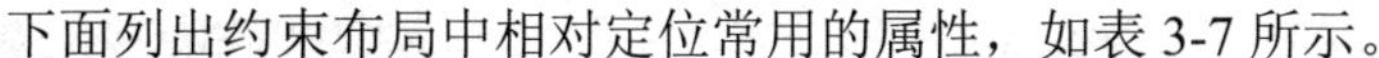

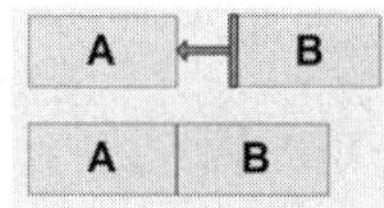

图 3-18　相对定位

下面列出约束布局中相对定位常用的属性，如表 3-7 所示。

表 3-7　约束布局中相对定位常用属性

控件属性	功能描述
app:layout_constraintLeft_toLeftOf	控件与某控件或父容器的左侧对齐
app:layout_constraintLeft_toRightOf	控件在某控件的右侧
app:layout_constraintRight_toLeftOf	控件在某控件的左侧
app:layout_constraintRight_toRightOf	控件与某控件或父容器的右侧对齐
app:layout_constraintTop_toTopOf	控件与某控件或父容器的顶部对齐
app:layout_constraintTop_toBottomOf	控件在某控件的下方
app:layout_constraintBottom_toTopOf	控件在某控件的上方
app:layout_constraintBottom_toBottomOf	控件与某控件或父容器的底部对齐
app:layout_constraintBaseline_toBaselineOf	控件与某控件文本基线对齐

接下来通过一个例子学习约束布局中相对定位的使用，具体代码如下：

```
<android.support.constraint.ConstraintLayout xmlns:android="http://schemas.
android.com/apk/res/android"
    xmlns:app="http://schemas.android.com/apk/res-auto"
    android:layout_width="match_parent"
    android:layout_height="match_parent">

    <TextView
        android:id="@+id/tv1"
        android:layout_width="150dp"
        android:layout_height="150dp"
        android:background="#e7e9e5"
        android:gravity="center"
        android:text="TextView1"
        android:textSize="25sp" />

    <TextView
        android:id="@+id/tv2"
        android:layout_width="wrap_content"
        android:layout_height="wrap_content"
```

```
        android:background="#8f8c8c"
        app:layout_constraintBaseline_toBaselineOf="@id/tv1"
        android:text="TextView2"
        android:textSize="25sp"
        app:layout_constraintLeft_toRightOf="@+id/tv1" />

    <TextView
        android:layout_width="wrap_content"
        android:layout_height="wrap_content"
        android:textSize="25sp"
        app:layout_constraintTop_toBottomOf="@id/tv1"
        app:layout_constraintRight_toRightOf="@id/tv1"
        android:background="#8f8c8c"
        android:text="TextView3"/>
</android.support.constraint.ConstraintLayout>
```

在上述代码中设置了 TextView2 显示在 TextView1 的右侧，并且文本基线对齐。TextView3 在 TextView 的下方，并且右侧对齐，相对定位预览效果如图 3-19 所示。

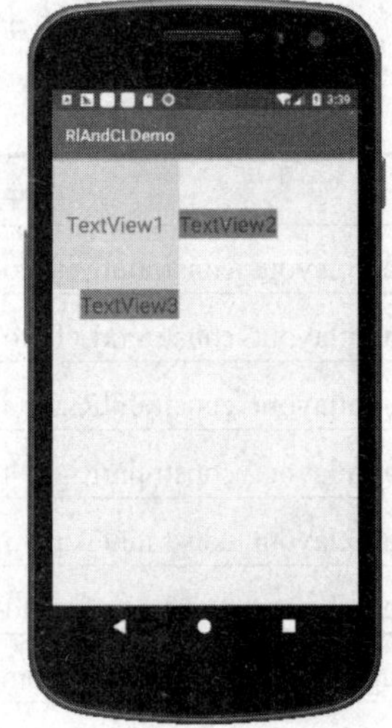

图 3-19　相对定位效果

多学一招：约束布局的边距

约束布局的边距常用属性与其他布局名称一样，但是需注意约束布局在设置边距前要指定约束对象。例如，控件距离父布局左边 30dp，要约束控件在 parent 的左边，具体代码如下：

```
<Button
    android:layout_width="wrap_content"
    android:layout_height="wrap_content"
    android:layout_marginLeft="30dp"
    app:layout_constraintLeft_toLeftOf="parent"
    android:text="BUTTON"/>
```

2)　Circular positioning(圆形定位)

圆形定位指的是可以用一个角度和一个半径来约束两个控件的中心，如图 3-20 所示。

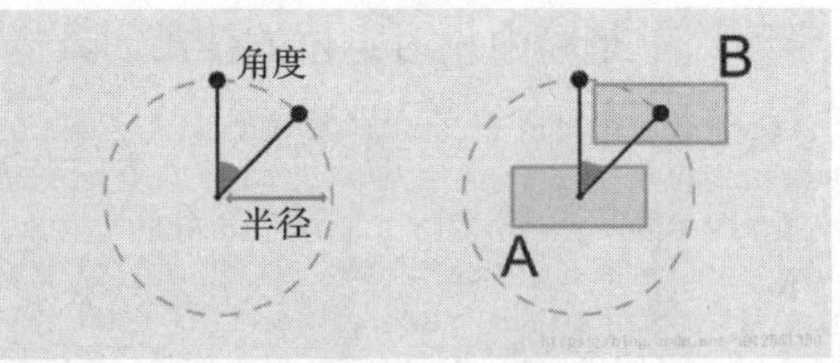

图 3-20　圆形定位

下面列出约束布局中圆形定位使用的属性，如表 3-8 所示。

表 3-8　约束布局中圆形定位属性

控件属性	功能描述
app:layout_constraintCircle	设置约束的圆心位置
app:layout_constraintCircleAngle	设置约束角度
app:layout_constraintCircleRadius	设置约束半径

接下来通过一个例子学习圆形定位的使用，具体代码如下：

```
<android.support.constraint.ConstraintLayout xmlns:android="http://schemas.android.com/apk/res/android"
    android:layout_width="match_parent"
    android:layout_height="match_parent"
    xmlns:app="http://schemas.android.com/apk/res-auto">

    <Button
        android:id="@+id/button1"
        android:layout_width="wrap_content"
        android:layout_height="wrap_content"
        android:text="BUTTON1"/>

    <Button
        android:id="@+id/button2"
        android:layout_width="wrap_content"
        android:layout_height="wrap_content"
        app:layout_constraintCircle="@id/button1"
        app:layout_constraintCircleAngle="120"
        app:layout_constraintCircleRadius="150dp"
        android:text="BUTTON2"/>
</android.support.constraint.ConstraintLayout>
```

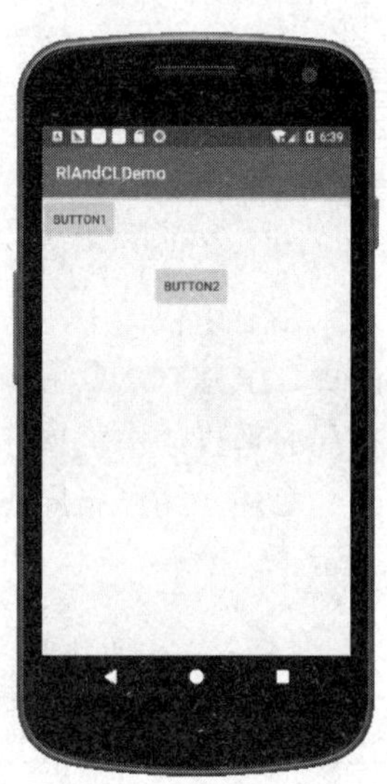

图 3-21 圆形定位效果

在上述代码中，BUTTON2 位置以 BUTTON1 为圆心，正上方沿顺时针方向旋转 120° 并距离 BUTTON1 中心 150dp 位置，圆形定位预览效果如图 3-21 所示。

3) Centering positioning and bias(居中以及设置偏差)

在相对布局中，把控件放在布局中央的方法是把 layout_centerInParent 属性值设为 true。而在约束布局中，把控件放在布局中央，需要设置 4 个方向的约束，写法如下：

```
app:layout_constraintTop_toTopOf="parent"
app:layout_constraintBottom_toBottomOf="parent"
app:layout_constraintLeft_toLeftOf="parent"
app:layout_constraintRight_toRightOf="parent"
```

同理，如果要设置控件在布局的水平居中，需要设置控件左侧约束在父容器的左侧为 app:layout_constraintLeft_toLeftOf="parent"，控件右侧约束在父容器的右侧为 app:layout_constraintRight_toRightOf="parent"。设置控件垂直居中，只需要设置上下方向的约束即可。接下来通过一个例子学习约束布局设置控件的居中方式，具体代码如下：

```
<android.support.constraint.ConstraintLayout xmlns:android="http://schemas.android.com/apk/res/android"
    android:layout_width="match_parent"
    android:layout_height="match_parent"
    xmlns:app="http://schemas.android.com/apk/res-auto">

    <Button
        android:id="@+id/button1"
        app:layout_constraintLeft_toLeftOf="parent"
        app:layout_constraintRight_toRightOf="parent"
        app:layout_constraintTop_toTopOf="parent"
        app:layout_constraintBottom_toBottomOf="parent"
```

```
        android:layout_width="wrap_content"
        android:layout_height="wrap_content"
        android:text="BUTTON1"/>

    <Button
        android:layout_width="wrap_content"
        android:layout_height="wrap_content"
        app:layout_constraintLeft_toLeftOf="parent"
        app:layout_constraintRight_toRightOf="parent"
        android:text="BUTTON2"/>

    <Button
        android:layout_width="wrap_content"
        android:layout_height="wrap_content"
        app:layout_constraintTop_toTopOf="parent"
        app:layout_constraintBottom_toBottomOf="parent"
        android:text="BUTTON3"/>
</android.support.constraint.ConstraintLayout>
```

在上述代码中，我们设置了 BUTTON1 显示在布局的中央，BUTTON2 在布局的水平居中位置，BUTTON3 在布局的垂直居中位置，预览效果如图 3-22 所示。

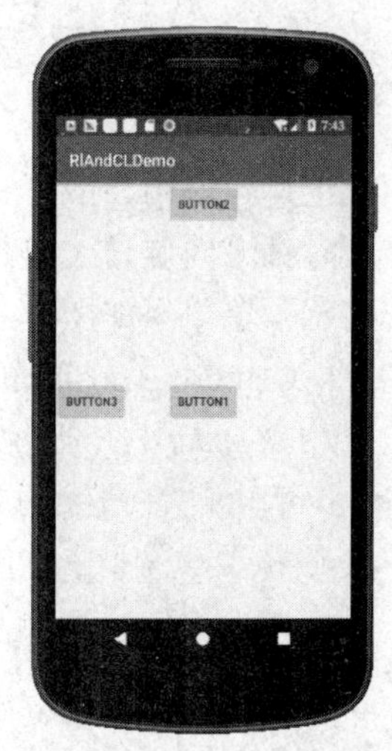

图 3-22　约束布局设置居中位置

此外，约束布局还提供了设置偏移的属性，如表 3-9 所示。

表 3-9　约束布局中的偏移属性

控件属性	功能描述
app:layout_constraintHorizontal_bias	水平偏移
app:layout_constraintVertical_bias	垂直偏移

偏移属性的取值范围在 0～1 之间，0 表示控件在布局的最左侧，1 表示控件在布局的最右侧，0～1 之间的浮点值表示偏移量。接下来通过一个例子学习偏移的使用，具体代码如下：

```
<android.support.constraint.ConstraintLayout xmlns:android="http://schemas.android.com/apk/res/android"
    xmlns:app="http://schemas.android.com/apk/res-auto"
    android:layout_width="match_parent"
    android:layout_height="match_parent">

    <Button
        android:layout_width="wrap_content"
        android:layout_height="wrap_content"
        android:text="BUTTON1"
        app:layout_constraintHorizontal_bias="0.3"
        app:layout_constraintLeft_toLeftOf="parent"
        app:layout_constraintRight_toRightOf="parent" />

    <Button
        android:layout_width="wrap_content"
```

```
        android:layout_height="wrap_content"
        android:text="BUTTON2"
        app:layout_constraintBottom_toBottomOf="parent"
        app:layout_constraintTop_toTopOf="parent"
        app:layout_constraintVertical_bias="0.3" />
</android.support.constraint.ConstraintLayout>
```

在上述代码中，给 BUTTON1 设置了水平偏移 0.3，那么它将显示在布局水平居中偏左的位置。同理，BUTTON2 设置了垂直偏移 0.3，那么它将显示在布局垂直居中偏上的位置。预览效果如图 3-23 所示。

4) Chains(链式约束)

链式约束，即一组控件通过一个双向的约束关系链接起来。链式约束能够对一组在水平或竖直方向互相关联控件的属性进行统一管理，如图 3-24 所示。

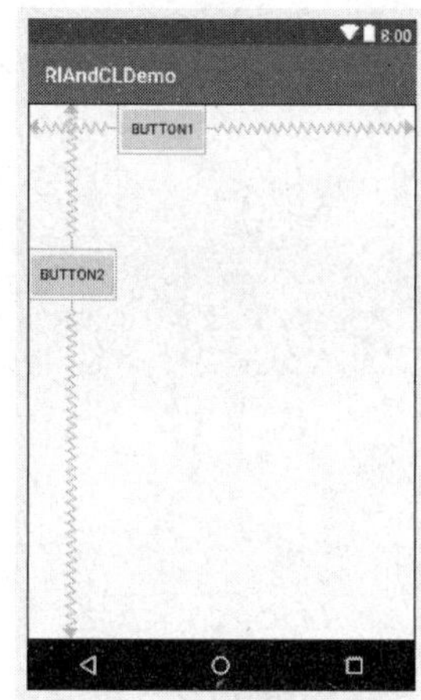

图 3-23 约束布局中的偏移

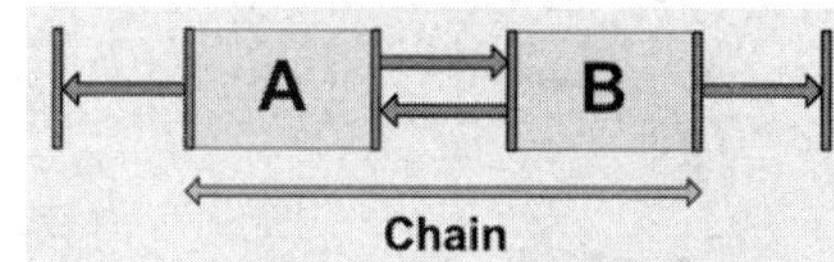

图 3-24 链式约束

接下来通过一个例子学习链式约束的简单使用，具体代码如下：

```
<android.support.constraint.ConstraintLayout xmlns:android="http://schemas.android.com/apk/res/android"
    android:layout_width="match_parent"
    android:layout_height="match_parent"
    xmlns:app="http://schemas.android.com/apk/res-auto">

    <Button
        android:id="@+id/btnA"
        android:layout_width="wrap_content"
        android:layout_height="wrap_content"
        app:layout_constraintLeft_toLeftOf="parent"
        app:layout_constraintRight_toLeftOf="@id/btnB"
        android:text="A"/>
    <Button
        android:id="@+id/btnB"
        android:layout_width="wrap_content"
        android:layout_height="wrap_content"
        app:layout_constraintLeft_toRightOf="@id/btnA"
        app:layout_constraintRight_toLeftOf="@id/btnC"
        android:text="B"/>
```

```
    <Button
        android:id="@+id/btnC"
        android:layout_width="wrap_content"
        android:layout_height="wrap_content"
        app:layout_constraintLeft_toRightOf="@id/btnB"
        app:layout_constraintRight_toRightOf="parent"
        android:text="C"/>
</android.support.constraint.ConstraintLayout>
```

在上述代码中，按钮 A 左侧约束在父容器的左侧，其右侧约束在按钮 B 的左侧。按钮 B 的左侧约束在按钮 A 的右侧，其右侧约束在按钮 C 的左侧。按钮 C 的左侧约束在按钮 B 的右侧，其右侧约束在父容器的右侧。这样，按钮 A、按钮 B、按钮 C 在父容器的范围两两之间形成约束，组成一条链。预览效果如图 3-25 所示。

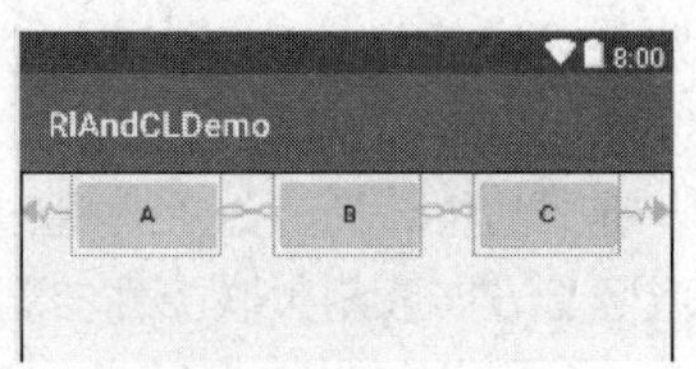

图 3-25　链式约束默认样式

链式约束除了上述的默认样式外，还提供了其他一些样式可供选择，如图 3-26 所示。

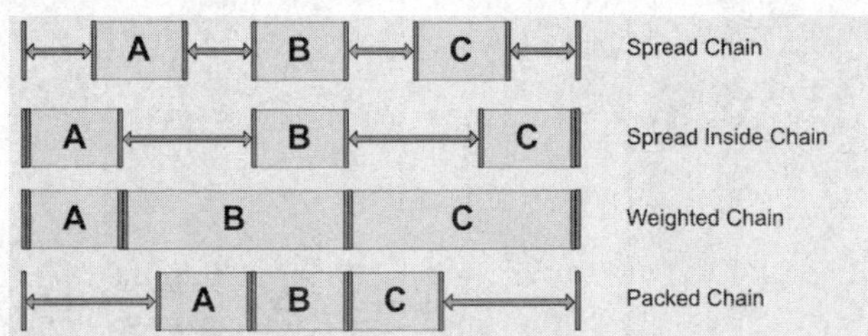

图 3-26　链式约束样式

一条链的第一个控件是这条链的链头，可以在链头中设置 layout_constraintHorizontal_chainStyle 来改变整条链的样式，该属性提供了 3 种样式，分别如下。

- CHAIN_SPREAD —— 展开元素 (默认)。
- CHAIN_SPREAD_INSIDE —— 展开元素，但链的两端贴近 parent。
- CHAIN_PACKED —— 链的元素将被打包在一起。

除了上述在链头设置链的样式外，还可以创建一个权重链。若需设置水平方向的权重链，则每个控件的宽度需要设置为 0dp，并加入属性 layout_constraintHorizontal_weight 设置宽度的权重。同理，垂直方向的权重链，每个控件高度需设置为 0dp，并加入属性 layout_constraintVertical_weight 设置高度的权重。

【任务实施】

1. 任务分析

本任务要求完成园区监控系统界面的开发(标题栏未实现)，界面的设计分析如下。

(1) 外层布局使用线性布局，设置背景为图片 bg_out.png，布局的内容居中显示。

(2) 右侧方向控制使用相对布局或约束布局(本次任务实现使用相对布局)，其余控件使用线性或相对布局都可以，完成效果如图 3-27 所示。

图 3-27　园区监控系统界面效果图

2. 任务实现

新建工程 Project3_2，修改 activity_main.xml 中的布局代码，具体代码如下：

```
<LinearLayout xmlns:android="http://schemas.android.com/apk/res/android"
    android:layout_width="match_parent"
    android:layout_height="match_parent"
    android:background="@drawable/bg_out"
    android:orientation="horizontal">

    <LinearLayout
        android:layout_width="0dp"
        android:layout_height="match_parent"
        android:layout_weight="3"></LinearLayout>

    <LinearLayout
        android:layout_width="0dp"
        android:layout_height="match_parent"
        android:layout_weight="1"
        android:background="#e0000000"
        android:gravity="center_horizontal"
        android:orientation="vertical">

        <RelativeLayout
            android:layout_width="180dp"
            android:layout_height="180dp"
            android:layout_marginTop="20dp"
            android:background="@drawable/bg_direction">

            <ImageView
                android:id="@+id/ivLeft"
                android:layout_width="wrap_content"
                android:layout_height="wrap_content"
                android:layout_alignParentLeft="true"
                android:layout_centerVertical="true"
                android:layout_marginLeft="10dp"
                android:src="@drawable/arrow_left_white" />
```

```

            <ImageView
                android:id="@+id/ivRight"
                android:layout_width="wrap_content"
                android:layout_height="wrap_content"
                android:layout_alignParentRight="true"
                android:layout_centerVertical="true"
                android:layout_marginRight="10dp"
                android:src="@drawable/arrow_right_white" />

            <ImageView
                android:id="@+id/ivTop"
                android:layout_width="wrap_content"
                android:layout_height="wrap_content"
                android:layout_alignParentTop="true"
                android:layout_centerHorizontal="true"
                android:layout_marginTop="10dp"
                android:src="@drawable/arrow_up_white" />

            <ImageView
                android:id="@+id/ivBottom"
                android:layout_width="wrap_content"
                android:layout_height="wrap_content"
                android:layout_alignParentBottom="true"
                android:layout_centerHorizontal="true"
                android:layout_marginBottom="10dp"
                android:src="@drawable/arrow_down_white" />
        </RelativeLayout>


        <Button
            android:id="@+id/btnOpenOrClose"
            android:layout_width="180dp"
            android:layout_height="wrap_content"
            android:layout_marginTop="20dp"
            android:background="#266fde"
            android:button="@null"
            android:drawableLeft="@drawable/ico_start"
            android:padding="10dp"
            android:text="开始监控"
            android:textColor="#fff" />

        <Button
            android:id="@+id/btnPeople"
            android:layout_width="180dp"
            android:layout_height="wrap_content"
            android:layout_marginTop="40dp"
            android:background="#d64646"
            android:button="@null"
            android:drawableLeft="@drawable/pic_icon_smoke"
            android:padding="10dp"
            android:text="无人"
            android:textColor="#fff" />

        <CheckBox
            android:id="@+id/cb_red"
            android:layout_width="120dp"
            android:layout_height="120dp"
```

```
                android:layout_marginTop="20dp"
                android:background="@drawable/lamp_off"
                android:button="@null"
                android:clickable="false" />

            <TextView
                android:layout_width="wrap_content"
                android:layout_height="wrap_content"
                android:layout_marginTop="20dp"
                android:text="三色灯-红灯"
                android:textColor="#fff" />
        </LinearLayout>
</LinearLayout>
```

3. 运行结果

将应用程序在工控平板上运行，运行结果如图 3-28 所示。

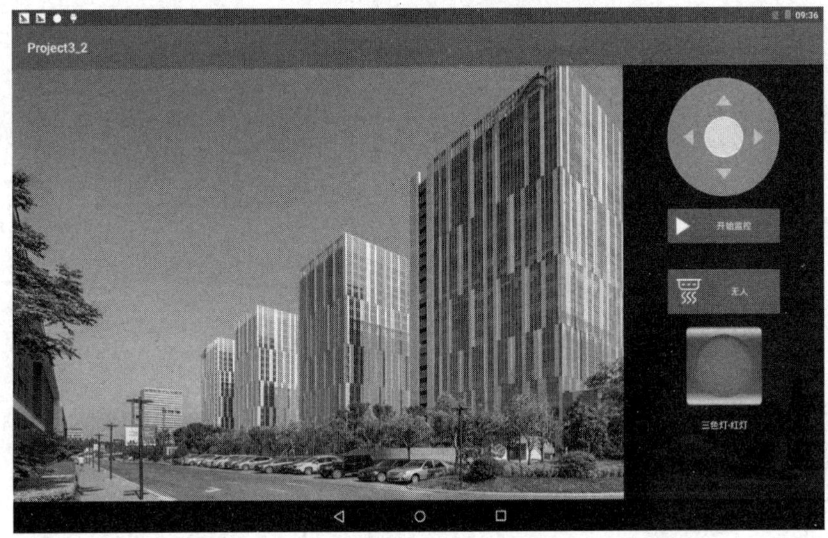

图 3-28　园区监控系统界面

任务 3　环境采集系统阈值设置对话框开发

【任务描述】

本任务要求使用自定义对话框以及 SeekBar 完成智慧园区环境采集系统阈值设置对话框的开发。在此对话框中能够通过拖动 SeekBar 实现进度值的改变。任务清单如表 3-10 所示。

扫码观看视频讲解

表 3-10　任务清单

任务课时	4 课时	任务组员数量	建议 1 人
任务组采用设备	PC 一台 模拟器或真机或工控平板		

【知识解析】

1. 常见对话框的简单使用

在 Android 的界面中，有一种以小窗口的形式显示在 Activity(活动界面)上的组件，称其为 Dialog(对话框)。对话框是程序与用户交互的一种重要方式，可以使用 Dialog 的子类 AlertDialog 来创建一些简单的对话框，如提示信息对话框、单选对话框、多选对话框等。也可以使用自定义类继承 Dialog 来设计自定义风格的对话框。接下来通过一个例子来学习对话框的简单使用。

首先修改 activity_main.xml 文件的布局代码，具体代码如下：

```
<LinearLayout xmlns:android="http://schemas.android.com/apk/res/android"
    android:layout_width="match_parent"
    android:layout_height="match_parent"
    android:orientation="vertical">
    <Button
        android:id="@+id/btnShowMsgDlg"
        android:layout_width="match_parent"
        android:layout_height="wrap_content"
        android:text="提示信息对话框"/>

    <Button
        android:id="@+id/btnShowSingleDlg"
        android:layout_width="match_parent"
        android:layout_height="wrap_content"
        android:text="单选对话框"/>

    <Button
        android:id="@+id/btnShowMultiDlg"
        android:layout_width="match_parent"
        android:layout_height="wrap_content"
        android:text="多选对话框"/>

    <Button
        android:id="@+id/btnShowCustomDlg"
        android:layout_width="match_parent"
        android:layout_height="wrap_content"
        android:text="自定义对话框"/>
</LinearLayout>
```

接着在 MainActivity 中分别对 4 个按钮注册点击监听器，具体代码如下：

```
public class MainActivity extends AppCompatActivity implements View.OnClickListener {
    //显示提示信息对话框按钮
    private Button btnShowMsgDlg;
    //显示单选对话框按钮
    private Button btnShowSingleDlg;
    //显示多选对话框按钮
    private Button btnShowMultiDlg;
    //显示自定义对话框按钮
    private Button btnShowCustomDlg;

    @Override
    protected void onCreate(Bundle savedInstanceState) {
        super.onCreate(savedInstanceState);
        setContentView(R.layout.activity_main);

```

```
            //初始化视图以及监听器
            initView();
        }

        //初始化视图以及监听器
        private void initView() {
            //初始化控件
            btnShowMsgDlg = findViewById(R.id.btnShowMsgDlg);
            btnShowSingleDlg = findViewById(R.id.btnShowSingleDlg);
            btnShowMultiDlg = findViewById(R.id.btnShowMultiDlg);
            btnShowCustomDlg = findViewById(R.id.btnShowCustomDlg);

            //注册按钮监听器
            btnShowMsgDlg.setOnClickListener(this);
            btnShowSingleDlg.setOnClickListener(this);
            btnShowMultiDlg.setOnClickListener(this);
            btnShowCustomDlg.setOnClickListener(this);
        }


        @Override
        public void onClick(View v) {
            switch (v.getId()) {
                case R.id.btnShowMsgDlg: //显示提示信息对话框按钮
                    showMsgDialog();
                    break;
                case R.id.btnShowSingleDlg: //显示单选对话框按钮
                    showSingleChoiceDialog();
                    break;
                case R.id.btnShowMultiDlg:  //显示多选对话框按钮
                    showMultiChoiceDialog();
                    break;
                case R.id.btnShowCustomDlg: //显示自定义对话框按钮
                    showCustomDialog();
                    break;
            }
        }
        //显示提示消息对话框
        private void showMsgDialog() {
        }
        //显示单选对话框
        private void showSingleChoiceDialog() {
        }
        //显示多选对话框
        private void showMultiChoiceDialog() {
        }
        //显示自定义对话框
        private void showCustomDialog() {
        }
}
```

1) 提示信息对话框

提示信息对话框主要是用来显示提示信息的，并通常具有“确定”和“取消”按钮，接下来在 MainActivity 类中的 showMsgDialog()方法实现提示对话框的显示，具体代码如下：

```
//显示提示消息对话框
private void showMsgDialog() {
    //创建 AlertDialog 构造器 Builder 对象，AlertDialog 建议使用 android.support.v7.app 包下的
    AlertDialog.Builder builder = new AlertDialog.Builder(this);
```

```
        //设置对话框标题
        builder.setTitle("提示信息对话框");
        //设置提示信息
        builder.setMessage("是否确定退出！");
        //设置对话框图标
        builder.setIcon(R.mipmap.ic_launcher);
        //添加确定按钮
        builder.setPositiveButton("确定", new DialogInterface.OnClickListener() {
            @Override
            public void onClick(DialogInterface dialog, int which) {
                //添加确定按钮点击的处理代码
                Toast.makeText(MainActivity.this, "点击了确定！",
Toast.LENGTH_SHORT).show();
            }
        });
        //添加取消按钮
        builder.setNegativeButton("取消",null);
        //创建并显示对话框
        builder.show();
}
```

在上述代码中，首先创建了 AlertDialog 的构造器 Builder，并通过 Builder 设置对话框的标题、提示消息、图标、“确定”按钮以及“取消”按钮，最后调用 Builder 对象的 show()方法创建并显示对话框。需要注意的是，添加对话框确认按钮的方法 setPositiveButton()，第一个参数为按钮显示的信息，第二个参数为按钮的监听器，如果不需要处理，可以填 null。同理，setNegativeButton()方法的设置也一样。运行应用程序，并点击提示信息对话框按钮，程序运行效果如图 3-29 所示。

图 3-29　提示信息对话框

2)　单选对话框

单选对话框一般用于从选项中选择一项，它是通过 AlertDialog 的构造器 Builder 对象调用 setSingleChoiceItems()方法设置的，接下来在 MainActivity 类中的 showSingleChoiceDialog()方法实现单选对话框的显示，具体代码如下：

```
//显示单选对话框
private void showSingleChoiceDialog() {
    //创建AlertDialog构造器Builder对象，AlertDialog建议使用android.support.v7.app包下的
    AlertDialog.Builder builder = new AlertDialog.Builder(this);
    //设置对话框标题
    builder.setTitle("请选择性别");
    //设置对话框图标
    builder.setIcon(R.mipmap.ic_launcher);
    final String[] sexs = new String[]{"男", "女"};
    //设置单选选项
    builder.setSingleChoiceItems(sexs, 0, new DialogInterface.OnClickListener() {
        @Override
        public void onClick(DialogInterface dialog, int which) {
            Toast.makeText(MainActivity.this, "您选择了："+sexs[which],
Toast.LENGTH_SHORT).show();
        }
    });
```

```
        //添加确定按钮
        builder.setPositiveButton("确定", null);
        //创建并显示对话框
        builder.show();
}
```

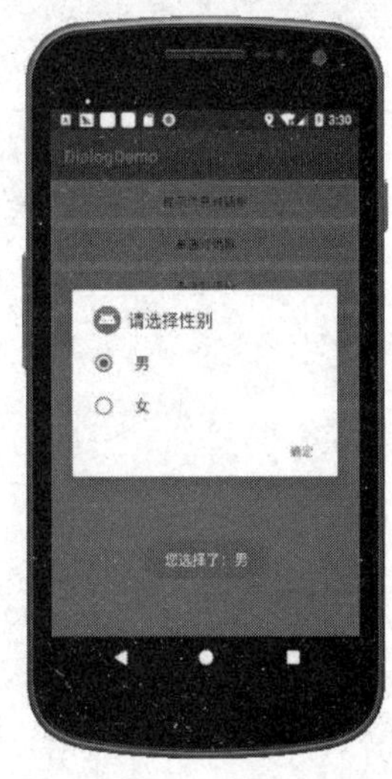

图 3-30　单选对话框

在上述代码中，setSingleChoiceItems()方法需要传入 3 个参数，第一个参数是选项数组，用于显示选项内容；第二个参数是设置默认选中的选项，0 表示选中第一个选项，如果没有默认选中的选项，参数填-1；第三个参数是选项点击的监听器。运行程序，并点击单选对话框按钮，运行效果如图 3-30 所示。

3)　多选对话框

多选对话框是用作从选项中选择多项，它是通过 AlertDialog 的构造器 Builder 对象调用 setMultiChoiceItems()方法设置的，接下来在 MainActivity 类中的 showMultiChoiceDialog()方法实现多选对话框的显示，具体代码如下：

```
//显示多选对话框
private void showMultiChoiceDialog() {
    //创建 AlertDialog 构造器 Builder 对象，AlertDialog 建议使用 android.support.v7.app 包下的
    AlertDialog.Builder builder = new AlertDialog.Builder(this);
    //设置对话框标题
    builder.setTitle("请选择传感器");
    //设置对话框图标
    builder.setIcon(R.mipmap.ic_launcher);
    final String[] sensors = new String[]{"温湿度传感器", "光照传感器","CO2 传感器", "风速传感器"};
    //设置多选选项
    builder.setMultiChoiceItems(sensors, new boolean[]{false,true,true,false}, new DialogInterface.OnMultiChoiceClickListener() {
        @Override
        public void onClick(DialogInterface dialog, int which, boolean isChecked) {

        }
    });
    //添加确定按钮
    builder.setPositiveButton("确定", null);
    //创建并显示对话框
    builder.show();
}
```

图 3-31　多选对话框

在上述代码中，setMultiChoiceItems()方法需要传入 3 个参数：①选项数组，用于显示选项内容；②boolean 数组，用于设置默认选中的选项，new booelan[]{false,true,true,false}表示默认选中第二个与第三个选项，如果没有默认选中的选项，参数填 null；③选项点击的监听器。运行程序，并点击多选对话框按钮，运行效果如图 3-31 所示。

4)　自定义对话框

在 Android 项目中，为了提高用户体验，当系统提供的对话框已不能满足使用要求时，可以根据项目需求自定义对话框

样式。接下来通过一个例子演示自定义对话框的使用。

(1) 创建一个自定义对话框的布局，这里命名为 my_dialog，布局代码如下：

```
<LinearLayout xmlns:android="http://schemas.android.com/apk/res/android"
    android:layout_width="300dp"
    android:layout_height="wrap_content"
    android:orientation="vertical">

    <TextView
        android:id="@+id/tvTitle"
        android:layout_width="match_parent"
        android:layout_height="wrap_content"
        android:background="#2372c1"
        android:gravity="center"
        android:text="提示"
        android:padding="5dp"
        android:textColor="#fff"
        android:textSize="25sp" />

    <TextView
        android:id="@+id/tvContent"
        android:layout_width="match_parent"
        android:layout_height="200dp"
        android:gravity="center"
        android:text="自定义对话框内容" />

    <LinearLayout
        android:layout_width="match_parent"
        android:layout_height="wrap_content"
        android:background="#c0c0c0"
        android:gravity="center"
        android:orientation="horizontal">

        <Button
            android:id="@+id/btnOk"
            android:layout_width="wrap_content"
            android:layout_height="wrap_content"
            android:text="确定" />

        <Button
            android:id="@+id/btnCancel"
            android:layout_width="wrap_content"
            android:layout_height="wrap_content"
            android:layout_marginLeft="20dp"
            android:text="取消" />
    </LinearLayout>
</LinearLayout>
```

(2) 在 Java 目录的包名下创建一个 MyDialog 类继承 android.app.Dialog，主要用于初始化自定义对话框中的控件以及注册按钮的监听器，具体代码如下：

```
public class MyDialog extends Dialog {
    private String title;
    private String content;
    private TextView tvTitle;
    private TextView tvContent;
    private Button btnOk;
```

```
    private Button btnCancel;

    public MyDialog(Context context) {
        super(context);
    }

    @Override
    protected void onCreate(Bundle savedInstanceState) {
        super.onCreate(savedInstanceState);
        //去除标题
        requestWindowFeature(Window.FEATURE_NO_TITLE);
        //引入自定义对话框布局
        setContentView(R.layout.my_dialog);
        //初始化控件
        initView();
        //设置标题
        tvTitle.setText(title);
        //设置内容
        tvContent.setText(content);
        //注册确认按钮监听器
        btnOk.setOnClickListener(new View.OnClickListener() {
            @Override
            public void onClick(View v) {
                //点击确认时的操作
            }
        });
        //注册取消按钮监听器
        btnCancel.setOnClickListener(new View.OnClickListener() {
            @Override
            public void onClick(View v) {
                //关闭对话框
                dismiss();
            }
        });
    }

    //初始化控件
    private void initView() {
        tvTitle = findViewById(R.id.tvTitle);
        tvContent = findViewById(R.id.tvContent);
        btnOk = findViewById(R.id.btnOk);
        btnCancel = findViewById(R.id.btnCancel);
    }

    public void setTitle(String title) {
        this.title = title;
    }

    public void setContent(String content) {
        this.content = content;
    }
}
```

(3) 在 MainActivity 中对 showCustomDialog()方法进行实现，显示一个自定义对话框，具体代码如下：

```
//显示自定义对话框
private void showCustomDialog() {
    MyDialog dialog = new MyDialog(this);
    dialog.setTitle("自定义对话框");
    dialog.setContent("你好!这里是自定义对话框! ");
    dialog.show();
}
```

(4) 运行程序，点击自定义对话框按钮，运行效果如图 3-32 所示。

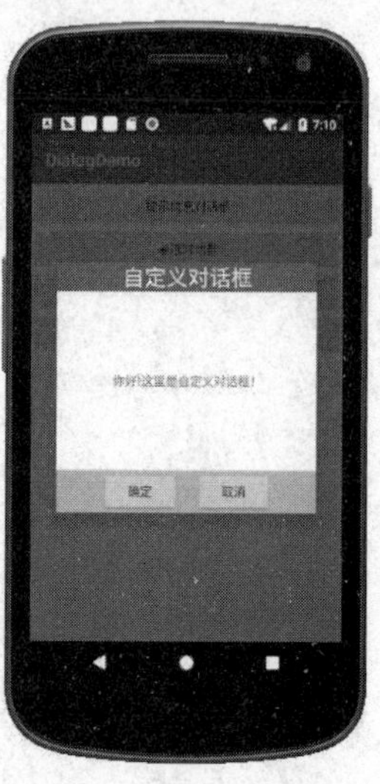

图 3-32　自定义对话框

2. ProgressBar(进度条)的使用

ProgressBar 用于在界面上显示一个进度条。ProgressBar 进度条有很多种风格，如圆环形进度条、水平显示进度的进度条等。接下来通过一个例子学习进度条 ProgressBar 的使用，具体代码如下：

```
<LinearLayout xmlns:android="http://schemas.android.com/apk/res/android"
    android:layout_width="match_parent"
    android:layout_height="match_parent"
    android:orientation="vertical">

    <ProgressBar
        android:id="@+id/pbCircle"
        android:layout_width="match_parent"
        android:layout_height="wrap_content"/>

    <ProgressBar
        android:id="@+id/pbHorizontal"
        android:layout_width="match_parent"
        android:layout_height="wrap_content"
        style="@style/Widget.AppCompat.ProgressBar.Horizontal"/>

    <Button
        android:id="@+id/button"
        android:layout_width="match_parent"
        android:layout_height="wrap_content"
        android:text="点我"/>
</LinearLayout>
```

上述代码中，在布局中加入了一个圆环形进度条(默认)和一个水平进度条(通过 style 样式可以设置的进度条风格)，并加入一个按钮，用于点击按钮增加进度条的进度。在 MainActivity 中添加代码，具体代码如下：

```
public class MainActivity extends AppCompatActivity {

    private ProgressBar pbHorizontal;
    private Button button;

    @Override
    protected void onCreate(Bundle savedInstanceState) {
        super.onCreate(savedInstanceState);
        setContentView(R.layout.activity_main);
        //初始化控件
        pbHorizontal = findViewById(R.id.pbHorizontal);
```

```
            button = findViewById(R.id.button);
            //注册按钮监听器
            button.setOnClickListener(new View.OnClickListener() {
                @Override
                public void onClick(View v) {
                    //setProgress()设置进度条进度，getProgress()获取进度条进度
                    pbHorizontal.setProgress
(pbHorizontal.getProgress()+10);
                }
            });
        }
    }
```

在上述代码中，使用 setProgress()方法设置进度条的进度，使用 getProgress()方法获取进度条的当前进度。程序运行效果如图 3-33 所示。

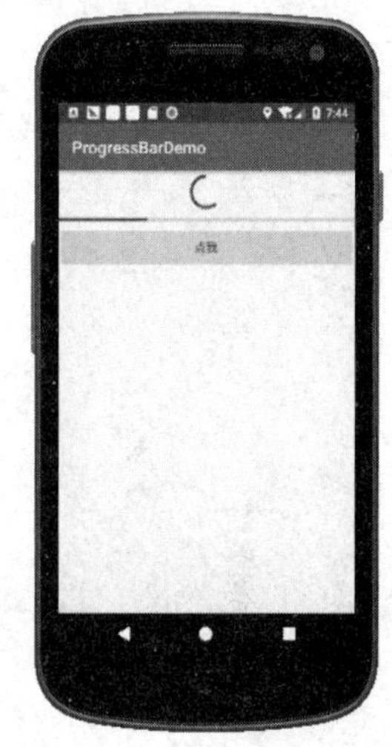

图 3-33　ProgressBar 控件

3. SeekBar(拖动条)的使用

SeekBar 是 ProgressBar 的子类，ProgressBar 主要用来显示进度，但是不能和用户进行交互，而 SeekBar 可以供用户进行拖动改变进度值。接下来通过一个例子演示 SeekBar 的使用。

在布局文件中加入 SeekBar，具体代码如下：

```
<LinearLayout xmlns:android="http://schemas.android.com/apk/res/android"
    android:layout_width="match_parent"
    android:layout_height="match_parent"
    android:orientation="vertical">

    <SeekBar
        android:id="@+id/seekBar"
        android:layout_width="match_parent"
        android:layout_height="wrap_content"
        android:layout_marginTop="20dp"
        android:progress="30" />

    <TextView
        android:id="@+id/tvProgress"
        android:layout_width="match_parent"
        android:layout_height="wrap_content"
        android:gravity="center"
        android:text="数值范围 0~100 之间，当前值：30"
        android:textSize="20sp" />
</LinearLayout>
```

上述代码中，在布局文件中加入 SeekBar 控件，并设置 SeekBar 的进度值为 30(SeekBar 默认最小进度为 0，最大进度为 100)。在后面还加入了一个 TextView，用于显示 SeekBar 的进度值。接下来修改 MainActivity 的代码，设置 SeekBar 进度值的显示，具体代码如下：

```
public class MainActivity extends AppCompatActivity {
    private SeekBar seekBar;
    private TextView tvProgress;

    @Override
```

```
    protected void onCreate(Bundle savedInstanceState) {
        super.onCreate(savedInstanceState);
        setContentView(R.layout.activity_main);

        //初始化控件
        seekBar = findViewById(R.id.seekBar);
        tvProgress = findViewById(R.id.tvProgress);
        //注册拖动条监听器
        seekBar.setOnSeekBarChangeListener(new SeekBar.OnSeekBarChangeListener() {
            @Override
            public void onProgressChanged(SeekBar seekBar, int progress, boolean fromUser) {
                //将 SeekBar 的进度设置到 TextView 上
                tvProgress.setText("数值范围 0~100 之间，当前值："+progress);
            }

            @Override
            public void onStartTrackingTouch(SeekBar seekBar) {

            }

            @Override
            public void onStopTrackingTouch(SeekBar seekBar) {

            }
        });
    }
}
```

在上述代码中，使用 setOnSeekBarChangeListener()方法为 SeekBar 注册进度值变化的监听器，其中 onProgressChanged()方法为当 SeekBar 进度改变时会调用该方法，方法的第一个参数为 SeekBar 控件，第二个参数为当前进度值，第三个参数为是否为用户触发；onStartTrackingTouch()方法用于通知用户已经开始一个触摸拖动手势；onStopTrackingTouch()方法用于通知用户触摸手势已经结束。运行程序，效果如图 3-34 所示。

图 3-34　SeekBar 控件

【任务实施】

1. 任务分析

本任务要求完成园区环境采集系统阈值设置对话框的开发(标题栏与界面控件未实现)，自定义阈值对话框过程分析如下。

(1) 自定义对话框布局。

(2) 创建自定义阈值对话框继承 android.app.Dialog，并完成对话框控件的初始化与功能。

(3) 在 MainActivity 使用创建阈值对话框并显示。

完成的效果如图 3-35 所示。

图 3-35 设置阈值对话框效果图

2. 任务实现

(1) 在 activity_main.xml 布局文件中设置背景图片为 bg_inner，并创建阈值对话框布局。在 res/values 目录下的 style.xml 文件中加入样式，代码如下：

```
!--设置阈值对话框主题-->
<style name="Dialog" parent="Theme.AppCompat.Light.Dialog">
</style>
```

(2) 自定义阈值对话框类，具体代码如下：

```
/**
 * 设置阈值对话框
 */
public class SettingThresholdDialog extends Dialog {

    private TextView tvTempValue,tvHumiValue,tvLightValue;
    private Button btnCancel;
    private Button btnConfirm;
    private SeekBar sbTemp,sbHumi,sbLight;


    public SettingThresholdDialog(@NonNull Context context) {
        super(context,R.style.Dialog);
        //关联布局文件
        this.setContentView(R.layout.dialog_setting_threshold);
        //初始化组件
        initView();
        addListener();
    }

    private void initView() {
        sbTemp=findViewById(R.id.sb_temp);
        sbHumi=findViewById(R.id.sb_humi);
        sbLight=findViewById(R.id.sb_light);
```

```
        tvTempValue=findViewById(R.id.tv_tempValue);
        tvHumiValue=findViewById(R.id.tv_humiValue);
        tvLightValue=findViewById(R.id.tv_lightValue);
        btnCancel=findViewById(R.id.btn_cancel);
        btnConfirm=findViewById(R.id.btn_confirm);
    }

    private void addListener() {
        //温度 SeekBar 状态改变监听
        sbTemp.setOnSeekBarChangeListener(new SeekBar.OnSeekBarChangeListener() {
            @Override
            public void onProgressChanged(SeekBar seekBar, int progress, boolean fromUser) {
                //SeekBar 进度显示到 TextView 上
                tvTempValue.setText(String.valueOf(progress));
            }
            @Override
            public void onStartTrackingTouch(SeekBar seekBar) {

            }
            @Override
            public void onStopTrackingTouch(SeekBar seekBar) {

            }
        });
        //湿度 SeekBar 状态改变监听
        sbHumi.setOnSeekBarChangeListener(new SeekBar.OnSeekBarChangeListener() {
            @Override
            public void onProgressChanged(SeekBar seekBar, int progress, boolean fromUser) {
                //SeekBar 进度显示到 TextView 上
                tvHumiValue.setText(String.valueOf(progress));
            }
            @Override
            public void onStartTrackingTouch(SeekBar seekBar) {

            }
            @Override
            public void onStopTrackingTouch(SeekBar seekBar) {

            }
        });
        //光照 SeekBar 状态改变监听
        sbLight.setOnSeekBarChangeListener(new SeekBar.OnSeekBarChangeListener() {
            @Override
            public void onProgressChanged(SeekBar seekBar, int progress, boolean fromUser) {
                //SeekBar 进度显示到 TextView 上
                tvLightValue.setText(String.valueOf(progress));
            }
            @Override
            public void onStartTrackingTouch(SeekBar seekBar) {

            }
            @Override
            public void onStopTrackingTouch(SeekBar seekBar) {

            }
```

```
            });
            //设置确定点击事件
            btnConfirm.setOnClickListener(new View.OnClickListener() {
                @Override
                public void onClick(View v) {
                    //对话框消失
                    SettingThresholdDialog.this.dismiss();
                }
            });
            //设置取消点击事件
            btnCancel.setOnClickListener(new View.OnClickListener() {
                @Override
                public void onClick(View v) {
                    //对话框消失
                    SettingThresholdDialog.this.dismiss();
                }
            });
        }
    }
```

(3) 在 MainActivity 创建阈值对话框并显示，具体代码如下：

```
public class MainActivity extends AppCompatActivity {

    @Override
    protected void onCreate(Bundle savedInstanceState) {
        super.onCreate(savedInstanceState);
        setContentView(R.layout.activity_main);

        SettingThresholdDialog dialog = new SettingThresholdDialog(this);
        dialog.show();
    }
}
```

3. 运行结果

将应用程序在工控平板上运行，运行结果如图 3-36 所示。

图 3-36　设置阈值对话框

任务 4　图片预览界面开发

【任务描述】

本任务要求使用 RecyclerView 完成图片预览界面的开发。在此界面上能够显示日期以及拍照的图片。任务清单如表 3-11 所示。

表 3-11　任务清单

任务课时	4 课时	任务组员数量	建议 1 人
任务组采用设备	PC 一台 模拟器或真机或工控平板		

【知识解析】

下面我们来介绍一下 RecyclerView 的简单使用。

从 Android 5.0 开始，Google 公司推出了一个用于大量数据展示的新控件 RecylerView，其功能更加强大和灵活。RecyclerView 是 support-v7 包中的新组件，是一个强大的滑动组件，与经典的 ListView 相比，同样拥有 item 回收复用的功能，这一点从它的名字 RecyclerView(即回收 View)也可以看出。

接下来演示使用 RecyclerView 完成一个瀑布流的显示效果。

首先，在 build.gradle 文件中引入 RecyclerView，如图 3-37 所示。

```
dependencies {
    implementation fileTree(dir: 'libs', include: ['*.jar'])
    implementation 'com.android.support:appcompat-v7:28.0.0'
    implementation 'com.android.support.constraint:constraint-layout:1.1.3'
    implementation 'com.android.support:recyclerview-v7:28.0.0'
    testImplementation 'junit:junit:4.12'
    androidTestImplementation 'com.android.support.test:runner:1.0.2'
    androidTestImplementation 'com.android.support.test.espresso:espresso-core:3.0.2'
}
```

图 3-37　引入 RecyclerView

添加完后，在右上角有一个同步 Sync Now 的提示，点击进行同步构建。接下来修改 activity_main.xml 中的代码，具体如下：

```
<LinearLayout xmlns:android="http://schemas.android.com/apk/res/android"
    android:layout_width="match_parent"
    android:layout_height="match_parent">

    <android.support.v7.widget.RecyclerView
        android:id="@+id/recyclerView"
        android:layout_width="match_parent"
        android:layout_height="match_parent"/>
</LinearLayout>
```

接下来，我们希望这个列表项的每一条信息能够显示图片和文字，为此需要准备一组图片放入 drawable 目录中。定义一个实体类，作为 RecyclerView 适配器的适配类型，新建

Vegie 类，具体代码如下：

```
/**
 * 蔬菜类
 */
public class Vegie {

    private String name;
    private int imageId;

    public Vegie(String name, int imageId) {
        this.name = name;
        this.imageId = imageId;
    }

    public String getName() {
        return name;
    }

    public int getImageId() {
        return imageId;
    }
}
```

在上述代码中，Vegie 类中有两个变量，name 表示蔬菜名称，imageId 表示蔬菜对应的图片资源 id。

然后，需要自定义一个列表项的布局，在 layout 目录下新建 vegie_item.xml，具体代码如下：

```
<LinearLayout xmlns:android="http://schemas.android.com/apk/res/android"
    android:layout_width="match_parent"
    android:layout_height="wrap_content"
    android:layout_margin="5dp"
    android:orientation="vertical">

    <ImageView
        android:id="@+id/img"
        android:layout_width="wrap_content"
        android:layout_height="wrap_content"
        android:layout_gravity="center_horizontal" />

    <TextView
        android:id="@+id/tv"
        android:layout_width="wrap_content"
        android:layout_height="wrap_content"
        android:layout_gravity="left" />
</LinearLayout>
```

在这个布局中定义了一个 ImageView 用来显示子项的蔬菜图片，TextView 用来显示蔬菜的名字。父布局使用线性布局，排列方向为 horizontal(水平)。

完成这些后，开始自定义适配器 VegieAdapter，这个适配器继承 RecyclerView.Adapter，并指定泛型 VegieAdapter.ViewHolder，其中 ViewHolder 是 VegieAdapter 中的一个内部类，具体代码如下：

```
public class VegieAdapter extends RecyclerView.Adapter<VegieAdapter.ViewHolder> {

    private List<Vegie> vegitList;
```

```

    static class ViewHolder extends RecyclerView.ViewHolder{
        ImageView vegieImg;
        TextView vegieName;
        View itemView;

        public ViewHolder(@NonNull View itemView) {
            super(itemView);
            this.itemView = itemView;
            vegieImg = itemView.findViewById(R.id.img);
            vegieName = itemView.findViewById(R.id.tv);
        }
    }

    public VegieAdapter(List<Vegie> vegitList) {
        this.vegitList = vegitList;
    }

    @Override
    public ViewHolder onCreateViewHolder(@NonNull final ViewGroup viewGroup, int i) {
        View view = LayoutInflater.from(viewGroup.getContext()).inflate
(R.layout.vegie_item,viewGroup,false);
        ViewHolder holder = new ViewHolder(view);
        //注册点击文字部分的监听器
        holder.itemView.setOnClickListener(new View.OnClickListener() {
            @Override
            public void onClick(View v) {
                Toast.makeText(viewGroup.getContext(),"你点击的是文字部分",
Toast.LENGTH_SHORT).show();
            }
        });
        //注册点击图片部分的监听器
        holder.vegieImg.setOnClickListener(new View.OnClickListener() {
            @Override
            public void onClick(View v) {
                Toast.makeText(viewGroup.getContext(),"你点击的是图片部分",
Toast.LENGTH_SHORT).show();
            }
        });
        return holder;
    }

    @Override
    public void onBindViewHolder(@NonNull ViewHolder viewHolder, int i) {
        Vegie vegie = vegitList.get(i);
        viewHolder.vegieImg.setImageResource(vegie.getImageId());
        viewHolder.vegieName.setText(vegie.getName());
    }

    @Override
    public int getItemCount() {
        return vegitList.size();
    }
}
```

适配器中的代码比较长，但还是比较容易理解。其中，ViewHolder 是一个内部类继承 RecyclerView.ViewHolder 类，在内部定义了一个构造方法并传入 View 参数，这个参数通常就是 RecyclerView 子项的最外层布局，那么就可以通过 findViewById()方法来获取布局中 ImageView 和 TextView 的实例了。

继续往下看，在 VegieAdapter 中也有一个构造方法，并传入 List，这个 List 就是适配器需要传入的数据。

再往下看，由于 VegieAdapter 继承自 RecyclerView.Adapter，那么就必须重写 onCreateViewHolder()、onBindViewHolder()和 getItemCount 这 3 个方法。onCreateViewHolder()是用于创建 ViewHolder 实例的，在这个方法中绑定子项的布局产生一个 View，并创建一个 ViewHolder 对象返回。onBindViewHolder()方法用于对 RecyclerView 指向的数据进行复制，可以通过第二个整型参数表示当前子项的索引(索引从 0 开始)。getItemCount()方法返回的是数据源的长度。

适配器写好后，就可以在 MainActivity 中使用 RecyclerView 了，修改 MainActivity 中的代码，具体代码如下：

```
public class MainActivity extends AppCompatActivity {

    private List<Vegie> vegieList = new ArrayList<>();

    @Override
    protected void onCreate(Bundle savedInstanceState) {
        super.onCreate(savedInstanceState);
        setContentView(R.layout.activity_main);
        initVegie();
        initView();
    }

    private void initVegie() {
        for (int i = 0; i < 3; i++) {
            Vegie artichoke = new Vegie(getRandomName("carrot"),R.drawable.carrot);
            vegieList.add(artichoke);
            Vegie carrot = new Vegie(getRandomName("chilli"),R.drawable.chilli);
            vegieList.add(carrot);
            Vegie chilli = new Vegie(getRandomName("mushroom"),R.drawable.mushroom);
            vegieList.add(chilli);
            Vegie leek = new Vegie(getRandomName("potato"),R.drawable.potato);
            vegieList.add(leek);
            Vegie cucumber = new Vegie(getRandomName("tomato"),R.drawable.tomato);
            vegieList.add(cucumber);
        }
    }

    private void initView() {
        RecyclerView recyclerView = findViewById(R.id.recyclerView);
        StaggeredGridLayoutManager layoutManager = new StaggeredGridLayoutManager(3,StaggeredGridLayoutManager.VERTICAL);
        recyclerView.setLayoutManager(layoutManager);
        VegieAdapter adapter = new VegieAdapter(vegieList);
        recyclerView.setAdapter(adapter);
    }

    //用于随机产生不同长度的蔬菜名称
    private String getRandomName(String name){
        Random random = new Random();
        int length = random.nextInt(20)+1;
        StringBuilder sb = new StringBuilder();
        for (int i = 0; i < length ; i++) {
            sb.append(name);
```

```
        }
        return sb.toString();
    }
}
```

在上面的代码中，initVegie()方法是用来模拟数据的，在这个方法中创建了 15 个 Vegie 实例，并保存在 vegieList 集合中。在 initView()方法中，首先获取 RecyclerView 实例，然后创建一个 StaggeredGridLayoutManager 对象，该对象用于指定 RecyclerView 使用瀑布流方式显示列表项。接下来创建 VegieAdapter 实例，并将 vegieList 集合的数据传入到 VeguiAdapter 的构造方法中。最后调用 RecyclerView 的 setAdapter()方法完成适配器设置。

最后运行程序，效果如图 3-38 所示。

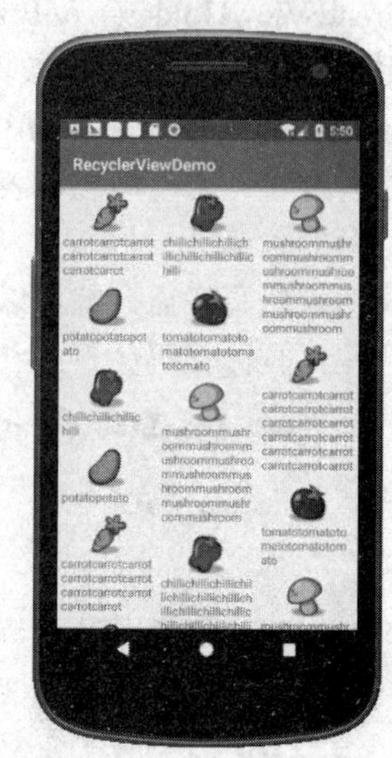

图 3-38　RecyclerView 控件

【任务实施】

1. 任务分析

本任务要求完成图片预览界面，图片预览界面使用 RecyclerView，过程分析如下。

(1) 在布局中加入 RecyclerView。

(2) 新建 RecyclerView 的列表项布局。

(3) 新建 RecyclerView 的数据适配器。

(4) 设置 RecyclerView 的数据适配器。

完成的效果如图 3-39 所示。

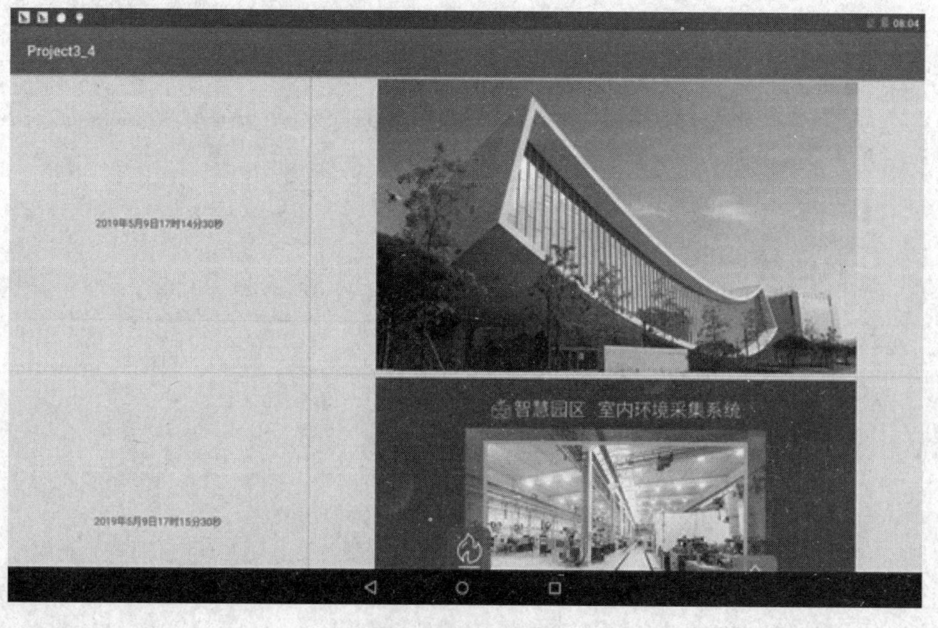

图 3-39　效果图

2. 任务实现

(1) 在 activity_main.xml 布局文件中加入 RecyclerView，具体代码如下：

```
<LinearLayout xmlns:android="http://schemas.android.com/apk/res/android"
    android:layout_width="match_parent"
    android:layout_height="match_parent"
    android:orientation="vertical">

    <android.support.v7.widget.RecyclerView
        android:id="@+id/recycler_view"
        android:layout_width="match_parent"
        android:layout_height="match_parent">
    </android.support.v7.widget.RecyclerView>
</LinearLayout>
```

(2) 创建 RecyclerView 的列表项布局，具体代码如下：

```
<LinearLayout xmlns:android="http://schemas.android.com/apk/res/android"
    android:layout_width="match_parent"
    android:layout_height="400dp"
    android:gravity="center">

    <TextView
        android:id="@+id/time"
        android:layout_width="0dp"
        android:layout_height="wrap_content"
        android:layout_weight="1"
        android:gravity="center" />

    <View
        android:layout_width="1dp"
        android:layout_height="fill_parent"
        android:background="#c0c0c0" />

    <ImageView
        android:id="@+id/image"
        android:layout_width="0dp"
        android:layout_height="match_parent"
        android:layout_margin="4dp"
        android:layout_weight="2"
        android:scaleType="fitCenter"
        android:src="@drawable/image1" />
</LinearLayout>
```

(3) 创建 RecyclerView 的数据适配器，具体代码如下：

```
public class MonitorAdapter extends RecyclerView.Adapter<MonitorAdapter.ViewHolder> {
    private String[] times = {"2019年5月9日17时14分30秒", "2019年5月9日17时15分30秒", "2019年5月9日18时30分30秒", "2019年5月9日18时40分30秒", "2019年5月10日17时14分30秒"};
    private int[] imageIds = {R.drawable.image1,R.drawable.image2,R.drawable.image3,R.drawable.image4,R.drawable.image5};

    @NonNull
    @Override
    public ViewHolder onCreateViewHolder(@NonNull ViewGroup parent, int viewType) {
        View view = LayoutInflater.from(parent.getContext()).inflate(R.layout.recyclerview_item, parent, false);
```

```
        ViewHolder holder = new ViewHolder(view);
        return holder;
    }

    @Override
    public void onBindViewHolder(@NonNull ViewHolder holder, int position) {
        holder.image.setImageResource(imageIds[position]);
        holder.time.setText(times[position]);
    }

    @Override
    public int getItemCount() {
        return times.length;
    }

    static class ViewHolder extends RecyclerView.ViewHolder {
        private ImageView image;//图片
        private TextView time;//时间

        public ViewHolder(View v) {
            super(v);
            image = v.findViewById(R.id.image);
            time = v.findViewById(R.id.time);
        }
    }
}
```

(4) 设置 RecyclerView 的适配器，具体代码如下：

```
public class MainActivity extends AppCompatActivity {
    private RecyclerView recyclerView;

    @Override
    protected void onCreate(Bundle savedInstanceState) {
        super.onCreate(savedInstanceState);
        setContentView(R.layout.activity_main);
        recyclerView = findViewById(R.id.recycler_view);
        //创建线性布局管理器，方向垂直
        LinearLayoutManager linearLayoutManager = new LinearLayoutManager(this, LinearLayoutManager.VERTICAL, false);
        //recyclerView 设置布局管理器
        recyclerView.setLayoutManager(linearLayoutManager);
        MonitorAdapter adapter = new MonitorAdapter();
        //添加 Android 自带的分割线
        recyclerView.addItemDecoration(new DividerItemDecoration(this, DividerItemDecoration.VERTICAL));
        //设置适配器
        recyclerView.setAdapter(adapter);
    }
}
```

3. 运行结果

将应用程序在工控平板上运行，运行结果如图 3-40 所示。

图 3-40 图片预览界面

项 目 总 结

本项目主要讲解了 Android 中 UI 开发，包括常见布局、常见控件、对话框、进度条、列表等，并使用 UI 的知识完成智慧园区部分界面的开发，如登录界面、园区监控系统界面、阈值设置对话框、图片预览界面等。除了本项目任务中界面外，读者还可以尝试开发项目中的其他界面。

思考与练习

3-1 简述 Android 中几种常见布局的区别。

3-2 简述 RecyclerView 的使用步骤。

项目 4 界面显示与切换

【项目描述】

通过项目 3 学习了 UI 的许多知识，读者已经有 UI 开发的基础了。本项目主要学习 Android 中的活动与碎片，实现引导页的功能以及侧滑菜单和标题栏的制作效果。学习完本项目能够让你对 Android 的界面显示和切换有更深的理解。

本项目通过 4 个任务的学习(图 4-1)，针对 Android 开发中必须掌握的 Activity、ViewPager、Fragment、DrawerLayout 和 Toolbar 技术进行讲解，加强读者 UI 开发的能力。具体任务列表如图 4-1 所示。

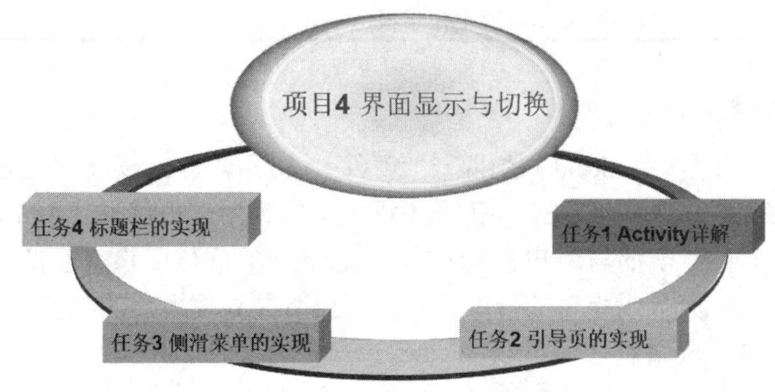

图 4-1　项目 4 任务列表

【学习目标】

知识目标：

- 会创建 Activity。
- 掌握 Activity 的生命周期和启动模式。
- 会 Fragment 的使用。
- 会 DrawerLayout 的使用。
- 会 Toolbar 的使用。

技能目标：

- 能实现页面跳转和传值。
- 能用 Activity 的 4 种启动模式启动 Activity。
- 能用 ViewPager 实现引导页。
- 能用 DrawerLayout 实现侧滑菜单。
- 能用 Fragment 实现页面切换。
- 能用 Toorbar 实现标题栏。

任务 1　Activity 详解

【任务描述】

本任务通过模拟用户的登录过程来讲解 Activity 的使用。任务清单如表 4-1 所示。

扫码观看视频讲解

表 4-1　任务清单

任务课时	10 课时	任务组员数量	建议 1 人
任务组采用设备	PC 一台 模拟器或真机或工控平板		

【知识解析】

Activity(活动)是 Android 的四大组件(Activity、Service、ContentProvider、BroadcastReceiver)之一，是用户操作的可视化界面，它为用户提供了一个完成操作指令的窗口。Activity 中所有操作都与用户密切相关，是一个负责与用户交互的组件，可以通过 setContentView(View)来显示指定的界面，也可以监听并处理用户的事件，同时做出响应。Activity 之间通过 Intent 进行通信。

1. 创建 Activity

(1) 新建工程 Demo4_1，右键单击 Android 工程的 java 代码目录，选择快捷菜单中的 New→Activity→Empty Activity 命令，如图 4-2 所示。弹出 Activity 配置窗口，在该窗口中可以修改 Activity 和布局文件的名称(Activity 的命名使用 XxxActivity 规则)，这里命名为 FirstActivity。默认勾选 Generate Layout File 复选框，如图 4-3 所示，这样在创建 Activity 时，会创建一个关联的布局文件。单击 Finish 按钮。

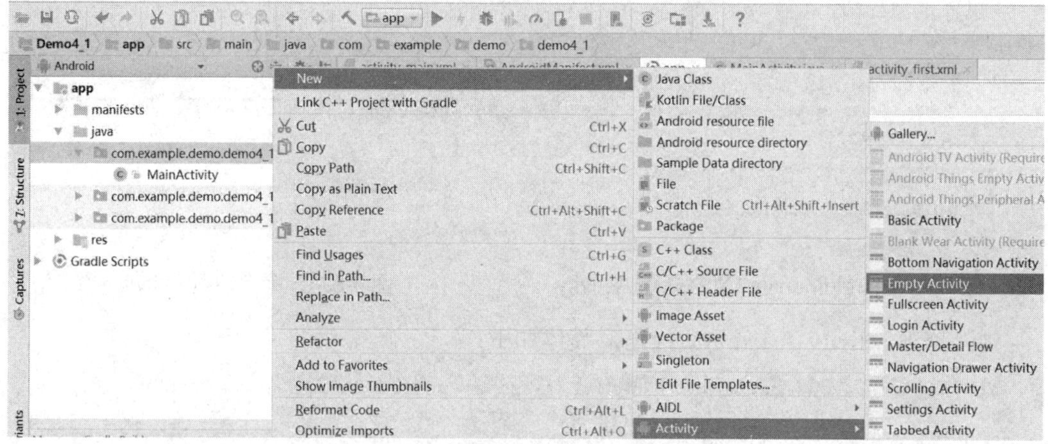

图 4-2　新建 Empty Activity

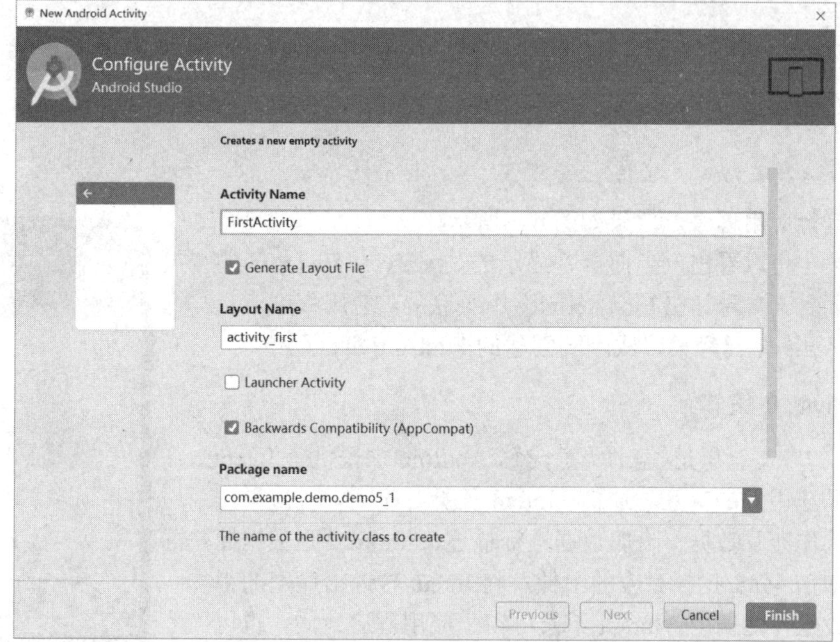

图 4-3　设置 Activity 及布局文件名称

(2) 成功创建后，打开 AndroidManifest.xml 文件，可以看到新增了一个名称为.FirstActivity 的 Activity 声明(每个 Activity 都要在 AndroidManifest.xml 文件中声明)，如图 4-4 所示。

(3) 在 AndroidManifest.xml 清单文件中，把 MainActivity 中的<intent-filter>标签，剪贴到声明 FirstActivity 的标签对中，如图 4-5 所示。

```
<application
    android:allowBackup="true"
    android:icon="@mipmap/ic_launcher"
    android:label="@string/app_name"
    android:roundIcon="@mipmap/ic_launcher_round"
    android:supportsRtl="true"
    android:theme="@style/AppTheme">
    <activity android:name=".MainActivity">
        <intent-filter>
            <action android:name="android.intent.action.MAIN"/>
            <category android:name="android.intent.category.LAUNCHER"/>
        </intent-filter>
    </activity>
    <activity android:name=".FirstActivity">
    </activity>
</application>
```

图 4-4　AndroidManifest 文件新增 Activity

```
<application
    android:allowBackup="true"
    android:icon="@mipmap/ic_launcher"
    android:label="@string/app_name"
    android:roundIcon="@mipmap/ic_launcher_round"
    android:supportsRtl="true"
    android:theme="@style/AppTheme">
    <activity android:name=".MainActivity">
    </activity>
    <activity android:name=".FirstActivity">
        <intent-filter>
            <action android:name="android.intent.action.MAIN"/>
            <category android:name="android.intent.category.LAUNCHER"/>
        </intent-filter>
    </activity>
</application>
```

图 4-5　设置启动项 Activity

(4)　修改 activity_fist.xml 的代码，具体如下：

```
<LinearLayout xmlns:android="http://schemas.android.com/apk/res/android"
              android:layout_width="match_parent"
              android:layout_height="match_parent"
              android:orientation="vertical">

    <TextView
        android:layout_width="match_parent"
        android:layout_height="wrap_content"
        android:gravity="center"
        android:text="这是FirstActivity绑定的布局" />

</LinearLayout>
```

(5)　运行应用程序，效果如图 4-6 所示。

图 4-6　运行效果

从结果中可以看出，当前屏幕显示的是 activity_first.xml 的布局效果，而该布局是在 FirstActivity 中关联的，应用程序运行的第一个 Activity 已经设置成刚刚创建的 FirstActivity 了。

2. Activity 的跳转

Intent 的中文意思是“意图，意向”，Android 中提供了 Intent 机制来协助应用间的交互与通信。Intent 负责对应用中一次操作的动作、动作涉及数据、附加数据进行描述，Android 则根据此 Intent 的描述，负责找到对应的组件，将 Intent 传递给调用的组件，并完成组件的调用。Intent 不仅可用于应用程序之间，也可用于应用程序内部的 Activity、Service 和 BroadcastReceiver 之间的交互。

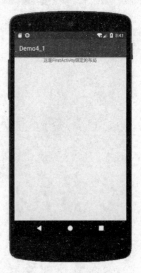

Intent 可以启动一个 Activity，也可以启动一个 Service，还可以发起一个广播 BroadcastReceiver。具体方法见表 4-2。

Intent 分为两种类型，即显式 Intent 和隐式 Intent。

1)　显式 Intent

按名称(完全限定类名)指定要启动的组件，创建显式 Intent 启动 Activity 或服务时，系统将立即启动 Intent 对象中指定的应用组件。

表 4-2 Intent 启动组件的方法

组件名称	方法名称	方法作用
Activity	startActivity()	跳转到新的(另一个)Activity
	startActivityForResult()	跳转到新的 Activity，新的 Activity 关闭后，可以得到返回数据
Service	startService()	启动服务
	bindService()	绑定服务
BroadcastReceiver	sendBroadcasts()	发送无序广播
	sendOrderdBroadcasts()	发送有序广播

通常，在自己的应用中使用显式 Intent 来启动组件，因为此时要启动的 Activity 或服务的类名是明确的。如启动新 Activity 以响应用户操作或者启动服务以在后台下载文件。

2) 隐式 Intent

隐式 Intent 不会指定特定的组件。

创建隐式 Intent 时，Android 系统通过将 Intent 的内容与在设备上其他应用的清单文件中声明的 Intent 过滤器(intent-filter)进行比较，从而找到要启动的相应组件。如果 Intent 与 Intent 过滤器匹配，系统将启动该组件，并向其传递 Intent 对象。如果多个 Intent 过滤器兼容，则系统会显示一个对话框，支持用户选取要使用的应用。

我们已经知道，Activity 是完成操作指令的窗口，这个窗口的显示界面由关联的布局文件决定。要实现 Activity 的跳转，需要有两个 Activity，两个布局文件，接下来通过显式 Intent 实现 Activity 之间的跳转。在上一个案例代码 Demo4_1 基础上实现 Activity 的跳转。

(1) 按上一个案例的方式创建第二个 Activity，命名为 SecondActivity，并修改 activity_second.xml 的内容，代码如下：

```
<LinearLayout xmlns:android="http://schemas.android.com/apk/res/android"
              android:layout_width="match_parent"
              android:layout_height="match_parent"
              android:orientation="vertical">

    <TextView
        android:layout_width="match_parent"
        android:layout_height="wrap_content"
        android:gravity="center"
        android:text="这是 SecondActivity 绑定的布局"/>
</LinearLayout>
```

(2) 在 FirstActivity 绑定的布局 activity_first.xml 代码中加入一个按钮，在该按钮的点击事件中创建一个 Intent 实例并启动，FirstActivity 的代码如下：

```
public class FirstActivity extends AppCompatActivity {
    private Button btnJump;
    @Override
    protected void onCreate(Bundle savedInstanceState) {
        super.onCreate(savedInstanceState);
        setContentView(R.layout.activity_first);
        btnJump=findViewById(R.id.btnJump);
        btnJump.setOnClickListener(new View.OnClickListener() {
            @Override
            public void onClick(View view) {
```

```
                //创建 Intent 实例，指定从 FirstActivity 跳转到 SecondActivity
                Intent intent=new Intent(FirstActivity.this,SecondActivity.class);
                //启动 Intent
                startActivity(intent);
            }
        });
    }
}
```

(3) 运行程序，效果如图 4-7 所示。

(4) 点击按钮将实现 Activity 的跳转，效果如图 4-8 所示。

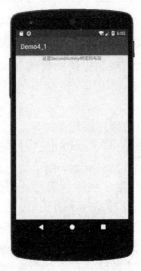

图 4-7　运行结果

图 4-8　跳转结果

3. Intent 传递数据

上一个案例已经实现了 Activity 之间的跳转，接下来在跳转过程中实现数据的传递，并在返回到前一个 Activity 的时候将数据返回。

1) 向下一个 Activity 传递数据

Intent 提供一系列的 putExtra()方法，可以把要传递的数据暂存在 Intent 中，启动另一个 Activity 后，在另一个 Activity 中就可以从 Intent 中取出这些数据。接下来实现点击按钮跳转并传递数据。

(1) 在布局文件 acticity_first.xml 中添加一个按钮，在 FirstActivity 中，添加该按钮的点击事件，代码如下：

```
btnRun.setOnClickListener(new View.OnClickListener() {
            @Override
            public void onClick(View view) {
                //创建 Intent 实例，指定从 FirstActivity 跳转到 ThirdActivity
                Intent intent=new Intent(FirstActivity.this,ThirdActivity.class);
                intent.putExtra("name","张三");//放入 String 类型的数据，key 为 name
                intent.putExtra("age",20);//放入 int 类型的数据，key 为 age
                //启动 Intent
                startActivity(intent);
            }
        });
```

在上面的代码第 6 行和第 7 行，调用了 putExtra()方法传递字符串，该方法接收 2 个参数：①键，用于后面从 Intent 中取值；②传递的数据。

(2) 接下来在 ThirdActivity 中取出数据，代码如下：

```
public class ThirdActivity extends AppCompatActivity {
    private TextView tvData;
    @Override
    protected void onCreate(Bundle savedInstanceState) {
        super.onCreate(savedInstanceState);
        setContentView(R.layout.activity_third);
        Intent intent = getIntent();
        String name = intent.getStringExtra("name");
        int age = intent.getIntExtra("age",0);//第二个参数表示 age 的默认值为 0
        tvData = findViewById(R.id.tvData);
        tvData.setText("接收到数据：\n 姓名："+name+"\n 年龄："+age);
    }
}
```

在 ThirdActivity 中，首先通过 getIntent()方法获取这个 Intent 实例，接着就可以通过这个 Intent 实例与相应的键取得相应的值。字符串类型的数据使用 getStringExtra()方法获取字符串的值，int 类型的数据使用 getIntExtra(String name, int defaultValue)方法获取 int 类型数据的值，第二个参数 defaultValue 为默认值。最后将收到的数据显示在文本上。

(3) 运行程序，跳转到 ThirdActivity，效果如图 4-9 所示。

图 4-9 ThirdActivity 界面效果

2) 返回数据给上一个 Activity

Activity 之间跳转时可以传递数据到下一个 Activity，那么能否返回数据到上一个 Activity 呢？答案是肯定的。这时启动 Intent 需要使用另一个方法 startActivityForResult (Intent intent, int requestCode)，该方法提供 2 个参数：①Intent 实例；②int 类型的请求码，要求不小于 0，用于区分数据的来源。

下面实现返回数据给上一个 Activity。

(1) 修改 FirstActivity 中的代码，具体如下：

```
    private static final int REQUEST_CODE= 1;
    @Override
    protected void onCreate(Bundle savedInstanceState) {
        super.onCreate(savedInstanceState);
        setContentView(R.layout.activity_first);
        btnReturn=findViewById(R.id.btnReturrn);
        btnReturn.setOnClickListener(new View.OnClickListener() {
            @Override
            public void onClick(View view) {
                //创建 Intent 实例，指定从 FirstActivity 跳转到 FourthActivity
                Intent intent=new Intent(FirstActivity.this,FourthActivity.class);
                intent.putExtra("name","张三");//放入 String 类型的数据，key 为 name
                intent.putExtra("age",20);//放入 int 类型的数据，key 为 age
                //REQUEST_CODE 是定义的 int 类型的常量，值为 1
               startActivityForResult(intent,REQUEST_CODE);
            }
        });
```

在上面的代码中，定义了一个整型常量 REQUEST_CODE，它的值为整数 1。一般开发中很少用具体的数字当作区分的标识，而是使用有名字的常量。这一点在后面的开发中很重要。

(2) 接下来在 FourthActivity 中添加返回数据的代码，具体如下：

```
Intent intent1 = new Intent();
intent1.putExtra("returnData","Hello FirstActivity");
//设置返回数据，第一个参数为返回的结果码，用于区分到底是谁的返回数据，第二个为 Intent 实例
setResult(RESULT_OK,intent1);
```

在上面的代码中，创建了一个 Intent 实例，放入了一个关键字为 returnData 的字符串数据。使用 setResult()方法设置返回结果，setResult()方法提供 2 个参数：①返回的结果码，一般使用 RESULT_OK 或 RESULT_CANCELED 这两个值；②Intent 实例，用于携带数据返回。

(3) 最后，返回 FirstActivity，在 FirstActivity 中需要添加 onActivityResult()方法进行处理，代码如下：

```
@Override
  protected void onActivityResult(int requestCode, int resultCode, Intent data) {
      super.onActivityResult(requestCode, resultCode, data);
      //如果请求码是常量 REQUEST_CODE，返回码是常量 RESULT_OK，弹出返回的数据提示
      if(requestCode==REQUEST_CODE||resultCode==RESULT_OK){
          String returnData = data.getStringExtra("returnData");
          Toast.makeText(this,"接收到 FourthActivity 返回的数据："+returnData,
Toast.LENGTH_LONG).show();
      }
  }
```

在上面的代码中，onActivityResult()方法在由跳转的 FourthActivity 返回到 FirstActivity 时会被调用，该方法提供 3 个参数：①请求码，请求码用于判断请求的来源；②结果码，结果码在返回数据时有传入；③Intent 的实例，返回的数据就在这个 Intent 实例中。

(4) 运行程序，出现图 4-10 所示的界面，点击箭头所示的按钮，跳转到图 4-11 所示的界面，此时按模拟器底部的返回键，回到 FirstActivity 效果，如图 4-12 所示。

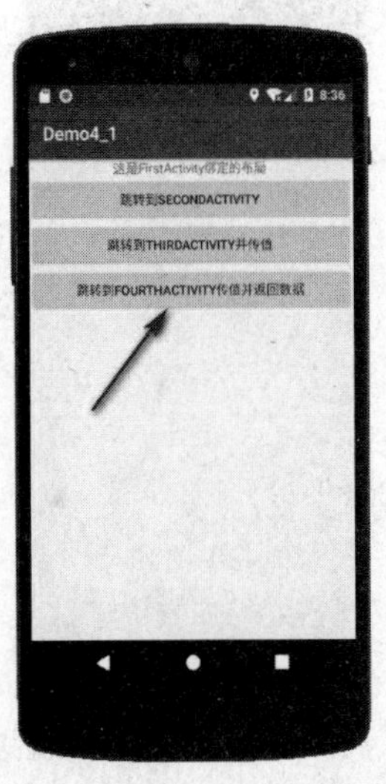

图 4-10　FirstActivity 界面

图 4-11　FourthActivity 界面

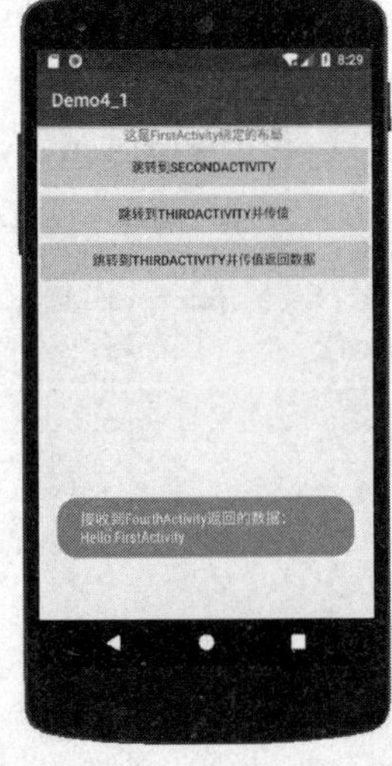

图 4-12　FirstActivity 界面
(FourthActivity 返回)

4. Activity 的生命周期

介绍生命周期之前，要先讲解一下“任务栈”的概念。每个应用都有一个任务栈，是用来存放 Activity 的实例，功能类似于函数调用的栈，任务栈具有先进后出的特点。例如，Activity 实例的入栈顺序是 Activity A→Activity B→Activity C，则任务栈如图 4-13 所示。

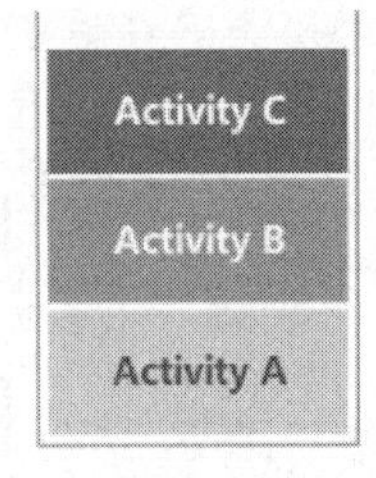

图 4-13　存放 Activity 的任务栈

1)　Activity 的生命周期

在 Android 中会维持一个任务栈，当创建一个新的 Activity 实例时，它就会放到栈顶，这个 Activity 实例就处于运行状态。当再有一个新的 Activity 实例被创建后，会重新压入栈顶，而之前的 Activity 实例则会在这个新的 Activity 实例底下，就像枪梭压入子弹一样，而且之前的 Activity 实例就会进入后台。Activity 整个生命周期有 4 种状态、7 个重要方法。

2)　Activity 的 4 种状态

每一个 Activity 在其生命周期中最多可能会有 4 种状态。

(1)　运行状态。当 Activity 运行在屏幕前台(处于当前任务栈的最上面)，此时它获取了焦点能响应用户的操作，属于运行状态，某一时刻只会有一个 Activity 处于活动(Active)或运行状态。

(2)　暂停状态。在该状态下，Activity 不能与用户进行交互，但 Activity 仍然可见。例如，显示一个对话框时，对话框只占用部分的屏幕，这时 Activity 处于暂停状态。

(3)　停止状态。当 Activity 处于完全不可见时，就会进入停止状态。停止状态下的 Activity 可能还在内存中，但这是不可靠的，随时都有可能被系统回收。

(4)　销毁状态。当 Activity 从任务栈中移除后就变成了销毁状态。

3)　Activity 生命周期的 7 个方法

Activity 类定义了 7 个回调方法，覆盖了其生命周期的每一个环节，下面对这 7 个方法进行介绍。

(1)　onCreate()。onCreate()方法是 Activity 第一次被启动时执行的，在这个方法中需要完成所有的正常静态设置，如创建一个视图(view)、绑定列表的数据等。

(2)　onStart()。onStart()方法是 Activity 界面被显示出来时执行的。

(3)　onResume()。当前 Activity 由被覆盖状态回到前台或解锁屏时系统会调用 onResume()方法，再次进入运行状态。

(4)　onPause()。当前 Activity 被其他 Activity 覆盖其上或被锁屏，系统会调用 onPause()方法，暂停当前 Activity 的执行。

(5)　onStop()。当前 Activity 转到新的 Activity 界面或按 Home 键回到主屏，自身退居后台时系统会先调用 onPause()方法，然后调用 onStop()方法，进入停滞状态。

(6)　onRestart()。onRestart()是当前 Activity 重新被启动时调用的。

(7)　onDestroy()。在 Activity 被销毁前会调用该方法。这是 Activity 能接收到的最后一个调用。可能会因为有人调用了 finish()方法使得当前 Activity 关闭，或系统为了保护内存临时释放这个 Activity 的实例，而调用该方法。

Android 官方提供一张 Activity 生命周期的示意图，如图 4-14 所示。

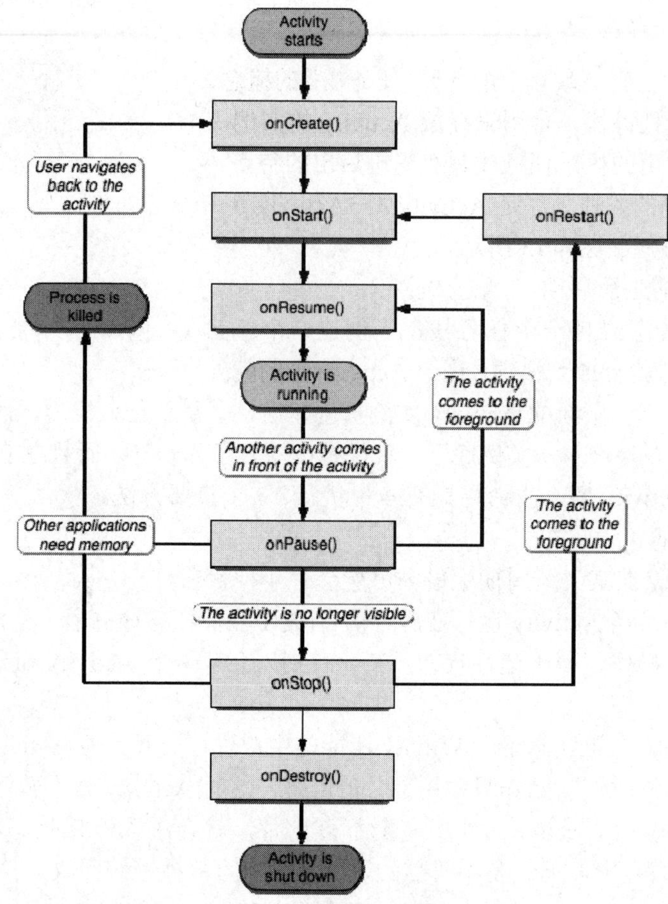

图 4-14　Activity 生命周期

在 Demo4_1 中实现 Activity 的生命周期的案例。

(1) 创建一个 Activity，命名为 LifeCycleActivity.java，布局命名为 activity_life_cycle.xml，并把 LifeCycleActivity 设置为启动应用程序显示的第一个 Activity。AndroidManifest.xml 文件配置如下：

```
<activity android:name=".LifeCycleActivity">
            <intent-filter>
                <action android:name="android.intent.action.MAIN"/>

                <category android:name="android.intent.category.LAUNCHER"/>
            </intent-filter>
        </activity>
```

(2) 重写 LifeCycleActivity 中的 onCreate()、onStart()、onResume()、onPause()、onStop()、onDestroy()、onRestart()方法，在这几个生命周期的方法中使用 Log 打印相关的日志。LifeCycleActivity.java 代码如下：

```
public class LifeCycleActivity extends AppCompatActivity {

    private static final String TAG = "LifeCycle";
    @Override
    protected void onCreate(Bundle savedInstanceState) {
```

```
        super.onCreate(savedInstanceState);
        setContentView(R.layout.activity_life_cycle);
        Log.d(TAG,"onCreate()方法被调用了...");
    }

    @Override
    protected void onStart() {
        super.onStart();
        Log.d(TAG,"onStart()方法被调用了...");
    }

    @Override
    protected void onResume() {
        super.onResume();
        Log.d(TAG,"onResume()方法被调用了...");
    }

    @Override
    protected void onPause() {
        super.onPause();
        Log.d(TAG,"onPause()方法被调用了...");
    }

    @Override
    protected void onStop() {
        super.onStop();
        Log.d(TAG,"onStop()方法被调用了...");
    }

    @Override
    protected void onDestroy() {
        super.onDestroy();
        Log.d(TAG,"onDestroy()方法被调用了...");
    }

    @Override
    protected void onRestart() {
        super.onRestart();
        Log.d(TAG,"onRestart()方法被调用了...");
    }
}
```

(3) 运行程序，打开 LogCat，并使用过滤器只查看 LifeCycle 的日志，结果如图 4-15 所示。

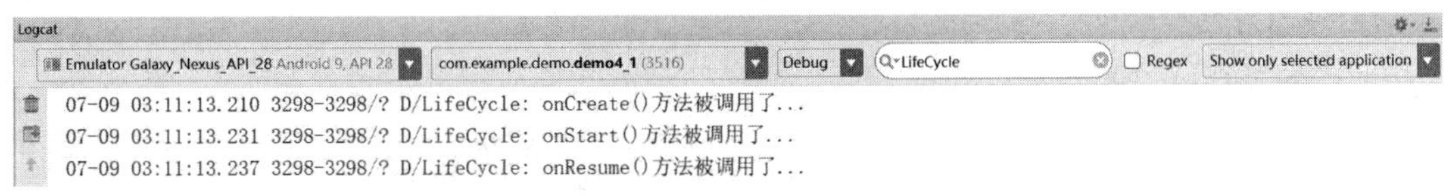

图 4-15 运行日志结果

可以看到 LifeCycleActivity 第一次显示呈现在用户界面上时，调用了 onCreate()、onStart()、onResume()方法。因为 Activity 第一次显示，会调用 onCreate()方法，紧接着执行 onStart()方法，直到调用 onResume()方法，Activity 才可以和用户交互。

(4) 清空 LogCat，按 Home 键返回到桌面，在 LogCat 中查看日志，结果如图 4-16 所示。

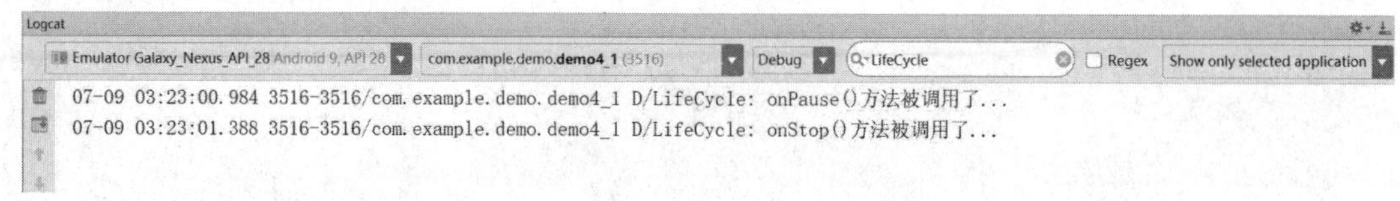

图 4-16　按 Home 键返回日志结果

可以看出 LifeCycleActivity 的 onPause()方法和 onStop()方法被调用了，因为 Activity 进入后台，不再可见，所以会执行 onPause()、onStop()方法。

(5) 清空 LogCat，再次打开 Demo4_1，在 LogCat 中查看日志，结果如图 4-17 所示。

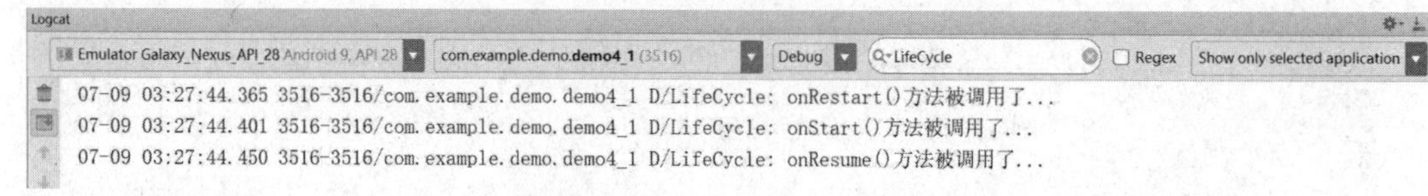

图 4-17　重新进入 Activity 日志结果

可以看出 LifeCycleActivity 的 onRestart()、onStart()、onResume()方法被调用了。这是因为 LifeCycleActivity 从后台重新变为可见，还是会执行 onRestart()方法，接着执行 onStart()、onResume()方法。onCreate()不会再执行，因为 Activity 的实例还没消亡，不会重新执行 onCreate()方法。

(6) 清空 LogCat，按返回键退出应用程序，查看日志，结果如图 4-18 所示。

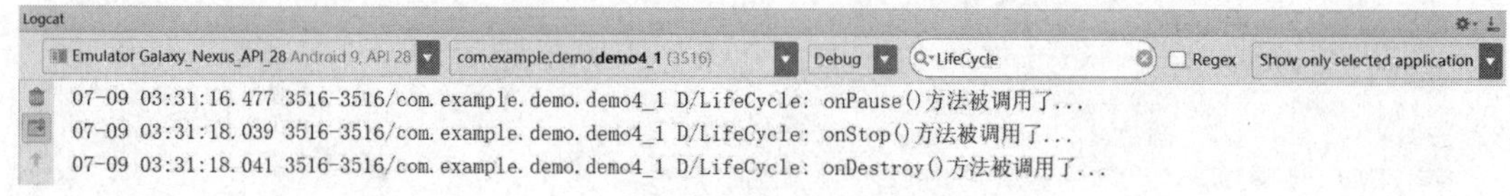

图 4-18　按返回键日志结果

可以看出 LifeCycleActivity 的 onPause()、onStop()、onDestroy()方法被调用了。这是因为当退出应用程序时，LifeCycleActivity 从运行状态到销毁状态，首先要执行 onPause()和 onStop()方法，接着执行 onDestroy()方法，此时 LifeCycleActivity 消亡。

【任务实施】

1. 任务分析

本任务模拟用户的登录过程。

(1) 新建工程 Project4_1，以创建 Activity 的方式创建 LoginActivity 及其布局 activity_login，在布局中添加一个登录按钮，并设置点击事件。

(2) 在 LoginActivity 登录按钮的点击事件中，将用户名 admin 和密码 123456 放到 Intent 中，并跳转到 MainActivity。

(3) 在 MainActivity 中接收传递过来的数据，判断用户名和密码是否正确，如果正确，则弹出欢迎提示。

(4) 将 LoginActivity 设置成第一个启动的 Activity。

2. 任务实现

(1) 新建工程 Project4_1，以创建 Activity 的方式创建 LoginActivity 及其布局 activity_login，在布局中添加一个登录按钮，并设置点击事件，布局代码如下：

```
<?xml version="1.0" encoding="utf-8"?>
<android.support.constraint.ConstraintLayout
    xmlns:android="http://schemas.android.com/apk/res/android"
    xmlns:app="http://schemas.android.com/apk/res-auto"
    xmlns:tools="http://schemas.android.com/tools"
    android:layout_width="match_parent"
    android:layout_height="match_parent"
    tools:context="com.example.demo.project4_1.LoginActivity">
<Button
    android:layout_width="wrap_content"
    android:layout_height="wrap_content"
    android:text="登录"
    android:onClick="login"
    />
</android.support.constraint.ConstraintLayout>
```

(2) 在 LoginActivity 登录按钮的点击事件中，将用户名 admin 和密码 123456 放到 Intent 中，并跳转到 MainActivity，LoginActivity 代码如下：

```
     public class LoginActivity extends AppCompatActivity {
    @Override
    protected void onCreate(Bundle savedInstanceState) {
        super.onCreate(savedInstanceState);
        setContentView(R.layout.activity_login);
    }
    public void login(View view) {
        String userName = "admin";
        String password = "123456";
        Intent intent = new Intent(this,MainActivity.class);
        intent.putExtra("username",userName);
        intent.putExtra("password",password);
        startActivity(intent);

    }
}
```

(3) 在 MainActivity 中接收传递过来的数据，判断用户名和密码是否正确，如果正确则弹出欢迎提示。MainActivity 代码如下：

```
public class MainActivity extends AppCompatActivity {

    @Override
    protected void onCreate(Bundle savedInstanceState) {
        super.onCreate(savedInstanceState);
        setContentView(R.layout.activity_main);
        //获取 Intent 实例
        Intent intent = getIntent();
        //通过 intent 实例的键取得数据
        String userName = intent.getStringExtra("username");
        String password = intent.getStringExtra("password");
        if(userName.equals("admin")&&password.equals("123456")){
            Toast.makeText(this, "欢迎 admin", Toast.LENGTH_LONG).show();
        }
    }
}
```

(4) 将 LoginActivity 设置成第一个启动的 Activity，AndroidManifest.xml 代码如下：

```
<?xml version="1.0" encoding="utf-8"?>
<manifest xmlns:android="http://schemas.android.com/apk/res/android"
        package="com.example.demo.project4_1">
    <application
        android:allowBackup="true"
        android:icon="@mipmap/ic_launcher"
        android:label="@string/app_name"
        android:roundIcon="@mipmap/ic_launcher_round"
        android:supportsRtl="true"
        android:theme="@style/AppTheme">
        <activity android:name=".MainActivity">
        </activity>
        <activity android:name=".LoginActivity">
            <intent-filter>
                <action android:name="android.intent.action.MAIN"/>

                <category android:name="android.intent.category.LAUNCHER"/>
            </intent-filter>
        </activity>
    </application>
</manifest>
```

3. 运行结果

运行程序，结果如图 4-19 和图 4-20 所示。

图 4-19　登录页面

图 4-20　主页面

(1) 运行程序，出现登录页面。

(2) 单击“登录”按钮，登录成功，出现主界面，弹出欢迎提示。

任务 2 引导页的实现

【任务描述】

扫码观看视频讲解

现在的大多数应用都会有一个引导页面，引导页面由几张图片组成，通过左右滑动来告知用户一些功能特性。这个引导页就是使用 ViewPager 来实现的。本任务通过 ViewPager 来实现引导页的效果，引导页由 4 张图片组成，如图 4-21 所示。滑动到最后一张图片时，界面上出现“开启应用”按钮，单击后跳转到 MainActivity。任务清单如表 4-3 所示。

图 4-21 引导页

表 4-3 任务清单

任务课时	2 课时	任务组员数量	建议 1 人
任务组采用设备	PC 一台 模拟器或真机或工控平板		

【知识解析】

ViewPager 是 android-support-v4 中提供的类，它是一个容器类，常用于页面之间的切换。android-support-v4 最低支持 Android1.6(API Level 4)，在 support-v4 包中，它所拥有的类有很多，主要包含对应用组件的支持、用户交互体验的一些工具类以及一些数据网络方面的工具类。

ViewPager 使用的适配器是 PagerAdapter。所以，编写 ViewPager 的适配器时要继承 PagerAdapter 类，并且要重写里面的方法。在每次创建 ViewPager 或滑动过程中，以下 4 个方法都会被调用。

① public abstract int getCount()，这个方法用于获取当前窗体界面数。

② public abstract boolean isViewFromObject(View arg0, Object arg1)，这个方法用于判断是否由对象生成界面。

③ public Object instantiateItem(View container, int position)，这个方法返回当前的view，功能是往 PageView 里添加自己需要的 page。

④ public void destroyItem(ViewGroup container, int position，Object object)，这个方法是从 ViewGroup 中移除当前 View。

多学一招：

(1) ViewPager 里面对每个页面的管理是 key-value 形式的，也就是说，每个 page 都有一个对应的 id(id 是 object 类型)，需要对 page 操作时都是通过 id 来完成的。instantiateItem(View container, int position)方法的返回值就是 page 对应的 id。

(2) public abstract boolean isViewFromObject (View view, Object object)，这个方法就是用来告诉框架 view 的 id 是不是这个 object。Google 官方推荐把 view 当 id 用，所以常规的 instantiateItem()函数的返回值是你自己定义的 view，而 isViewFromObject()的返回值是 view == object。

【任务实施】

1. 任务分析

本任务通过 ViewPager 实现引导页。

(1) 新建工程，将引导页的图片复制到工程中，编写引导页对应的布局。

(2) 实现 ViewPager 的适配器。

(3) 新建引导页对应的 Activity-GuideActivity，并声明同时设置成第一个启动的Activity，判断是否为第一次进入，如果不是则直接跳转到 MainActivity。

(4) 将引导页的图片添加到图片集合中，然后设置 ViewPager 的适配器。

(5) 设置 ViewPager 的滑动监听，如果滑动到最后一张引导页，显示“开启应用”按钮；否则隐藏。单击“开启应用”按钮可进入 MainActivity。

2. 任务实现

(1) 新建工程 Project4_2，将随书资料中的 pic_guide_1、pic_guide_2、pic_guide_3 和pic_guide_4 复制到 mipmap 文件夹下，编写引导页对应的布局文件 activity_guide.xml，该布局由 ViewPager 和一个按钮组成，代码如下：

```
<?xml version="1.0" encoding="utf-8"?>
<RelativeLayout xmlns:android="http://schemas.android.com/apk/res/android"
    android:layout_width="match_parent"
    android:layout_height="match_parent">
    <!--ViewPager 填充整个布局，用于向导页面图片展示-->
    <android.support.v4.view.ViewPager
        android:id="@+id/vp"
        android:layout_width="match_parent"
        android:layout_height="match_parent"></android.support.v4.view.ViewPager>
    <!--Button 覆盖在 ViewPager 上，默认隐藏，在向导页面最后一页，通过代码控制让其显示-->
    <Button
        android:id="@+id/btn_enter_login"
        android:text="开启应用"
```

```
            android:onClick="enterhome"
            android:paddingLeft="40dp"
            android:paddingRight="40dp"
            android:background="#fff15d"
            android:textSize="20sp"
            android:textStyle="bold"
            android:textColor="#cccccc"
            android:layout_alignParentBottom="true"
            android:layout_centerHorizontal="true"
            android:layout_marginBottom="130dp"
            android:layout_width="wrap_content"
            android:layout_height="wrap_content"
            android:visibility="gone"/>
</RelativeLayout>
```

(2) 接下来实现 ViewPager 的适配器 MyPagerAdapter 类。

① 新建 MyPagerAdapter 类继承 PagerAdapter，并定义 ImageView 集合用于存放 ViewPager 显示的图片，在构造方法中将传进来的 ImageView 集合赋值给定义的 ImageView 列表，代码如下：

```
public class MyPagerAdapter extends PagerAdapter {
    private ArrayList<ImageView> list;
    public MyPagerAdapter(ArrayList<ImageView> list) {
        this.list = list;
    }
```

② 重写 PagerAdapter 中的 4 个方法，代码如下：

```
    //PagerAdapter 适配器选择图片对应的 View 显示在当前的 ViewPager 中
    @Override
    public Object instantiateItem(ViewGroup container, int position) {
        View view = list.get(position);
        container.addView(view);
        return view;
    }
    //从 ViewGroup 中移除当前图片对应的 View
    @Override
    public void destroyItem(ViewGroup container, int position, Object object) {
        View view = list.get(position);
        container.removeView(view);
    }
    //当前窗体数目即 list 的个数
    @Override
    public int getCount() {
        return list.size();
    }
    //判断一个视图是否与一个给定的 key 相对应
    @Override
    public boolean isViewFromObject(View view, Object object) {
        //返回值固定为 view==object
        return view == object;
    }
```

(3) 新建 GuideActivity，在 AndroidManifest.xml 中声明并设置为第一个启动的 Activity，在 GuideActivity 中定义变量并初始化。GuideActivity 代码如下：

```
public class GuideActivity extends Activity {
    //引导页的四张图片
    private int imgs[] = {R.mipmap.pic_guide_1, R.mipmap.pic_guide_2,
R.mipmap.pic_guide_3,R.mipmap.pic_guide_4};
```

```
    private ViewPager vp;
    private ArrayList<ImageView> list;
    private Button btnEnterLogin;
    private SharedPreferences sp;
    private SharedPreferences.Editor editor;
    @Override
    protected void onCreate(Bundle savedInstanceState) {
        super.onCreate(savedInstanceState);
        setContentView(R.layout.activity_guide);
        //初始化组件
        initView();
        //检查是否进入过引导页面
        checkIsEnter();
        //加载引导页的图片
        loadData();
        //监听
        addListener();
    }
    public void initView() {
        vp = (ViewPager) findViewById(R.id.vp);
        btnEnterLogin = (Button) findViewById(R.id.btn_enter_login);
        sp=getSharedPreferences("MySp",Activity.MODE_PRIVATE);
        editor=sp.edit();
    }
```

(4) 在 GuideActivity 中添加 checkIsEnter()方法用于判断是否为第一次进入 App，如果不是就直接跳转到 MainActivity，代码如下：

```
  /**
     * 检查是否进入过引导页面
     */
    private void checkIsEnter(){
        //获取是否进入过向导页面
        boolean isFirst = sp.getBoolean("isFirst",true);
        //如果不是第一次打开 App，直接跳转到登录页面
        if (!isFirst){
            //进入 MainActivity
            enter();
        }
    }
```

enter()方法代码如下：

```
  /**
     * 进入 MainActivity
     */
    private void enter(){
        Intent intent = new Intent(this,MainActivity.class);
        startActivity(intent);
        this.finish();
    }
```

(5) 在 GudieActivity 的 loadData()方法中将引导页的图片放到 List 集合中，然后设置 ViewPager 的适配器，代码如下：

```
   /**
     * 加载引导页的图片
     */
    public void loadData() {
        //加载图片，将图片放在 ImageView 中，然后将 ImageView 加载到 ViewPager 中
```

```
        //定义 ImageView 列表
        list = new ArrayList<ImageView>();
        //遍历 imgs
        for (int resId : imgs) {
            //实例化 ImageView
            ImageView iv = new ImageView(this);
            //将 ImageView 的背景设为引导页的图片
            iv.setBackgroundResource(resId);
            //添加 ImagView 到 list 中
            list.add(iv);
        }
        //设置 ViewPaper 的适配器
        vp.setAdapter(new MyPagerAdapter(list));
    }
```

(6) 添加监听的方法 addListener()。

① 设置 ViewPager 的滑动监听，如果滑动到最后一张引导页，显示进入按钮，否则隐藏，代码如下：

```
    public void addListener() {
        //设置 ViewPager 滑动监听
        vp.addOnPageChangeListener(new ViewPager.OnPageChangeListener() {
            @Override
            public void onPageScrolled(int position, float positionOffset, int positionOffsetPixels) {
            }
            @Override
            public void onPageSelected(int position) {
                //如果当前在最后一张引导页，显示进入按钮，否则隐藏
                //因为 position 的初始值为 0，所以最后一张引导页等于 img.length-1
                if (position == imgs.length - 1) {
                    btnEnterLogin.setVisibility(View.VISIBLE);
                } else {
                    btnEnterLogin.setVisibility(View.GONE);
                }
            }
            @Override
            public void onPageScrollStateChanged(int state) {
            }
        });
```

② 单击“开启应用”按钮，进入 MainActivity，并在 Shareperferences 中设置 isFirst 为 false，表示不是第一次打开应用程序，代码如下：

```
//开始体验按钮，单击进入 MainActivity 并记录
        btnEnterLogin.setOnClickListener(new View.OnClickListener() {
            @Override
            public void onClick(View v) {
                //进入 MainActivity
                enter();
                //记录已经进入过向导页面了
                editor.putBoolean("isFirst",false);
                editor.commit();
            }
        });
    }
```

3. 运行结果

(1) 运行程序，出现引导页，向左滑动到最后一张图片，如图 4-22 所示。

图 4-22 引导页

(2) 单击“开启应用”按钮，进入 MainActivity 界面，如图 4-23 所示。
(3) 再次运行程序，则会直接进入 MainActivity 界面。

图 4-23 MainActivity 界面

任务 3 侧滑菜单的实现

【任务描述】

本任务通过 DrawerLayout 实现侧滑菜单，效果如图 4-24 和图 4-25 所示，默认进入室内环境采集系统界面。滑动出侧滑菜单后，单击“室内

扫码观看视频讲解

环境”可进入室内环境采集系统界面，单击“园区监控”可进入园区监控系统。任务清单如表 4-4 所示。

图 4-24　室内环境采集系统界面

图 4-25　园区监控系统界面

表 4-4　任务清单

任务课时	4 课时	任务组员数量	建议 1 人
任务组采用设备	PC 一台 模拟器或真机或工控平板		

【知识解析】

1. Fragment 介绍

1)　Fragment 概述

Fragment 是一种可以嵌入在活动中的 UI 片段，能够让程序更加合理和充分利用大屏幕的空间，出现的初衷是为了适应大屏幕的平板电脑，可以将其看成一个小型 Activity，又称为 Activity 片段。

Fragment 允许这样的一种设计，而不需要开发者亲自来管理 UI 布局的复杂变化。例如，新闻应用可以在屏幕左侧使用 Fragment 来展示一个文章的列表，然后在屏幕右侧使用另一个 Fragment 来展示一篇文章。两个 Fragment 并排显示在同一个 Activity 中。每个 Fragment 拥有它自己的一套生命周期回调方法，并能处理 Fragment 的用户输入事件，如图 4-26 所示。

Fragment 在应用中表现为一个模块化和可重用的组件，即在 Fragment 定义了它自己的布局，以及通过它自己的生命周期回调方法定义了它自己的行为，在使用时可以将 Fragment 包含到多个 Activity 中。

2)　Fragment 的常用方法

Fragment 的常用方法见表 4-5。

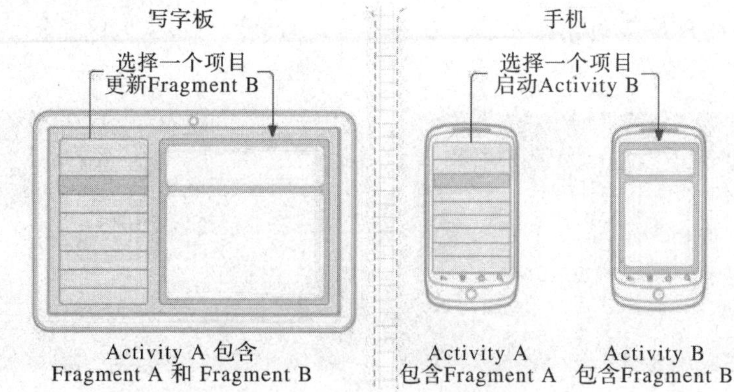

图 4-26　Fragment 模块管理图

表 4-5　Fragment 常用方法

方法名称	功能说明
getSupportFragmentManager()	获取 FragmentManager 对象，该对象用于在 Activity 中操作 Fragment
FragmentManager.beginTransaction()	获取 FragementTransaction 对象，该对象为 Fragment 中的事务，可通过 FragementTransaction 进行 Fragment 的相关操作
FragmentTransaction.add()	添加 Fragment
FragmentTransaction.remove()	去除 Fragment
FragmentTransaction.replace()	替换 Fragment
FragmentTransaction.commit()	提交更改 Fragment 的事务

2. Fragment 实现页面切换

使用 Fragment 实现页面切换功能，单击页面上的“按钮 1”按钮和“按钮 2”按钮，分别跳转到 FragmentA 和 FragmentB 页面，页面效果如图 4-27 所示。

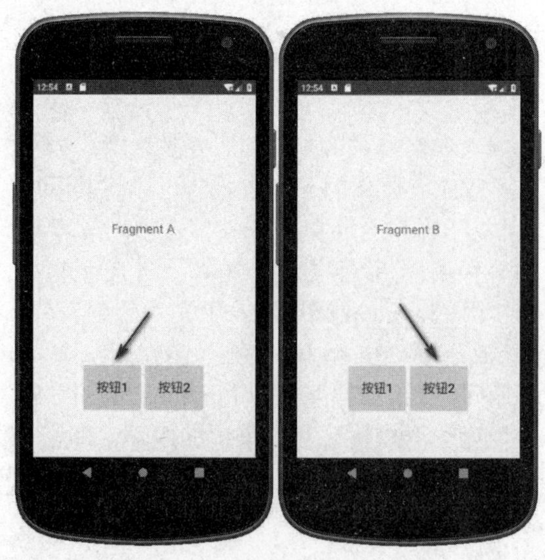

图 4-27　Fragment 案例效果

(1) 新建工程 Demo4-3。要实现图 4-27 所示的显示效果，需要有 3 个布局文件。第一个布局文件为主页面的布局，还有 FragmenA 和 FragmentB 的布局。主页面布局底部是两个导航按钮，两个导航按钮上方是一个空白的布局，主页面布局 activity_main.xml 代码如下：

```
<RelativeLayout xmlns:android="http://schemas.android.com/apk/res/android"
    android:layout_width="match_parent"
    android:layout_height="match_parent"
    android:orientation="vertical">
<!--第一部分：Fragment 的内容显示区域-->
    <LinearLayout
        android:id="@+id/line"
        android:layout_width="fill_parent"
        android:layout_height="fill_parent"
        android:layout_above="@+id/layout"
        android:gravity="center"
        android:orientation="vertical" >
    </LinearLayout>
<!--第二部分：底部导航按钮-->
    <LinearLayout
        android:id="@+id/layout"
        android:layout_width="match_parent"
        android:layout_height="wrap_content"
        android:layout_alignParentBottom="true"
        android:layout_marginBottom="70dp"
        android:orientation="horizontal"
        android:layout_marginLeft="80dp"
        android:layout_marginRight="80dp">
        <Button
            android:id="@+id/button1"
            android:layout_width="0dp"
            android:layout_height="80dp"
            android:layout_weight="1"
            android:text="按钮 1"
            android:textSize="20sp" />
        <Button
            android:id="@+id/button2"
            android:layout_width="0dp"
            android:layout_height="80dp"
            android:layout_weight="1"
            android:text="按钮 2"
            android:textSize="20sp" />
    </LinearLayout>
</RelativeLayout>
```

(2) FragemntB 与 FragmentA 的布局类似，只是显示的文字不同，FragmentA 对应布局 fragment_a 代码如下：

```
<?xml version="1.0" encoding="utf-8"?>
<RelativeLayout xmlns:android="http://schemas.android.com/apk/res/android"
    android:layout_width="match_parent"
    android:layout_height="match_parent">
    <TextView
        android:id="@+id/textb"
        android:layout_width="wrap_content"
        android:layout_height="wrap_content"
        android:layout_centerInParent="true"
        android:text="Fragment A"
        android:textSize="20sp" />
</RelativeLayout>
```

(3) 在项目中建立 FragmentA.java 文件。该文件继承 Fragment 类，并关联布局文件，代码如下：

```
1.      public class FragmentA extends Fragment{
2.      @Override
3.      public View onCreateView(LayoutInflater inflater, ViewGroup container,
4.              Bundle savedInstanceState) {
5.          //把布局转换成对应的 View 对象
6.          View view=inflater.inflate(R.layout.fragment_a, null);
7.          return view;
8.      }
9.
10. }
```

第 7 行返回的是要显示的视图，在用 Fragment 时需要在 onCreateView()方法中返回要显示的视图，相当于 Activity 里面的 setContentView()，作用是关联布局文件。FragmentB 与 FragmentA 的代码基本一致，只是知识关联的布局不同而已。

(4) MainActivity 类继承 FragmentActivity(继承之后才能管理 Fragment)，加载 FragmentA 页面，并对按钮进行监听，代码如下：

```
public class MainActivity extends FragmentActivity implements View.OnClickListener{
    private Button button1;
    private Button button2;
    //Fragment 的事务
    private FragmentTransaction transaction;
    FragmentA fragmentA;
    FragmentB fragmentB;
    //Fragment 管理者
    private FragmentManager manager;
    @Override
    protected void onCreate(Bundle savedInstanceState) {
        super.onCreate(savedInstanceState);
        setContentView(R.layout.activity_main);
        button1=(Button)findViewById(R.id.button1);
        button2=(Button)findViewById(R.id.button2);
        button1.setOnClickListener(this);
        button2.setOnClickListener(this);
        //获得 Fragment 实例
        fragmentA=new FragmentA();
        fragmentB=new FragmentB();
        //得到 Fragment 的管理器
        manager=getSupportFragmentManager();
        //得到 Fragment 的事务
        transaction=manager.beginTransaction();
        transaction.add(R.id.line,fragmentA);
        //提交事务
        transaction.commit();
    }
```

(5) 在按钮的监听事件中实现单击“按钮 1”按钮，跳转到 FragmentA 页面，单击“按钮 2”按钮跳转到 FragmentB 页面的功能，代码如下：

```
@Override
    public void onClick(View view) {
        switch (view.getId()) {
            case R.id.button1:
                //得到 Fragment 的事务
                transaction=manager.beginTransaction();
```

```
                transaction.replace(R.id.line,fragmentA);
                //提交事务
                transaction.commit();
                break;
            case R.id.button2:
                //得到 Fragment 的事务
                transaction=manager.beginTransaction();
                transaction.replace(R.id.line,fragmentB);
                //提交事务
                transaction.commit();
                break;
        }
    }
```

3. DrawerLayout 简介

DrawerLayout 是 Google 官方提供的一个侧滑菜单控件。DrawerLayout 分为侧边菜单和主内容区两部分，侧边菜单可以根据手势展开与隐藏，主内容区的内容可以随着菜单的单击而变化。

使用 DrawerLayout 有以下几个注意事项。

① 主内容视图一定为 DrawerLayout 的第一个子视图。

② 主内容视图宽度和高度需要 match_parent。也就是说，当抽屉隐藏时，要让用户看到主视图的全部内容。

③ 必须显示指定侧滑视图的 android:layout_gravity 属性。

android:layout_gravity = "start"时，从左向右滑出菜单；

android:layout_gravity = "end"时，从右向左滑出菜单。

④ 侧滑视图的宽度以 dp 为单位，不建议超过 320dp(为了总能看到一些主内容视图)。

【任务实施】

1. 任务分析

本任务通过 DrawerLayout 实现侧滑菜单。

(1) 新建工程 Project4_3，将任务需要的图片复制到工程下。

(2) 编写侧滑菜单对应的布局 home_drawerlayout.xml。

(3) 在 activity_main.xml 中引入侧滑菜单。

(4) 编写室内环境采集系统对应的 Fragment 和布局文件。

(5) 编写园区监控系统对应的 Fragment 和布局文件。

(6) 在 MainActivity 中实现侧滑菜单的功能。

(7) 将 MainActivity 的屏幕方向设成水平。

2. 任务实现

(1) 新建工程 Project4_3，将随书资料中项目四任务 3 的图片复制到 mipmap-hdpi 文件夹下，将 radio_button_gray_selector.xml 文件复制到 drawable 文件夹下。

(2) 编写侧滑菜单对应的布局 home_drawerlayout.xml，该布局最外层是一个 DrawerLayout 组件，内层是帧布局，用于显示 Fragment 的内容。代码如下：

```
1.  <?xml version="1.0" encoding="utf-8"?>
2.  <android.support.v4.widget.DrawerLayout xmlns:android="http://schemas.android.com/apk/res/android"
3.      android:id="@+id/dl_left"
4.      android:layout_width="match_parent"
5.      android:layout_height="match_parent">
6.      <!--内容占位，后面通过代码替换-->
7.      <FrameLayout
8.          android:id="@+id/fg_content"
9.          android:layout_width="match_parent"
10.         android:layout_height="match_parent"></FrameLayout>
11.     <!--侧滑菜单-->
12.     <LinearLayout
13.         android:layout_gravity="left"
14.         android:background="#242B3E"
15.         android:layout_width="200dp"
16.         android:layout_height="match_parent">
17.         <RadioGroup
18.             android:layout_marginTop="10dp"
19.             android:id="@+id/rg"
20.             android:gravity="center_horizontal"
21.             android:layout_width="match_parent"
22.             android:layout_height="match_parent">
23.             <!--引用 RadioButton 样式-->
24.             <RadioButton
25.                 android:checked="true"
26.                 android:id="@+id/rb_environment"
27.                 android:drawableTop="@mipmap/icon_inner"
28.                 style="@style/home_RadioButtonStyle" />
29.
30.             <RadioButton
31.                 android:id="@+id/rb_monitoring"
32.                 style="@style/home_RadioButtonStyle"
33.                 android:drawableTop="@mipmap/icon_camera" />
34.         </RadioGroup>
35.     </LinearLayout>
36. </android.support.v4.widget.DrawerLayout>
```

第 28 行 RadioButton 引用了 styles.xml 文件中的样式，代码如下：

```
<style name="home_RadioButtonStyle">
        <item name="android:layout_width">150dp</item>
        <item name="android:layout_height">150dp</item>
        <item name="android:button">@null</item>
        <item name="android:paddingTop">20dp</item>
        <item name="android:paddingRight">10dp</item>
        <item name="android:paddingLeft">10dp</item>
        <item name="android:background">@drawable/radio_button_gray_selector</item>
    </style>
```

(3) 在 activity_main.xml 中引入侧滑菜单，activity_main.xml 代码如下：

```
<?xml version="1.0" encoding="utf-8"?>
<LinearLayout
    xmlns:android="http://schemas.android.com/apk/res/android"
    xmlns:tools="http://schemas.android.com/tools"
    android:layout_width="match_parent"
    android:layout_height="match_parent"
    android:orientation="vertical">
    <!--侧滑菜单-->
    <include layout="@layout/home_drawerlayout"/>

</LinearLayout>
```

(4) 编写室内环境采集系统对应的 Fragment 和布局文件，代码如下：

EnvironmentFragment.java

```
public class EnvironmentFragment extends Fragment{
    @Override
    public View onCreateView(LayoutInflater inflater, ViewGroup container,
            Bundle savedInstanceState) {
        //把布局转换成对应的 View 对象
        View view=inflater.inflate(R.layout.fragment_environment, null);
        return view;
    }

}
```

fragment_environment.xml

```
<?xml version="1.0" encoding="utf-8"?>
<RelativeLayout
    xmlns:android="http://schemas.android.com/apk/res/android"
    android:layout_width="match_parent"
    android:layout_height="match_parent"
    android:background="@mipmap/bg_out"
    android:orientation="horizontal">

    <TextView
        android:layout_width="wrap_content"
        android:layout_height="wrap_content"
        android:layout_centerHorizontal="true"
        android:text="室内环境采集系统"
        android:textSize="20sp"/>
</RelativeLayout>
```

(5) 编写园区监控系统对应的 Fragment 和布局文件，代码如下：

MonitoringFragment.java

```
public class MonitoringFragment extends Fragment{
    @Override
    public View onCreateView(LayoutInflater inflater, ViewGroup container,
            Bundle savedInstanceState) {
        //把布局转换成对应的 View 对象
        View view=inflater.inflate(R.layout.fragment_monitoring, null);
        return view;
    }
}
```

fragment_monitoring.xml

```
<?xml version="1.0" encoding="utf-8"?>
<RelativeLayout
    xmlns:android="http://schemas.android.com/apk/res/android"
    android:layout_width="match_parent"
    android:layout_height="match_parent"
    android:background="@mipmap/bg_out"
    android:orientation="horizontal">

    <TextView
        android:layout_width="wrap_content"
        android:layout_height="wrap_content"
        android:layout_centerHorizontal="true"
        android:text="园区监控系统"
        android:textSize="20sp"/>
</RelativeLayout>
```

(6) 在 MainActivity 中实现侧滑菜单的功能，代码如下：

```
public class MainActivity extends AppCompatActivity implements RadioGroup.OnCheckedChangeListener{
    private DrawerLayout mDrawerLayout;
    //侧滑菜单 RadioGroup
    private RadioGroup radioGroup;
    //侧滑菜单当前条目
    private int currentItem = 0;
    //Fragment 集合
    private ArrayList<Fragment> list;
    @Override
    protected void onCreate(Bundle savedInstanceState) {
        super.onCreate(savedInstanceState);
        setContentView(R.layout.activity_main);
        mDrawerLayout = (DrawerLayout) findViewById(R.id.dl_left);
        radioGroup = (RadioGroup) findViewById(R.id.rg);
        //设置侧滑菜单的监听
        radioGroup.setOnCheckedChangeListener(this);
        list = new ArrayList<Fragment>();
        //将 EnvironmentFragment 添加到 Fragment 集合中
        list.add(new EnvironmentFragment());
        //将 MonitoringFragment 添加到 Fragment 集合中
        list.add(new MonitoringFragment());
        //设置默认显示室内环境采集系统 Fragment 并提交事务
        getSupportFragmentManager().beginTransaction().add(R.id.fg_content, list.get(0)).commit();
    }

    @Override
    public void onCheckedChanged(RadioGroup radioGroup, int checkedId) {
        try {
            //侧滑菜单选中条目，对应界面切换，更改标题栏
            switch (checkedId) {
                case R.id.rb_environment:
                    currentItem = 0;
                    break;
                case R.id.rb_monitoring:
                    currentItem = 1;
                    break;
            }
            //显示选中的 Fragment 并提交事务
            getSupportFragmentManager().beginTransaction().replace(R.id.fg_content, list.get(currentItem)).commit();
            //重绘菜单选项
            invalidateOptionsMenu();
            //显示或隐藏侧滑菜单
            drawerToggle();
        } catch (Exception e) {
            e.printStackTrace();
        }
    }
    /**
     * 切换侧滑菜单状态
     */
    public void drawerToggle() {
        //如果侧滑菜单当前有显示，关闭侧滑菜单；否则显示侧滑菜单
        if (mDrawerLayout.isDrawerOpen(Gravity.START)) {
            mDrawerLayout.closeDrawers();
```

```
            } else {
                mDrawerLayout.openDrawer(Gravity.START);
            }
        }

}
```

(7) 将 MainActivity 的屏幕方向设置成水平，在 AndroidManifest.xml 中修改代码如下：

```
<activity android:name=".MainActivity"
            android:screenOrientation="landscape">
            <intent-filter>
                <action android:name="android.intent.action.MAIN"/>

                <category android:name="android.intent.category.LAUNCHER"/>
            </intent-filter>
        </activity>
```

3. 运行结果

(1) 运行程序，出现室内环境采集系统界面，在屏幕最左侧向右滑动，默认选中的是室内环境菜单，如图 4-28 所示。

(2) 在屏幕最左侧向右滑动，选中“园区监控”，出现园区监控系统界面，如图 4-29 所示。

图 4-28 室内环境采集系统界面

图 4-29 园区监控系统界面

任务 4 标题栏的实现

【任务描述】

扫码观看视频讲解

在开发中，如果有多个页面标题类似，往往会统一编写一个标题栏，以便统一管理和处理。本任务使用 Toolbar 实现标题栏的效果，如图 4-30

和图 4-31 所示。标题栏左侧是 Logo 图标，中间是标题，右侧是设置菜单按钮，在室内环境采集系统界面单击“设置”按钮，弹出“设置阈值”菜单项；在园区监控系统单击“设置”按钮，弹出“设置摄像头地址”和“查看截图”菜单项，单击菜单项弹出对应的提示信息。任务清单如表 4-6 所示。

图 4-30　室内环境采集系统标题栏效果

图 4-31　园区监控系统标题栏效果

表 4-6　任务清单

任务课时	4 课时	任务组员数量	建议 1 人
任务组采用设备	PC 一台 模拟器或真机或工控平板		

【知识解析】

1. Toolbar 简介

Toolbar 是在 Android 5.0(API 21)版本开始推出的，它的出现规范了 Android 开发者 App 标题栏的设计风格，极大地提高了开发效率，Google 非常推荐大家使用 Toolbar 作为 Android 客户端的导航栏，以此来取代之前的 Actionbar。由于其高度的灵活性、可定制性，越来越多的应用出现了 Toolbar，很多 App 的标题栏使用的都是 Toolbar。Google 为了将这一设计向下兼容，推出了兼容版的 Toolbar，为此，需要在工程中引入 appcompat-v7 的兼容包，使用 android.support.v7.widget.Toolbar 进行开发，这样就能兼容 Android 5.0 以下的版本。图 4-32 是 Toolbar 效果示意图。

2. Toolbar 的使用

(1) 为了使 Toolbar 兼容 Android 5.0 以下的版本，项目的 build.gradle 文件中要引入 appcompat-v7 的兼容包，如图 4-33 所示。

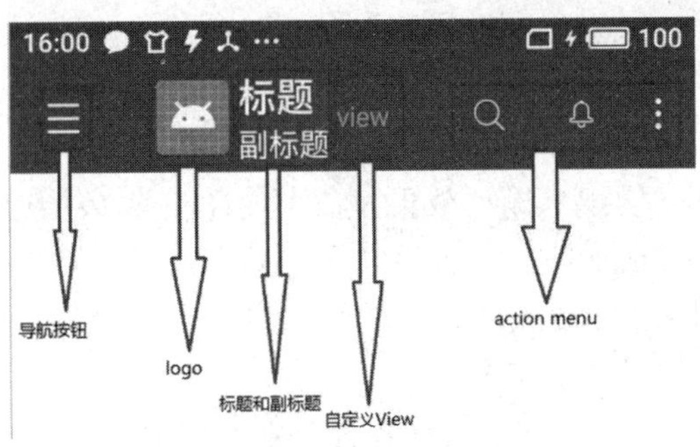

图 4-32　Toolbar 效果示意图

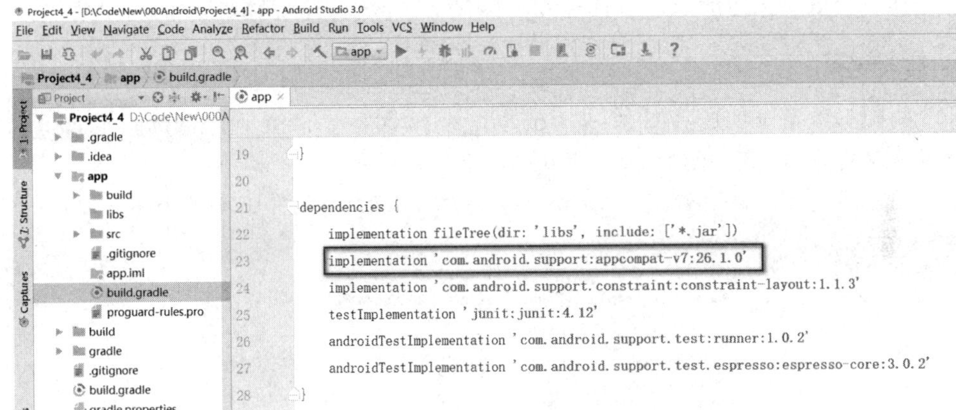

图 4-33　添加依赖界面

(2) 在布局文件中如何去使用Toolbar的自定义属性呢？在使用之前必须指定引用第三方控件的命名空间，在布局文件的根节点中添加以下一行代码：

```
xmlns:app=http://schemas.android.com/apk/res-auto
```

app 是第三方命名空间的名字，可以任意命名，在使用自定义属性时必须以它开头。Toolbar 中常用自定义属性的说明见表 4-7。

表 4-7　Toolbar 中常用自定义属性说明

变　量	说　明
app:logo	定义 Logo 图标
app:navigationIcon	定义导航按钮的图标
app:theme	定义 app 的主题
app:popupTheme	定义弹出菜单的主题
app:title	定义标题文字
app:subtitle	定义副标题文字
app:titleTextColor	定义标题文字的颜色

(3) Toolbar 弹出菜单的显示效果，需要在 styles.xml 样式文件中去设置。

【任务实施】

1. 任务分析

本任务使用 Toolbar 实现标题栏效果，在本项目任务 3 的代码 Project4_3 基础上实现，项目中已经引入了 appcompat-v7 的兼容包。

(1) 程序界面有用户名输入框、密码输入框、“记住密码”复选框和“立即登录”按钮。

(2) 在程序的 onCreate 方法中处理各种对象的初始化，并从 SharedPreferences 文件中取值回填到界面上。

(3) 在“登录”按钮的点击事件中处理保存记住密码的标志和对应的用户名和密码。

2. 任务实现

(1) 将随书资料中项目 4 任务 4 下的 btn_selected.png、btn_setting.png 和 icon_smart_park.png 放到 Project4_3 下的 mapmap-hdpi 文件夹下。

(2) 将系统自带的标题栏去掉。在 res→values-styles.xml 文件中添加去除标题栏的代码，如图 4-34 所示。

```
<style name="AppTheme" parent="Theme.AppCompat">
    <item name="colorPrimary">@color/colorPrimary</item>
    <item name="colorPrimaryDark">@color/colorPrimaryDark</item>
    <item name="colorAccent">@color/colorAccent</item>
    <item name="windowNoTitle">true</item>
</style>
```

图 4-34 设置 styles.xml 样式

(3) 在 styles.xml 样式文件中定义弹出菜单的样式。

① 定义弹出菜单的样式，样式名称为 OverflowMenuStyle，代码如下：

```
<!--弹出菜单样式-->
<style name="OverflowMenuStyle" parent="@style/Widget.AppCompat.PopupMenu.Overflow">
        <!-- 是否覆盖锚点，默认为 true，即盖住 Toolbar -->
        <item name="overlapAnchor">false</item>
        <!-- 弹出层背景颜色 -->
        <item name="android:popupBackground">#eeeeee</item>
        <!-- 弹出层垂直方向上的偏移,即在竖直方向上距Toolbar的距离,值为负则会盖住Toolbar -->
        <item name="android:dropDownVerticalOffset">0dp</item>
        <!-- 弹出层水平方向上的偏移，即距离屏幕左边的距离，负值会导致右边出现空隙 -->
        <item name="android:dropDownHorizontalOffset">10dp</item>
        <!-- 设置弹出菜单文字颜色 -->
        <item name="android:textColor">@android:color/black</item>
    </style>
```

在上面的 style 标签中，设置了 parent="@style/Widget.AppCompat.PopupMenu.Overflow"，是弹出菜单的样式，通过 parent 属性继承 Android 系统中的弹出菜单的样式。再通过 style 标签中 item 子节点来自定义弹出菜单的属性。

② 弹出菜单如果有多行文字，文字之间是没有分割线的，通过 style 样式来设置弹出菜单多行文字之间的分割线，代码如下：

```
<!--toolbar menu 分割线-->
<style name="AppDropDownListViewStyle" parent="Widget.AppCompat.ListView.DropDown">
    <!--设置分割线的颜色-->
    <item name="android:divider">#cccccc</item>
    <!--设置分割线的高度-->
    <item name="android:dividerHeight">1dp</item>
</style>
```

③ 定义一个 Toolbar 弹出菜单的主题，在这个 style 中，去引用这两个样式，代码如下：

```
<!--toolbar 弹出菜单主题-->
<style name="AppToolbarPopupTheme" parent="@style/ThemeOverlay.AppCompat.Light">
    <!--弹出菜单样式-->
    <item name="actionOverflowMenuStyle">@style/OverflowMenuStyle</item>
    <!--toolbar menu 分割线-->
    <item name="android:dropDownListViewStyle">@style/AppDropDownListViewStyle</item>
</style>
```

以上就完成了自定义弹出菜单 style 的编写，后续在设置 Toolbar 的时候可引用这个弹出菜单的主题。

(4) 菜单按钮默认是一 3 个点的图标。接下来通过自定义，设置菜单按钮的图标。

① 菜单按钮的图标是在 android:style/Widget.Holo.Light.ActionButton.Overflow 的 style 中设置。编写一个 style 继承该 style，设置菜单按钮的图标，代码如下：

```
<!--修改 toolbar menu 默认图标-->
<style name="ActionButton.Overflow.Setting" parent="android:style/Widget.Holo.Light.ActionButton.Overflow">
    <item name="android:src">@mipmap/btn_login</item>
</style>
```

② 直接引用上面 ActionButton.Overflow.Setting 的样式无法将菜单按钮的默认图标修改过来，必须把上面的样式设置到 Theme.AppCompat.Light.NoActionBar 样式的属性中才能修改默认图标，代码如下：

```
<!--toolbar 主题-->
<style name="AppToolBarTitleTheme" parent="Theme.AppCompat.Light.NoActionBar">
    <!-- actionOverflowButtonStyle，该属性为设置菜单按钮的样式的属性-->
    <item name="actionOverflowButtonStyle">@style/ActionButton.Overflow.Setting</item>
</style>
```

(5) 接下来编写 activity_main 布局文件。在布局文件中添加一个 Toolbar 组件，并自定义 Toolbar 的样式，代码如下：

```
<?xml version="1.0" encoding="utf-8"?>
<LinearLayout
    xmlns:android="http://schemas.android.com/apk/res/android"
    xmlns:tools="http://schemas.android.com/tools"
    android:layout_width="match_parent"
    android:layout_height="match_parent"
    android:orientation="vertical">
    <!--设置 toolbar 高度为 50dp，toolbar:logo 设置 Logo 图标，toolbar:theme 定义 Toolbar 的主题，修改菜单的默认图标，toolbar:popupTheme 定义弹出菜单的主题，设置弹出菜单的样式-->
    <android.support.v7.widget.Toolbar xmlns:android="http://schemas.android.com/apk/res/android"
                         xmlns:toolbar="http://schemas.android.com/apk/res-auto"
                         android:id="@+id/toolbar"
                         android:layout_width="match_parent"
                         android:layout_height="50dp"
```

```
                                android:background="#0189ff"
                                android:orientation="horizontal"
                                toolbar:logo="@mipmap/icon_smart_park"
                                toolbar:popupTheme="@style/AppToolbarPopupTheme"
                                  toolbar:theme="@style/AppToolBarTitleTheme"
                                  toolbar:title=" ">
            <TextView
                android:id="@+id/tv_subtitle"
                android:layout_width="wrap_content"
                android:layout_height="wrap_content"
                android:layout_gravity="center"
                android:textColor="#fff"
                android:textSize="20sp"/>
        </android.support.v7.widget.Toolbar>
        <!--侧滑菜单-->
        <include layout="@layout/home_drawerlayout"/>
</LinearLayout>
```

(6) 在 MainActivity 的 onCreate()方法中添加代码，绑定 Toolbar 标题栏标题的文本，实例化 Toolbar，并用 Toolbar 替换 ActionBar。代码如下：

```
    private TextView tvSubTitle;
@Override
protected void onCreate(Bundle savedInstanceState) {
    super.onCreate(savedInstanceState);
    setContentView(R.layout.activity_main);
    tvSubTitle = findViewById(R.id.tv_subtitle);
    Toolbar toolbar = (Toolbar) findViewById(R.id.toolbar);
    //使用 ToolBar 替换 Actionbar
    setSupportActionBar(toolbar);
```

(7) 设置标题栏菜单内容。右键单击 res 文件夹，新建 menu 资源文件夹，在 menu 文件夹中新建 park_toolbar_menu.xml 菜单文件，代码如下：

```
<?xml version="1.0" encoding="utf-8"?>
<menu xmlns:android="http://schemas.android.com/apk/res/android"
    xmlns:app="http://schemas.android.com/apk/res-auto">
    <item
        android:id="@+id/action_set_threshold"
        android:title="设置阈值"
        app:showAsAction="withText" />
    <item
        android:id="@+id/action_set_camera"
        android:title="设置摄像头地址"
        app:showAsAction="withText" />
    <item
        android:id="@+id/action_look_pic"
        android:title="查看截图"
        app:showAsAction="withText" />
</menu>
```

(8) 在 MainActivity 中重写创建菜单的 onCreateOptionsMenu()方法，设置 Toolbar 的弹出菜单，代码如下：

```
//创建标题栏菜单
    @Override
    public boolean onCreateOptionsMenu(Menu menu) {
        //将 park_toolbar_menu.xml 作为标题栏的菜单
        getMenuInflater().inflate(R.menu.park_toolbar_menu, menu);
```

```
            return super.onCreateOptionsMenu(menu);
    }
```

(9) 跳转到不同的 Fragment 页面显示对应的标题。在侧滑菜单选中条目的时候设置对应页面的标题。在 onCheckedChanged()方法中添加图 4-35 方框中的代码。

```
@Override
public void onCheckedChanged(RadioGroup radioGroup, int checkedId) {
    try {
        //侧滑菜单选中条目，对应界面切换，更改标题栏
        switch (checkedId) {
            case R.id.rb_environment:
                currentItem = 0;
                tvSubTitle.setText("园区环境采集系统");
                break;
            case R.id.rb_monitoring:
                currentItem = 1;
                tvSubTitle.setText("园区监控系统");
                break;
```

图 4-35　onCheckedChanged 代码

(10) 重写动态设置菜单的方法 onPrepareOptionsMenu(Menu menu)，分别设置各个页面弹出菜单的显示内容，代码如下：

```
//动态设置 ToolBar 弹出菜单
    @Override
    public boolean onPrepareOptionsMenu(Menu menu) {
        // 动态设置 ToolBar 状态
        switch (currentItem) {
            case 0:
                //显示设置阈值菜单
                menu.findItem(R.id.action_set_threshold).setVisible(true);
                menu.findItem(R.id.action_set_camera).setVisible(false);
                menu.findItem(R.id.action_look_pic).setVisible(false);
                break;
            case 1:
                //显示设置摄像头地址菜单和查看截图菜单
                menu.findItem(R.id.action_set_camera).setVisible(true);
                menu.findItem(R.id.action_look_pic).setVisible(true);
                menu.findItem(R.id.action_set_threshold).setVisible(false);
                break;
        }
        return super.onPrepareOptionsMenu(menu);
    }
```

每次单击菜单的时候，系统都会调用 onPrepareOptionsMenu(Menu menu)方法，切换侧滑菜单的时候已经对切换后的 Fragment 做了标记，所以在 onPrepareOptionsMenu(Menu menu)方法中可以根据这个标记判断当前是哪个 Fragment。如果是 0，就是室内环境采集系统对应的 Fragment，弹出菜单显示"设置阈值"菜单项；如果是 1，则是园区监控系统对应的 Fragment，弹出菜单显示"设置摄像头地址"和"查看截图"菜单项。

(11) 在 MainActivity 中重写菜单项的点击事件处理方法 onOptionsItemSelected (MenuItem item)，单击不同的菜单项会弹出对应的提示信息。

```
//标题栏菜单点击事件
    @Override
    public boolean onOptionsItemSelected(MenuItem item) {
        switch (item.getItemId()) {
            //设置传感器阈值菜单
            case R.id.action_set_threshold:
                Toast.makeText(this, "设置阈值", Toast.LENGTH_SHORT).show();
                break;
            //设置摄像头地址菜单
            case R.id.action_set_camera:
                Toast.makeText(this, "设置摄像头地址", Toast.LENGTH_SHORT).show();
                break;
            //查看截图菜单
            case R.id.action_look_pic:
                Toast.makeText(this, "查看截图", Toast.LENGTH_SHORT).show();
                break;
        }
        return true;
    }
```

3. 运行结果

(1) 运行程序，出现室内环境采集系统的界面，单击右上角的“设置”按钮，弹出“设置阈值”菜单项，如图 4-36 所示。

图 4-36　室内环境采集系统标题栏效果

(2) 运行程序，出现园区监控系统的界面，单击右上角的“设置”按钮，弹出“设置摄像头地址”和“查看截图”菜单项，如图 4-37 所示。

图 4-37　园区监控系统标题栏效果

项 目 总 结

本项目主要讲解了 Activity 的创建和简单使用 Fragment 实现页面切换，实现引导页的功能以及侧滑菜单和标题栏的效果。学习完本项目，能够让你对 Android 的界面显示和切换有更深的理解。希望读者能够通过 Android 的官方文档继续深入学习。

思考与练习

4-1 如果在 Activity 中可以跳转到不同的 Activity，那么怎么区分不同 Activity 返回的数据？

4-2 当你在使用手机应用时，突然接到了一个电话，请问这时这个活动的生命周期怎么变化？

项目5 数据存储的实现

【项目描述】

通过前面几个项目的学习，读者应该已能够开发出各种漂亮 Android 的界面。当我们平时在 App 上看新闻、与朋友微信聊天时都会产生数据，并且这些数据不会因为关了程序就丢失，那么 Android 是如何保存这些数据的呢？

本项目通过 3 个任务的学习(图 5-1)，针对 Android 开发中必须掌握的文件存储、SharedPreferences 和 SQLite 数据库等数据存储技术进行讲解，带领大家进入数据存储的世界。具体任务列表如图 5-1 所示。

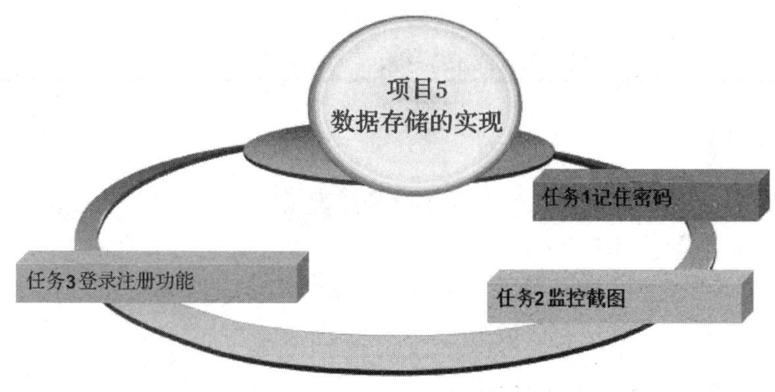

图 5-1　项目 5 任务列表

【学习目标】

知识目标：

- 会运用 SharedPreferences 存取数据。
- 会运用 openFileOutput 写数据。
- 会运用 openFileInput 读数据。
- 会读写 SD 卡的数据。
- 会处理运行时权限。
- 会在 SQLite 数据库中建库、建表，以及对数据进行添加、更新、删除、查询操作。

技能目标：

- 能用 SharedPreferences 实现记住密码的功能。
- 能把拍照的图片保存到 SD 卡上。
- 能把用户注册信息保存到 SQLite 数据库中，用户登录时从数据库中取数据进行验证。

任务 1 记住密码

【任务描述】

扫码观看视频讲解

本任务要求当登录界面中的“记住密码”复选框被选中时，把用户输入的用户名和密码保存起来。当用户再次登录时，能判断出是否记住用户名和密码，如果记住则取出用户名和密码回填到界面上。任务清单如表 5-1 所示。

表 5-1 任务清单

任务课时	2 课时	任务组员数量	建议 1 人
任务组采用设备	PC 一台 模拟器或真机或工控平板		

【知识解析】

用户名和密码这些具有键值对特征的数据存储可以用 SharedPreferences。SharedPreferences 存储适用于保存一些简单数据和键值对，它可以存储不同的数据类型，如果存储的是字符串，读取出来的也是字符串；如果存储的是整数，读取出来的也是整数。接下来讲解 SharedPreferences 存取数据的用法。

1. 用 SharedPreferences 存数据

用 SharedPreferences 存数据，得先获取 SharedPreferences 对象。SharedPreferences 对象可以通过 getSharedPreferences()方法获取。获取到 SharedPreferences 对象后，通过该对象的

edit()方法获取一个 Editor 对象，当要存数据时，Editor 对象提供一系列的 put 方法把键值对数据按类型进行添加，最后通过 commit 提交数据，完成存储，具体代码如下：

```
public void write()
{
    //1.获取 SharedPreferences 对象，参数一是文件名，参数二是操作权限
    SharedPreferences sp = getSharedPreferences("checkpass",MODE_PRIVATE);
   //2.获取 editor 对象，通过这个对象可以保存数据
    SharedPreferences.Editor editor = sp.edit();
  //3.往 editor 对象中添加数据，放的是键值对的数据
    editor.putString("name","newland");//存字符串类型的数据。键是 name，值是 newland
    editor.putBoolean("rememberpass", true);// 存布尔类型的数据。键是 rememberpass，值是 true
  //4.提交数据，完成存储
    editor.commit();
}
```

getSharedPreferences("checkpass",MODE_PRIVATE)方法需要 2 个参数：①文件名，第一次使用时文件名不存在则会创建一个，文件存放在/data/data/<应用程序包名>/shared_prefs 目录下；②操作模式，MODE_PRIVATE 表示只有当前应用程序才可以读写这个文件。

2. 取 SharedPreferences 中的数据

从 SharedPreferences 中取数据时要指明取什么类型的数据，SharedPreferences 对象提供了一系列的 get 方法用于取不同类型的数据。如果用 putString()存字符串，则用 getString()取出来；如果用 putBoolean()存布尔类型数据，则用 getBoolean()取出来。

get 方法需要填 2 个参数：①键值；②默认值。当传入的键值找不到对应的值时，就以默认值进行返回，代码如下：

```
public  void read()
{
    //1.获取 SharedPreferences 对象
    SharedPreferences sp=getSharedPreferences("checkpass",MODE_PRIVATE);
    //2.从 SharedPreferences 中取数据
  String sname= sp.getString("name","");
          //取字符串类型的数据。键值是 name，取不到值，默认返回为""
  Boolean flag= sp.getBoolean("rememberpass",false);
          //取布尔类型的数据。键值是 remerberpass，取不到值默认返回 0
  System.out.println("sname="+sname+",flag="+flag);
}
```

新建工程 Demo5_1，主 Activity 为 SharedActivity，在 onCreate()方法中先执行上述的 write()，发布程序到设备并运行后，打开/data/data/包名/shared_prefs 文件夹，可以看到生成了文件 sheckpass.xml，如图 5-2 所示。

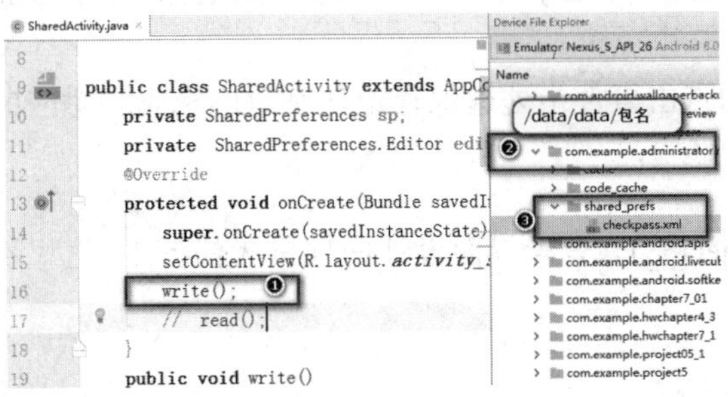

图 5-2 写键值对的结果

双击 checkpass.xml，可以看到保存好的数据，如图 5-3 所示。

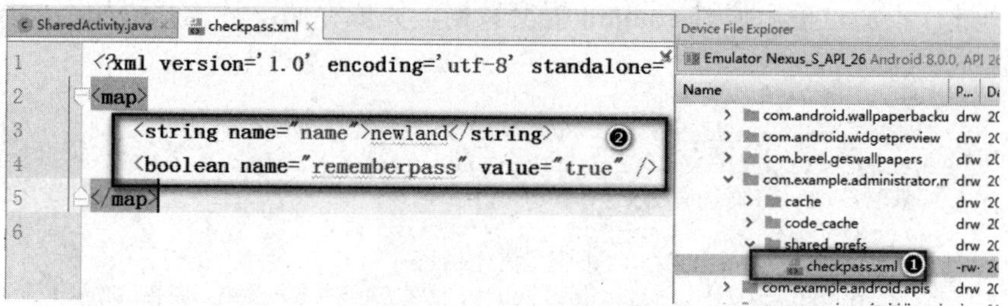

图 5-3　保存的键值对数据

接着屏蔽 onCreate()方法中的 write()方法，执行 read()方法，过滤日志，可以看到已经读取到文件 checkpass.xml 中的值，如图 5-4 所示。

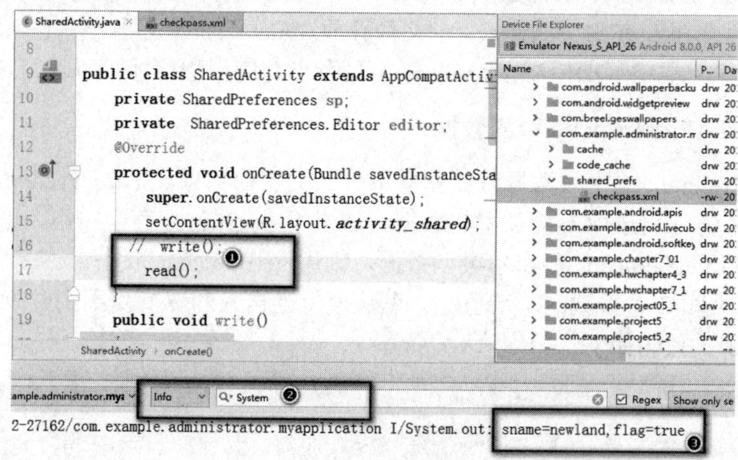

图 5-4　读取键值对的结果

【任务实施】

1. 任务分析

本任务要完成记住密码功能。

(1) 程序界面有用户名输入框、密码输入框、“记住密码”复选框和“立即登录”按钮。

(2) 在程序的 onCreate 方法中处理各种对象的初始化，并从 SharedPreferences 文件中取值回填到界面上。

(3) 在登录按钮的点击事件中处理保存记住密码的标志和对应的用户名和密码。

2. 任务实现

(1) 新建项目 Project5_1，登录界面设计同项目 3 任务 1，这里不再展开，效果如图 5-5 所示。

图 5-5 登录界面

(2) 在 LoginActivity.java 中，声明用户名和密码输入框对象、复选框对象、SharedPreferences 存储对象和 Editor 对象，并在 onCreate()方法中进行初始化。每次程序运行时，都要从 SharedPreferences 文件中把值取出来，检测勾选“记住密码”复选框的值，如果值为 true，则同时取出用户名和密码回填到界面上，复选框设置成被选中状态。如果值为 false，则无须处理。代码如下：

```
public class LoginActivity extends AppCompatActivity {
    private CheckBox cb_rem_pwd ;//记住密码复选框
    private EditText et_pwd,et_name;//用户名和密码输入框
    private SharedPreferences sp;//声明 SharedPreferences 存储对象
    private SharedPreferences.Editor editor;//声明 Editor 对象
   @Override
   protected void onCreate(Bundle savedInstanceState) {
       super.onCreate(savedInstanceState);
       setContentView(R.layout.activity_login);
       cb_rem_pwd = findViewById(R.id.cb_rem_pwd);
       et_pwd = findViewById(R.id.et_pwd);
       et_name = findViewById(R.id.et_name);
      //初始化 SharedPreferences 对象，保存数据的文件名为 check，权限仅当前程序可以访问
       sp =  getSharedPreferences("check", Context.MODE_PRIVATE);
      //初始化 Editor 对象
       editor = sp.edit();
       //从 check 文件中凭关键字 rememberpass 获取对应的值，如果获取不到，默认值返回为 false
      boolean flag= sp.getBoolean("rememberpass",false);
      //如果有勾选记住密码
      if(flag)
      {
          //从 check 文件中取出用户名和密码回填到输入框中，取不到就设置为""
          et_name.setText(sp.getString("name",""));
          et_pwd.setText(sp.getString("pass",""));
          //复选框勾上
          cb_rem_pwd.setChecked(true);
      }
   }
   //在这里写登录按钮的点击处理事件
}
```

(3) 当单击“登录”按钮时，正常是判断完用户名和密码是否正确后，要跳转到主页面，因为这里是讲解记住密码的功能，所以利用登录按钮的点击事件来实现这个功能，跳转方面就不展开讲解了。

在登录按钮的点击事件里，检测复选框是否为被选中状态，如果是，保存对应的值为 true，同时对输入的用户名和密码也一起保存。如果否，则保存对应的值为 false，同时用户名和密码的值保存为""。代码如下：

```
public void onLogin(View v)
{
    //获取输入的用户名和密码
    String sname=et_name.getText().toString().trim();
   String spass= et_pwd.getText().toString().trim();
   //如果用户名和密码输入为空
    if (TextUtils.isEmpty(sname) || TextUtils.isEmpty(spass)) {
        Toast.makeText(LoginActivity.this, "用户名或密码不能为空",Toast.LENGTH_LONG).show();
        return;
    }
   boolean flag= cb_rem_pwd.isChecked();
   //如果勾选了记住密码
   if(flag)
   {
       //把记住密码的标志置成 true
       editor.putBoolean("rememberpass",true);
       //保存用户名和密码
       editor.putString("name",sname);
       editor.putString("pass",spass);
       //提交数据
       editor.commit();
   }
   else
   {
       //如果没有勾选记住密码，则保存 false 和""
       editor.putBoolean("rememberpass",false);
       editor.putString("name","");
       editor.putString("pass","");
       editor.commit();
   }
    Toast.makeText(LoginActivity.this, "登录成功",Toast.LENGTH_LONG).show();
}
```

3. 运行结果

(1) 运行程序，输入用户名和密码，选中“记住密码”复选框后单击“立即登录”按钮，弹出消息提示“登录成功”，如图 5-6 所示。

(2) 查看 check.xml 文件内容，如图 5-7 所示。

(3) 验证是否实现了记住密码功能。

退出程序，重新运行程序(记住：不是重新发布程序)，可以看到用户名和密码已经回填了，“记住密码”复选框也被选中了，如图 5-8 所示。

把“记住密码”复选框的选中状态去掉，再次单击“立即登录”按钮。退出程序，再重新运行程序，可以看到用户名和密码没有被回填，“记住密码”复选框也未被选中。

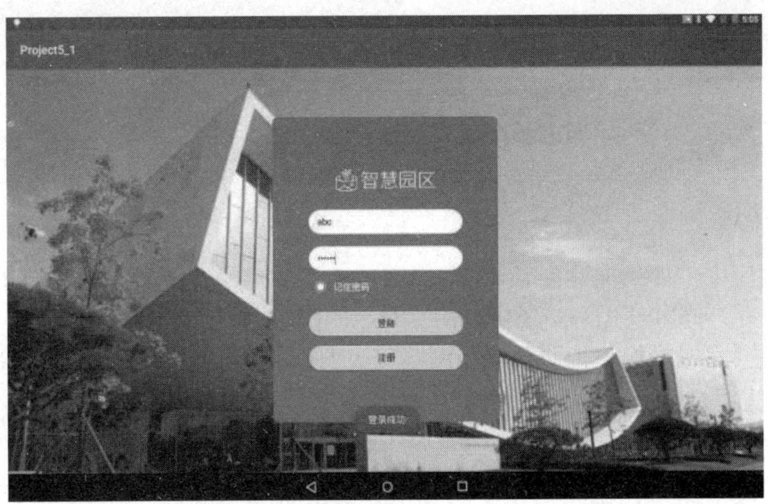

图 5-6 登录成功

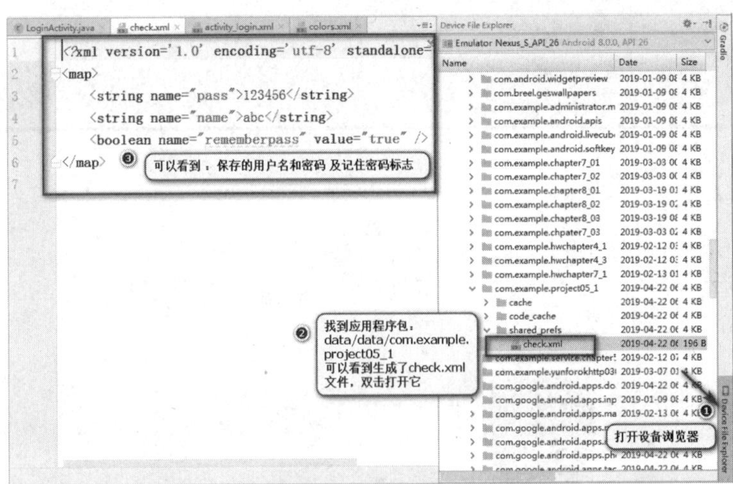

图 5-7 check.xml 文件

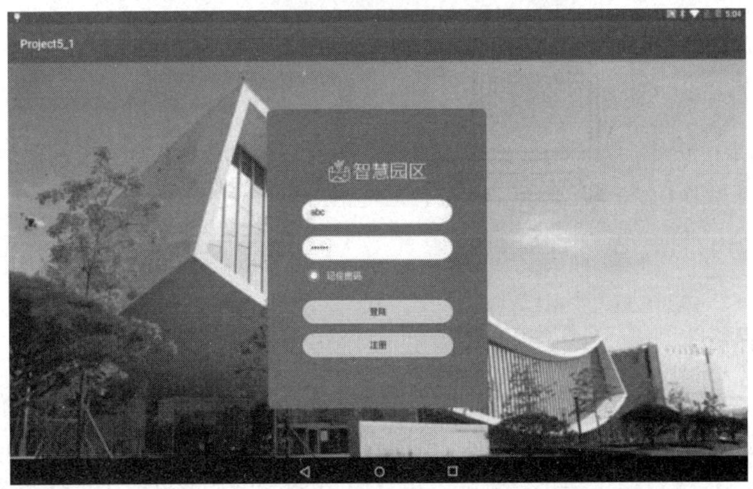

图 5-8 用户名和密码回填到了界面上

至此，记住密码功能已完成。具体代码可参考随书源代码 Project5_1。

任务 2　监 控 截 图

【任务描述】

在智慧园区中，如果检测到有非法入侵，会自动启动摄像头进行拍照，拍照后要对图片数据进行保存，同时安保人员还可以浏览对应的图片。

本任务主要完成摄像头拍照存图和浏览监控图片的功能，非法入侵自动拍照部分功能实现可参阅项目 11。任务清单如表 5-2 所示。

表 5-2　任务清单

任务课时	4 课时	任务组员数量	建议 3 人
任务组采用设备	PC 一台 摄像头一个 路由器一个 工控平板一个		

【拓扑图】

监控系统拓扑图如图 5-9 所示。

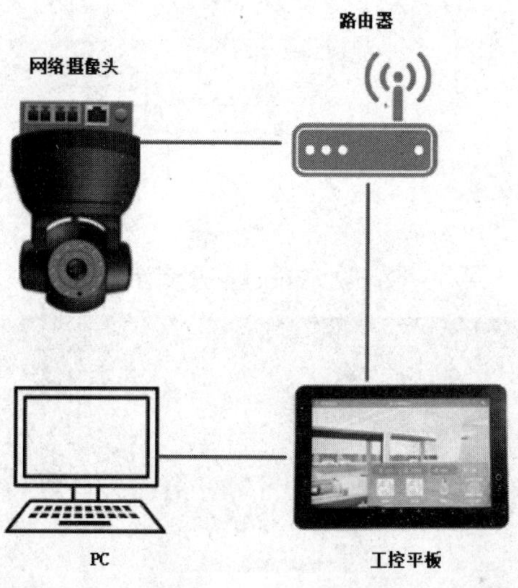

图 5-9　拓扑图

【知识解析】

每个发布运行在 Android 设备上的应用都会把数据存储在绝对路径/data/data/<应用程序包名>下，这些数据是应用程序私有的，除了本应用之外的其他程序都不能访问这些文件，当程序卸载了，这些数据也会跟着删除，可以用 openFileOutput()把这些私有数据存到文件中，再用 openFileInput()从文件中读取出来。

当文件比较大时，通常把数据存放在 SD 卡中。存放到 SD 卡中的这些数据应该是可以被所有应用程序所共享的，把数据写到 SD 卡要用 FileOutputStream 流，从 SD 卡中读取数据要用到 FileInputStream 流，而不能使用 openFileOutput 和 openFileInput。读写 SD 卡需要有读写 SD 卡的权限，同时 Android 6.0 以上还要处理 SD 卡的运行时权限。

1. 用 openFileOutput 保存数据

当要读写的数据不是 SD 中的数据时，可以使用 openFileOutput，它需要以下两个参数。

(1) 文件名，文件名不可以包含路径，因为所有的文件都是默认存储到/data/data/<包名>/files/下的。

(2) 操作模式，MODE_PRIVATE 是默认的操作模式，表示该文件是私有数据，只能被应用本身访问，如果文件已存在内容会覆盖。MODE_APPEND 表示如果文件已存在会追加内容。MODE_WORLD_READABLE 表示当前文件可被其他应用读取。MODE_WORLD_WRITEABLE 允许其他应用程序写入数据。

```
private void mysave() {
    //写数据
    String datas = " save...data...";//要保存的字符串数据
    FileOutputStream out = null;//
    try {
    //1.调用 openFileOutput 方法传入文件名和操作模式参数，返回 FOutputStreamile 对象 out
        out = openFileOutput("mydata", Context.MODE_PRIVATE);
    //2.调用 out 对象的 write()方法，将数据转为字节数组并写入到文件 mydata 中
        out.write(datas.getBytes());
    } catch (Exception e) {
        e.printStackTrace();
    } finally {
        try {
            if (out != null) {
     //3.调用 out 对象的 close()方法关闭输出流
                out.close();
            }
        } catch (IOException e) {
            e.printStackTrace();

        }
    }
}
```

新建工程 Demo5_2，在 MainActivity 的 onCreate()中调用 mysave()方法。

运行程序，打开 Device File Explorer(见图 5-10)，可以看到/data/data/包名/files/下产生了 mydata 文件。右击 mydata，选择快捷菜单中的 open 命令，可以看到数据被保存到文件中了(见图 5-11)。

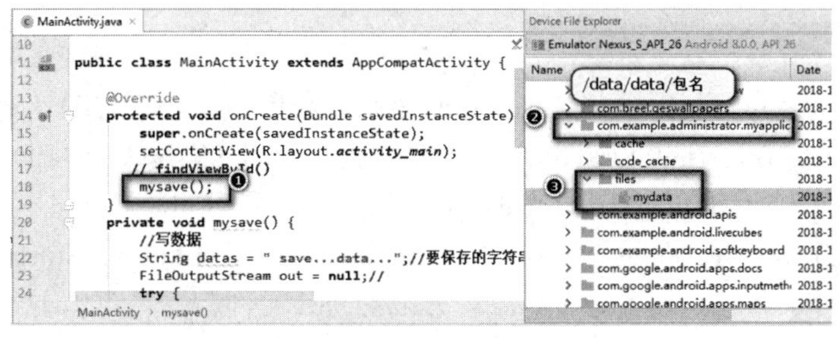

图 5-10 保存文件

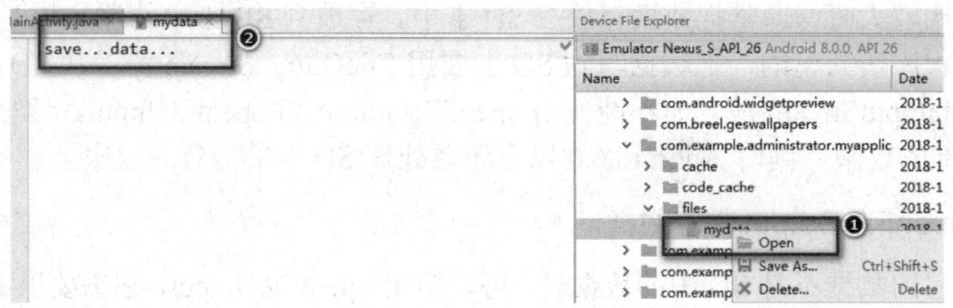

图 5-11　文件中的数据

2. 用 openFileInput 读数据

要把上面 mydata 文件中的数据读取出来，可以用 openFileInput，它只需要传入要读取的文件名参数即可。

```
private void myread() {
    //读取数据
    FileInputStream fin = null;
    try {
        byte[] buffer=new byte[1024];
        //1.调用 openFileInput()传入要读的文件名，返回一个 FileInputStream 对象 fin
        fin = openFileInput("mydata");
        //2.调用 fin 对象的 read()方法，读取到的字节数据放入 buffer 中
        int n=fin.read(buffer);
Log.i("TAG", "读到的数据=" + new String(buffer,0,n));

    } catch (Exception e) {
        e.printStackTrace();
    }finally{
      if(fin!=null)
        {
            try {
//3.关闭输入流
                fin.close();
            } catch (IOException e) {
                e.printStackTrace();
            }
        }
    }
}
```

在 MainActivity 的 onCreate()方法中调用 myread()方法，运行程序，可以看到文件中的数据被读取出来了，如图 5-12 所示。

3. SD 卡的数据存储

读写 SD 卡一般遵循以下步骤。

(1)　读写 SD 卡要用到环境变量 Environment 工具类。每次使用 SD 卡前，都要判断 SD 卡是否存在。如果 getExternalStorageState()的返回值是 MEDIA_MOUNTED，存在并可读写；如果返回值是 MEDIA_MOUNTED_READ_ONLY，则存在且可读，但不可写。

(2)　检查 SD 卡存在并可读写后，调用 Environment.getExtermalStorgeDirectory()可以获取到 SD 卡的目录路径，通常这个目录是/mnt/sdcard/文件名，也可以不用获取而直接写/mnt/sdcard/文件名。

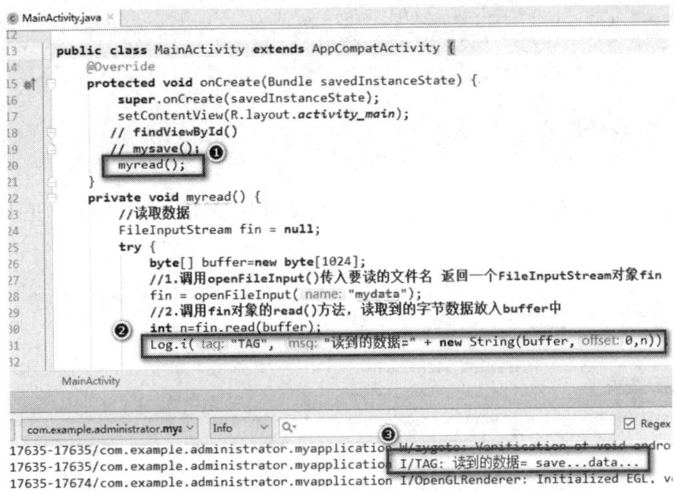

图 5-12 读取文件中的数据

(3) 获得文件路径后，就可以用 FileInputStream、FileOutputStream 操作字节数据的读写和用 FileReader、FileWriter 操作字符数据的读写。

```
public void saveToSD()
{
    //1.如果存在并有读写权限
    if (Environment.getExternalStorageState().equals(Environment.MEDIA_MOUNTED)) {
        Log.i("TAG", "SD 卡可读可写");
        //2.1 获取 SD 卡存储路径
        File path = Environment.getExternalStorageDirectory();
        //2.2 目的文件名
        String fileName =path+ "/data.txt";//要操作的文件
        File f=new File(fileName);
        if(!f.exists())
        {
            try {
                f.createNewFile();
            } catch (IOException e) {
                e.printStackTrace();
            }
        }
        //3.把数据写到 SD 卡中
        FileOutputStream out=null;
        try {
            out= new FileOutputStream(f);
         out.write("this is text".getBytes());//要把数据 this is text 写到文件 data.txt 中
        } catch (IOException e) {
            e.printStackTrace();
        } finally {
            try {
                if(out!=null) {
                    out.close();
                }
            } catch (IOException e) {
                e.printStackTrace();
            }
        }
    }
    //如果存在但只可以进行读的操作
```

```
    else if (Environment.getExternalStorageState().equals(Environment.MEDIA_
MOUNTED_READ_ONLY)) {
        Log.i("TAG", "SD 卡只读");
    }else
        Log.i("TAG", "SD 卡未挂载");
}
```

(4) 在 Android 中使用 SD 卡，需要在清单文件中添加读写 SD 卡的用户访问权限，如图 5-13 所示。

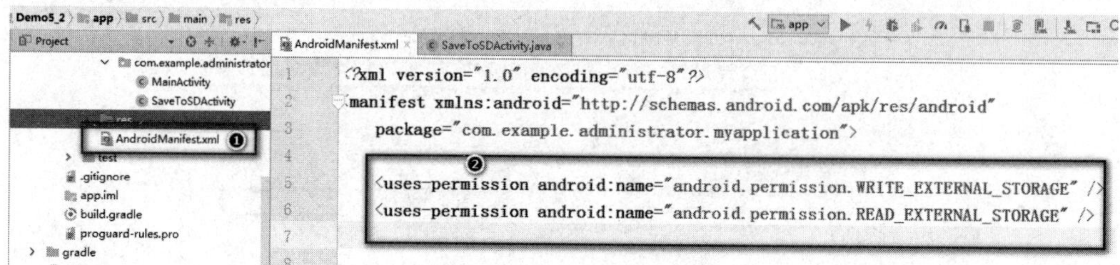

图 5-13　清单文件

在工程 Demo5_2 中添加 SaveToSDActivty，对应的布局文件是 activitysd.xml，布局中只有一个“保存”按钮，在按钮的点击事件中调用上面的 saveToSD()方法。

```
//保存按钮的点击事件处理
public void save(View v) {
    saveToSD();
}
```

在清单文件中把 SaveToSDActivty 改成主入口后，运行程序，会报错，如图 5-14 所示。

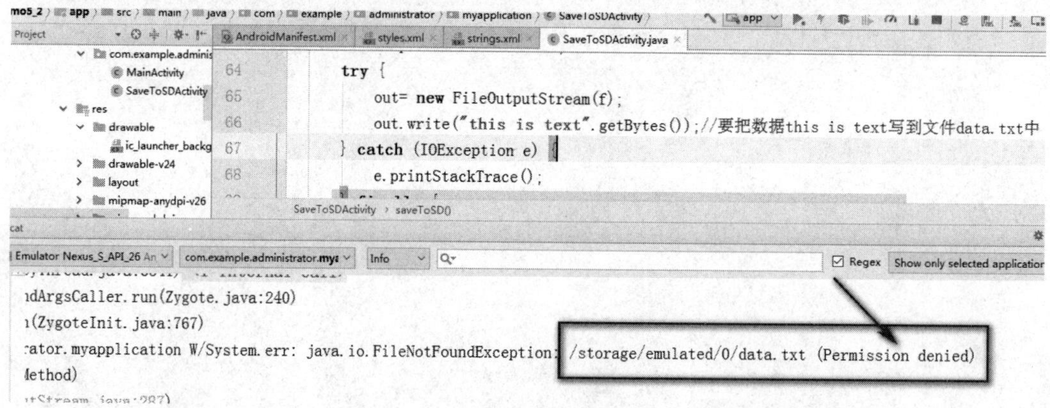

图 5-14　权限被拒绝

因为 Android 从 6.0 版本开始，访问 SD 卡需要处理运行时权限，没有访问权限无法打开 SD 卡的文件。

4. 运行时权限

Android 从 6.0 版本开始，为了保护用户隐私，将一些权限放在程序运行的时候去申请。运行时权限功能的核心就是在程序运行过程中由用户授权去执行一些危险操作。写 SD 卡需要运行时获取 WRITE_EXTERNAL_STORAGE 权限。把上面保存数据到 SD 卡的代码进行改写，写在 saveToSD2()中，在“保存”按钮的事件处理中先判断 SD 卡是否可读写，如果可读写，则进行 SD 卡的运行时权限并授权处理，代码如下：

```
//保存按钮的点击事件处理
public void save(View v) {

    //1.如果存在并有读写权限
    if (Environment.getExternalStorageState().equals(Environment.MEDIA_MOUNTED)) {
       //2.授权
        checkGrant();
    }
    //如果存在但只可以进行读的操作
    else if (Environment.getExternalStorageState().equals(Environment.MEDIA_MOUNTED_READ_ONLY)) {
        Log.i("TAG", "SD 卡只读");
    }else
        Log.i("TAG", "SD 卡未挂载");
}
@SuppressLint("NewApi")
private  void checkGrant() {
    //1.如果用户没有授权
    if (checkSelfPermission(Manifest.permission.WRITE_EXTERNAL_STORAGE)!= PackageManager.PERMISSION_GRANTED) {
        //2.请求授权：参数一要授权的权限放 String 数组中，参数二是请求码
        requestPermissions(new String[]{Manifest.permission.WRITE_EXTERNAL_STORAGE}, 1);
    } else {
        //有授权的话在这里处理业务 如读 SD 卡的内容
        //3.把数据保存到 SD 卡指定的文件中
        saveToSD2();
        Log.i("TAG", "SD 卡可读可写");
    }
}
//授权监听的返回结果  参数：  请求码   授权结果的数组   申请授权的数组
@Override
public void onRequestPermissionsResult(int requestCode, String[] permissions, int[] grantResults) {
    super.onRequestPermissionsResult(requestCode, permissions, grantResults);
    if (requestCode == 1  && grantResults[0]==PackageManager.PERMISSION_GRANTED) {
        //把数据保存到 SD 卡指定的文件中
        saveToSD2();
        Log.i("TAG", "请求授权成功");
    }
}
}
public void saveToSD2()
{
        //获取 SD 卡存储路径
        File path = Environment.getExternalStorageDirectory();
        //目的文件名
        String fileName =path+ "/data.txt";//要操作的文件
        File f=new File(fileName);
        if(!f.exists())
        {
            try {
                f.createNewFile();
            } catch (IOException e) {
                e.printStackTrace();
            }
```

```
            }
            //把数据写到 SD 卡中
            FileOutputStream out=null;
            try {
             out= new FileOutputStream(f);
             out.write("this is text".getBytes());//要把数据this is text写到文件data.txt中
            } catch (IOException e) {
                e.printStackTrace();
            } finally {
                try {
                    if(out!=null) {
                        out.close();
                    }
                } catch (IOException e) {
                    e.printStackTrace();
                }
            }
        }
```

从上面的程序可以看出，checkSelfPermission()方法的返回值与 PackageManager.PERMISSION_GRANTED 做比较，相等则说明已经授权。如果已授权就去处理业务，如果没有授权则要调用 requestPermissions()来向用户申请授权，申请授权方法有 2 个参数：①所有要申请的授权存放的字符串数组；②请求码。调用完 requestPermissions()后，系统会弹出一个权限申请的对话框，用户可以同意或拒绝。不论是哪种结果最终都会回调给 onRequestPermissionsResult()，授权的结果在 grantResults 中，通过它判断请求码是否是申请授权时的请求码，如果是则处理业务逻辑。

当处理了运行时权限后，程序运行时就会提示是否允许访问 SD 卡，允许可以继续访问，拒绝则结束访问，如图 5-15 所示。

图 5-15　处理运行时权限

【任务实施】

1. 任务分析

本案例要完成摄像头监控图像的保存和读取(因为工控平板是 Android 5.1 的，所以不需要处理 SD 卡的运行时权限)。

(1) 创建工程，添加摄像头的 jar 包。

(2) 程序界面比较复杂，左边是一个 FrameLayout，里面有 TextureView 和一个 View，当摄像头停止时加载 View 作背景，TextureView 用于承载摄像头图像。要求有 4 个用来输入连接摄像头的用户名、密码、IP、channel 号的 EditText。还有可以控制摄像头进行上、下、左、右转动和截图以及浏览监控图的按钮。

(3) 在 MainActivity 中实现摄像头的控制。

(4) 实现摄像头监控截图功能，在清单文件中添加访问 SD 卡的权限。

(5) 实现监控页面浏览功能。

2. 任务实现

(1) 新建工程 Project5_2，在 app 的 libs 下添加摄像头的库文件①，在 app build.gradle ②中的 android 闭包中添加 jni 的设置③，进行同步④，等待同步完成⑤，如图 5-16 所示。

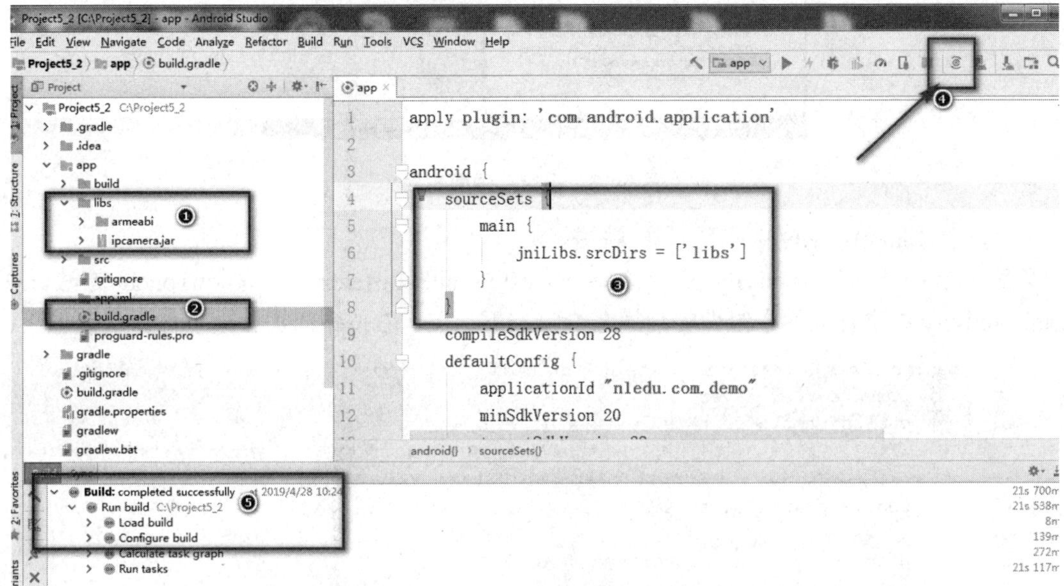

图 5-16 添加摄像头的库

(2) 在 MainActivity 对应的布局 activity_main 中添加对应的控件，界面效果如图 5-17 所示。

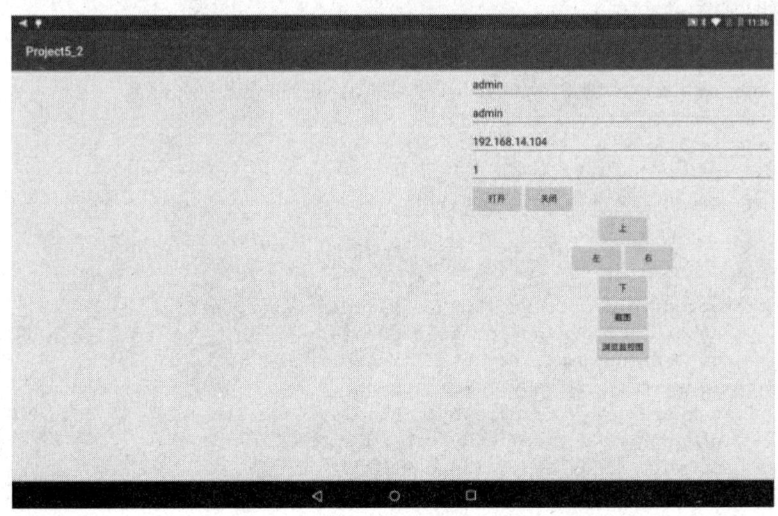

图 5-17 监控界面

其中，左边代码如图 5-18 所示，右边代码可参考随书源代码或自行设计。

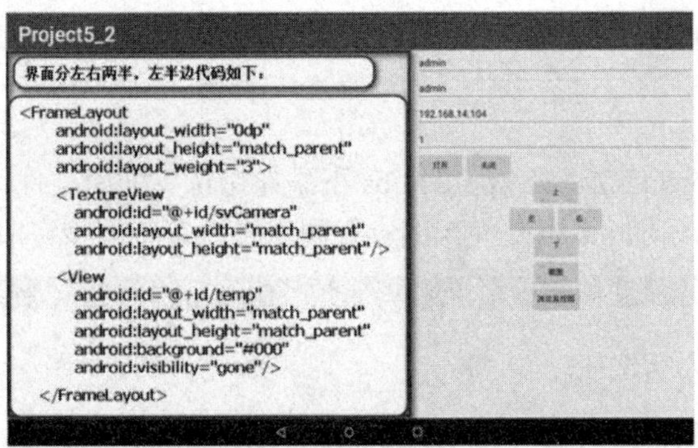

图 5-18　摄像头画面布局

(3) 在 MainActivity 中实现摄像头的控制。

第一步：让 MainActivity 实现接口 View.OnTouchListener，重写 onTouch 方法。在 MainActivity 中进行界面中组件的查找，给上、下、左、右按钮添加触摸事件。代码如下：

```
public class MainActivity extends AppCompatActivity implements View.OnTouchListener {
    private View layer;
    public static TextureView textureView;
    //提升为全局变量，保持持有一个TextureView的引用，如此一来该对象就不会被当作垃圾回收(从而被弃用)，
    //否则当跳转到下一个Activity时会报错："SurfaceTexture has been abandoned"
    public static CameraManager cameraManager;//摄像头管理工具类

    private EditText etUserName;//用户名
    private EditText etPwd;//密码
    private EditText etIP;//ip地址
    private EditText etChannel;//通道号

    public static List<String> imageList;  //保存拍照图片路径
    public static boolean isOpen;//摄像头是否已打开的标志
    Public static String userName;//获取输入的连接摄像头的用户名
    Public static String pwd;//获取输入的连接摄像头的密码
    Public static String ip;//获取输入的连接摄像头的IP
    Public static String channel;//获取输入的连接摄像头的信道号

    static {
        //从SD卡中取出监控截图
        imageList = getAllPics();
    }

    @Override
    protected void onCreate(Bundle savedInstanceState) {
        super.onCreate(savedInstanceState);
        setContentView(R.layout.activity_main);
        layer = findViewById(R.id.temp);
        textureView = findViewById(R.id.svCamera);
        etUserName = findViewById(R.id.etUserName);
        etPwd = findViewById(R.id.etPwd);
        etIP = findViewById(R.id.etIP);
        etChannel = findViewById(R.id.etChannel);
        //给上、下、左、右按钮添加触摸事件
        findViewById(R.id.up).setOnTouchListener(this);
        findViewById(R.id.left).setOnTouchListener(this);
        findViewById(R.id.right).setOnTouchListener(this);
        findViewById(R.id.down).setOnTouchListener(this);
```

```
            getCameraData();//获取连接摄像头的参数并打开摄像头
        }
    @Override
     public boolean onTouch(View arg0, MotionEvent arg1) {

        }
  }
```

注意 TextureView 要提升为全局变量，保持持有一个 TextureView 的引用，如此一来该对象就不会被当作垃圾回收(从而被弃用)；否则当跳转到下一个 Activity 时会报错："SurfaceTexture has been abandoned"。

第二步：在摄像头初始化方法 getCameraData()中获取输入的用户名、密码、IP 地址、Channel，调用 CameraManager.getInstance()获取摄像头管理对象 cameraManager，通过 cameraManager 的 setupInfo()方法进行初始化。代码如下：

```
private void getCameraData() {
    textureView = findViewById(R.id.svCamera);
     userName = etUserName.getText().toString();
     pwd = etPwd.getText().toString();
     ip = etIP.getText().toString();
     channel = etChannel.getText().toString();
     cameraManager = CameraManager.getInstance();
     cameraManager.setupInfo(textureView, userName, pwd, ip, channel);
}
```

注意：像一些 DVR 或者 NVR 的摄像头，接入了多路的视频流，当要访问的时候，应该指定是哪一路视频，也就是哪一个通道。本任务用的摄像头只有一路视频，所以通道参数设置为 1 即可。

第三步：处理打开摄像头的按钮事件 open，用 openCamera()方法打开摄像头。代码如下：

```
public void open(View view) {
     isOpen = true;
     layer.setVisibility(View.GONE);
     getCameraData();
     cameraManager.openCamera();
}
```

第四步：处理关闭摄像头的按钮事件 release，用 releaseCamera()方法关闭摄像头。当停止摄像头后，把摄像头的监控界面换成显示 layer，此时屏幕变成白的。代码如下：

```
public void release(View view) {
     isOpen = false;
     if (cameraManager != null) {
         cameraManager.releaseCamera();
     }
     layer.setVisibility(View.VISIBLE);
}
```

第五步：处理控制摄像头的触摸事件 onTouch，当手触摸屏幕上的上、下、左、右按钮时，分别记下对应的动作，当手抬起时动作为 stop，调用 cameraManager.controlDir(ptz)对摄像头进行方向控制。其中 PTZ 是摄像头 jar 包中提供的枚举对象，里面用来记录上、下、左、右、停的动作。代码如下：

```
@Override
public boolean onTouch(View arg0, MotionEvent arg1) {
    int action = arg1.getAction();
    PTZ ptz = null;
    if (action == MotionEvent.ACTION_CANCEL || action == MotionEvent.ACTION_UP) {
        ptz = PTZ.Stop;
    } else if (action == MotionEvent.ACTION_DOWN) {
        int viewId = arg0.getId();
        switch (viewId) {
            case R.id.up:
                ptz = PTZ.Up;
                break;
            case R.id.down:
                ptz = PTZ.Down;
                break;
            case R.id.left:
                ptz = PTZ.Left;
                break;
            case R.id.right:
                ptz = PTZ.Right;
                break;
        }
    }
    cameraManager.controlDir(ptz);
    return false;
}
```

因为要访问摄像头，所以要在清单文件中添加访问网络的权限。

```
<uses-permission android:name="android.permission.INTERNET" />
```

运行程序，应该可以控制摄像头的打开和关闭，并且可以控制上、下、左、右转动了。效果参考运行结果。

(4) 实现摄像头监控截图功能。

当要进行监控截图时，文件名保存为当前系统时间的年月日时分秒.png，可以用方法getPicName()进行处理。在截图的按钮事件 capture，先在 SD 卡中创建目录 pic，接着开启线程，通过 textureView.getBitmap()获取监控图像，用 FileOutputStream 流操作文件，用 bitmap.compress(Bitmap.CompressFormat.JPEG, 100, fos)把图像按格式保存到文件中。代码如下：

```
public String getPicName() {
    Date date = new Date();
    SimpleDateFormat sd = new SimpleDateFormat("yyyy年MM月dd日HH时mm分ss秒");
    return sd.format(date) + ".png";
}
String currPicName;//当前保存的摄像图片
public void capture(View view) {
    final File f = new File(Environment.getExternalStorageDirectory().getPath(), "pic");
    if (!f.exists()) {
        f.mkdirs();
    }
    currPicName = getPicName();

    if (this.textureView != null) {
        (new Thread() {
            public void run() {
                Bitmap bitmap = textureView.getBitmap();
```

```

                try {
                    File file = new File(f.getPath() + "/" + currPicName);
                    FileOutputStream fos = new FileOutputStream(file);
                    bitmap.compress(Bitmap.CompressFormat.JPEG, 100, fos);
                    fos.close();
                    imageList.add(currPicName);
                } catch (Exception var4) {
                    var4.printStackTrace();
                }
            }
        }).start();
    }
}
```

(5) 实现监控页面浏览监控图像功能。

第一步：读取所有监控图像。

在 MainActivity 类中的成员变量 imageList 用于保存所有监控图像的存放在 SD 卡中的文件名，写方法 getAllPics()用于读取已有的监控图像。在 MainActivity 类的静态代码块中进行调用。当类 MainActivity 被加载进内存时就进行图像的读取。代码如下：

```
static{
    //从 SD 卡中取出监控截图
    imageList= getAllPics();
}
public static List<String> getAllPics()
{
    List<String> list=new ArrayList();
    File f = new File(Environment.getExternalStorageDirectory().getPath(),"pic");
    String[] fileNames= f.list();

    for(String s:fileNames)
    {
        list.add(s);

    }
    return list;
}
```

因为截图后需要把图片存放到 SD 卡中，所以需要在清单文件中添加 SD 卡读写权限。代码如下：

```
<uses-permission android:name="android.permission.READ_EXTERNAL_STORAGE" />
<uses-permission android:name="android.permission.WRITE_EXTERNAL_STORAGE" />
```

当单击“截图”按钮时，在“截图”按钮点击事件方法 capture(View view)中的 imageList.add(currPicName)语句就把图片路径添加到集合中了。

运行程序，单击“截图”按钮后，打开 Device File Explorer 并找到 sdcard 下的 pic 文件夹，可以看到图片已经保存到 SD 卡中了，如图 5-19 所示。

第二步：添加 RecyclerView 的依赖包。

浏览页面一般有上下滚动功能，可以用控件 RecyclerView 实现。想要使用 RecyclerView，首先需要在 app 的 build.gradle 中添加相应的依赖库。

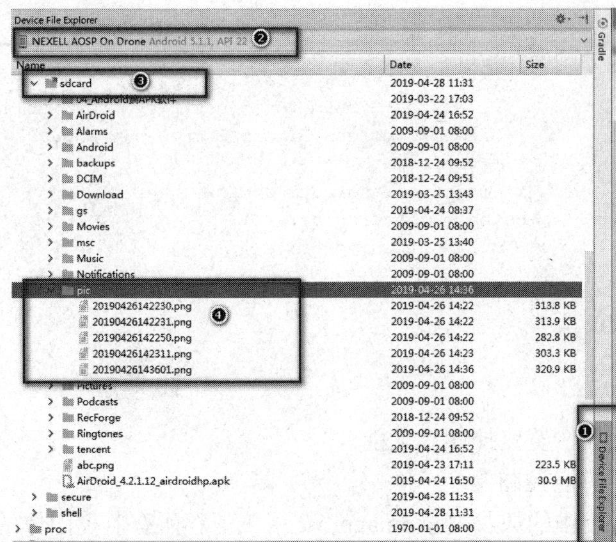

图 5-19　监控截图文件

添加 RecyclerView 的依赖包有以下两种方式。

① 通过搜索的方式添加，步骤如图 5-20 至图 5-23 所示。

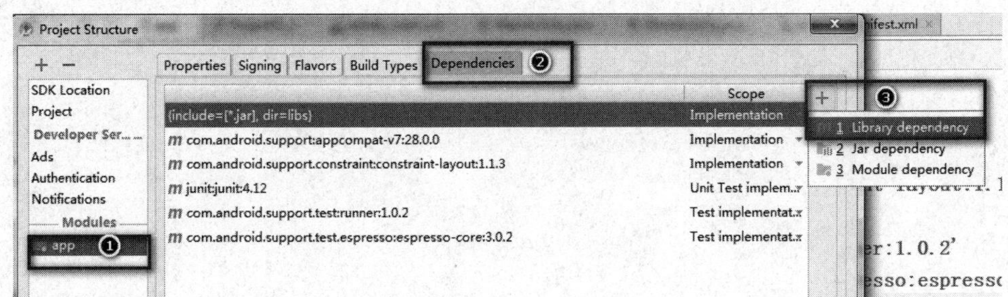

图 5-20　添加依赖库

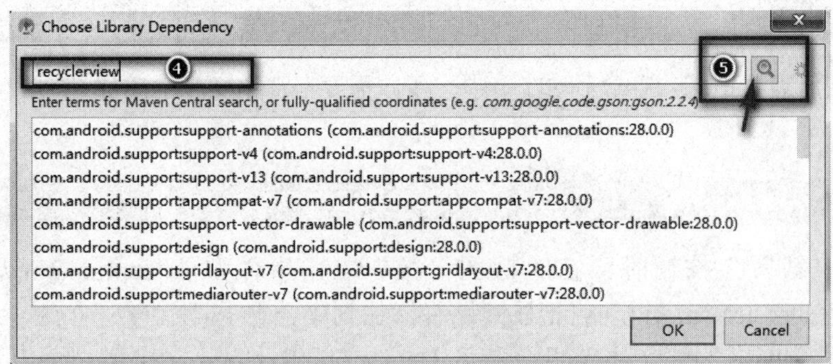

图 5-21　输入 recyclerview 并查找

查看 app 下的 build.grade，可以看到 recyclerview-v7:28.0.0 已经成功添加进来了，如图 5-24 所示。

② 直接在上述 app build.gradle 的 dependencies 闭包中添加，如图 5-24 所示，这种方式添加完后要单击 Sync Now 来进行同步。

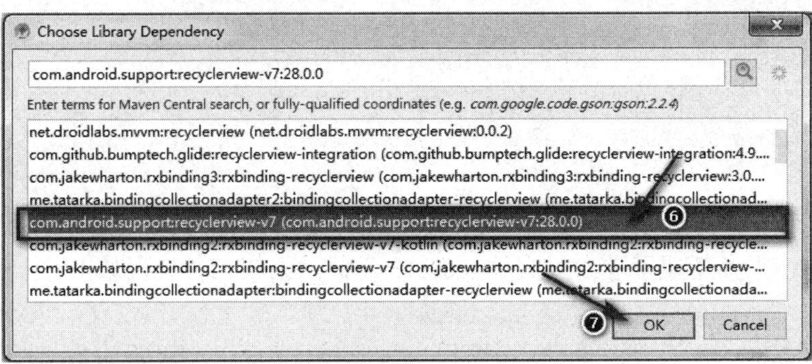

图 5-22 选中 recyclerview

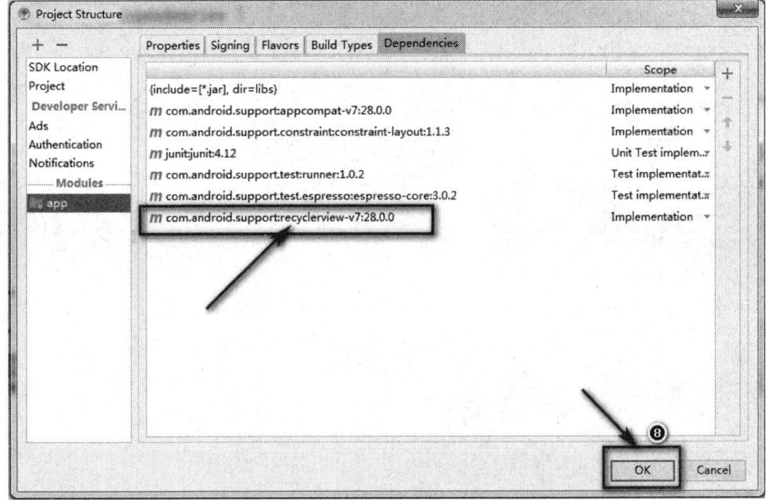

图 5-23 查看依赖添加结果

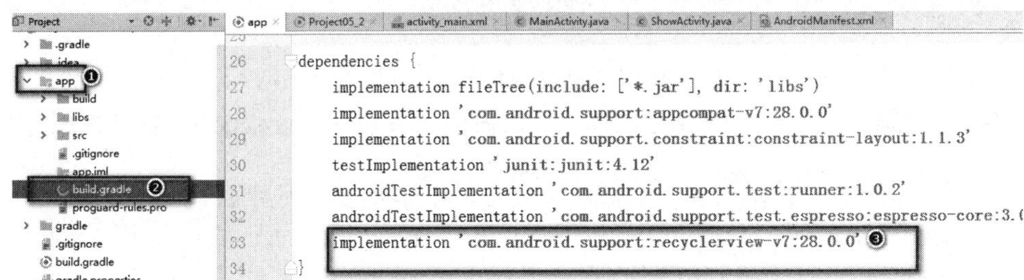

图 5-24 在 app 的 build.gradle 文件中查看依赖结果

第三步： 新建 ShowActivity 用于显示监控图像，布局为 activity_show.xml，用 RecyclerView 进行滚动控制。代码如下：

```
<?xml version="1.0" encoding="utf-8"?>
<LinearLayout xmlns:android="http://schemas.android.com/apk/res/android"
    xmlns:app="http://schemas.android.com/apk/res-auto"
    xmlns:tools="http://schemas.android.com/tools"
    android:layout_width="match_parent"
    android:layout_height="match_parent"
    android:orientation="vertical"
    tools:context=".ShowActivity">
  <android.support.v7.widget.RecyclerView
```

```
        android:layout_width="match_parent"
        android:layout_height="match_parent"
        android:id="@+id/recycler_view"
        >
    </android.support.v7.widget.RecyclerView>
</LinearLayout>
```

RecyclerView 中每个子项中都有 TextView 用来显示监控图像的日期，ImageView 用来显示监控图像，需要新建子项布局文件 recyclerview_item.xml。代码如下：

```
<?xml version="1.0" encoding="utf-8"?>
<LinearLayout xmlns:android="http://schemas.android.com/apk/res/android"
    android:layout_width="match_parent"
    android:layout_height="400dp"
    android:gravity="center"
    >
    <TextView
        android:id="@+id/time"
        android:layout_width="wrap_content"
        android:layout_weight="1"
        android:layout_height="wrap_content"
        android:gravity="center"
        />
    <View
        android:layout_width="1dp"
        android:layout_height="fill_parent"
        android:background="#cccccc"
        />
    <ImageView
        android:id="@+id/image"
        android:layout_width="0dp"
        android:layout_weight="2"
        android:scaleType="fitCenter"
        android:src="@mipmap/ic_launcher"
        android:layout_height="match_parent" />
</LinearLayout>
```

当想从 ShowActivity 跳转回到 MainActivity 时，可以在 ShowActivity 标题栏上显示返回箭头，要实现这个返回功能，可以在清单文件 ShowActivity 节点中添加代码：

```
<activity android:name=".ShowActivity"
    android:parentActivityName=".MainActivity">
</activity>
```

同时，利用 Activity 生命周期中的 onResume()方法，当从 ShowActivity 中跳转回到 MainActivity 时判断如果之前摄像头是打开的，则需要再次打开摄像头。代码如下：

```
@Override
public void onResume() {

    if (isOpen) {
        try {
            Thread.sleep(500);
        } catch (InterruptedException e) {
            e.printStackTrace();
        }
        new Thread(new Runnable() {
            @Override
```

```
                public void run() {
                    getCameraData();
                    //如果开启监控，0.3s 后打开摄像头
                    cameraManager = CameraManager.getInstance();
                    cameraManager.setupInfo(textureView, userName, pwd, ip, channel);
                    cameraManager.openCamera();
                }
            }).start();

        }
        super.onResume();
    }
```

第四步： 处理 RecyclerView 适配器。

定义 RecyclerView 适配器类 MonitorAdapter，该类继承自 RecyclerView.Adapter，并将泛型定义为 MonitorAdapter.ViewHolder。继承自 RecyclerView.Adapter 的类要重写 3 个方法，即 onCreateViewHolder()、onBindViewHolder()和 getItemCount()。

onCreateViewHolder()方法用于创建 ViewHolder 实例，并把加载出来的布局传入到构造方法中，最后将 ViewHolder 的实例返回。

onBindViewHolder()方法用于对 RecyclerView 子项的数据进行赋值，会在每个子项被滚动到屏幕内时执行，在这里通过 position 参数得到当前项的监控图像文件名，然后再将数据设置到 ViewHolder 的 time 和 image 中。

getItemCount()方法用于告诉 RecyclerView 共有多少子项，直接返回存放监控图像的集合长度即可。

ViewHolder 是在 MonitorAdapter 中定义的一个内部类。

在 MonitorAdapter 类中，通过构造方法把保存监控图像的集合传进来，在 onCreateViewHolder 中加载布局 recyclerview_item.xml，在 onBindViewHolder 中进行控件的绑定并取出日期和图片显示在控件上，取图片时利用 FileInputString 流加载数据，通过 BitmapFactory.decodeStream(input)转换成 Bitmap 后设置到 ImageView 控件上。代码如下：

```
public class MonitorAdapter extends RecyclerView.Adapter<MonitorAdapter.ViewHolder> {
    private List<String> monitorImageList;

    public MonitorAdapter(List<String> monitorImageList)
    {
        this.monitorImageList=monitorImageList;
    }

    @NonNull
    @Override
    public ViewHolder onCreateViewHolder(@NonNull ViewGroup parent, int viewType) {
        View view= LayoutInflater.from(parent.getContext()).inflate(R.layout.recyclerview_item,parent,false);
        ViewHolder holder = new ViewHolder(view);
        return holder;
    }

    @Override
    public void onBindViewHolder(@NonNull ViewHolder holder, int position) {
        String imagepath=monitorImageList.get(position);
```

```
            //取出拍照时间
            String time= imagepath.substring(0,imagepath.indexOf(".png"));
            InputStream input=null;
            Bitmap bitmap=null;
            try {
                input= new FileInputStream("/mnt/sdcard/pic/"+imagepath);
              bitmap=  BitmapFactory.decodeStream(input);

            } catch (FileNotFoundException e) {
                e.printStackTrace();
            }
            holder.image.setImageBitmap(bitmap);
            holder.time.setText(time);
        }

        @Override
        public int getItemCount() {
            return monitorImageList.size();
        }

        static class ViewHolder extends RecyclerView.ViewHolder{
            private ImageView image;//图片
            private TextView time;//时间
            public ViewHolder(View v)
            {
                super(v);
                image = v.findViewById(R.id.image);
                time = v.findViewById(R.id.time);
            }
        }
    }
```

第五步：在 ShowActivity 中加载 RecyclerView。代码如下：

```
    public class ShowActivity extends AppCompatActivity {

        @Override
        protected void onCreate(Bundle savedInstanceState) {
            super.onCreate(savedInstanceState);
            setContentView(R.layout.activity_show);

            RecyclerView recyclerView = findViewById(R.id.recycler_view);
            LinearLayoutManager linearLayoutManager = new LinearLayoutManager(ShowActivity.this);
            recyclerView.setLayoutManager(linearLayoutManager);
            MonitorAdapter adapter = new MonitorAdapter(MainActivity.imageList);
            //添加 Android 自带的分割线
            recyclerView.addItemDecoration(new DividerItemDecoration(this, DividerItemDecoration.VERTICAL));
            recyclerView.setAdapter(adapter);
        }
    }
```

在 onCreate()方法中先获取到 RecyclerView 的实例，然后创建 LinearLayoutManager 对象，并将它设置到 RecyclerView 中，接着创建适配器对象，调用 setAdaper()方法来完成适配器的设置，这样 RecyclerView 就和存放监控数据的集合建立了关联。

3. 运行结果

发布程序到工控平板上，运行结果如图 5-25 所示。

图 5-25 摄像头监控界面

连接路由器的用户名是 admin，密码是 admin，IP 地址填写配置给摄像头的真实地址，channel 通道填 1，确认上述信息无误后单击“打开”按钮，应该可以打开图 5-25 所示的摄像头监控界面。

单击“关闭”按钮，效果如图 5-25 所示，单击“上”“下”“左”“右”按钮可以控制摄像头的不同方向，单击“截图”按钮可以对监控界面进行图像保存。

单击“浏览监控图”按钮可以跳转到浏览界面，在浏览界面可以进行上下滚动以查看更多的监控图，单击左上角的◀按钮可以回到摄像头监控界面，如图 5-26 所示。

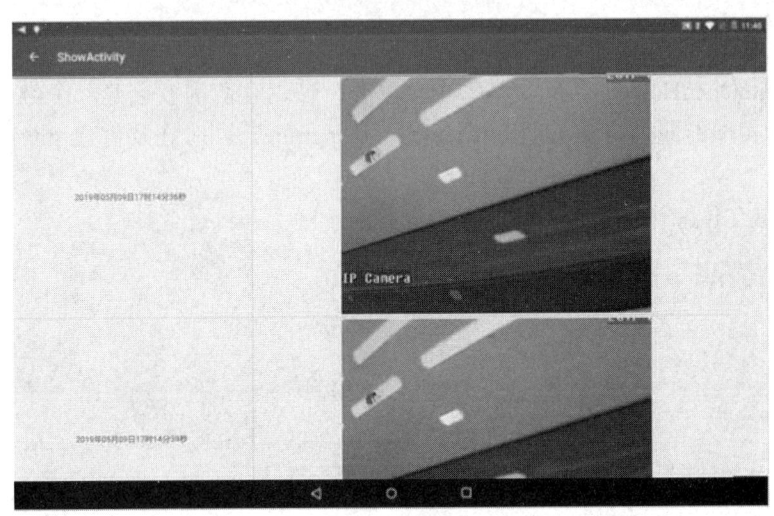

图 5-26 监控图像浏览界面

至此，任务 2 摄像头监控图片的存储功能就完成了。具体代码可参考随书源代码 Project5_2。

任务 3　登录注册功能

【任务描述】

扫码观看视频讲解

在使用 App 下载数据时，可以暂停下载，当下一次继续下载时会从断点处接着下载，这是怎么做到的呢？其实程序是把当前下载点实时保存在本地数据库中，当下载完毕才清除掉相关信息。

在 App 的注册和登录过程中，当用户注册时，可以把输入的用户名和密码保存到本地数据库中。当登录时，可以从本地数据库中查询用户相关信息并进行判断。本地数据库可以使用 Android 内置的 SQLite 数据库。

本任务要求用户注册时把用户信息保存到 SQLite 数据库中，当用户登录时到数据库中核查用户信息，如果正确则允许登录。任务清单如表 5-3 所示。

表 5-3　任务清单

任务课时	6 课时	任务组员数量	建议 1 人
任务组采用设备	PC 一台或模拟器或工控平板		

【知识解析】

SQLite 是一款 Android 系统内嵌的轻量级的关系数据库，支持标准的 SQL 语法，其运算速度快，占用资源少，一般一个数据库文件只有几百 KB 大小，所以适合在移动设备上使用。Android 提供了一个 SQLiteOpenHelper 帮助类，可以基于 SQLiteOpenHelper 帮助类对数据库进行操作。

1. 创建数据库

(1) SQLiteOpenHelper 帮助类是一个抽象类，使用它前需要创建一个类去继承它，继承后要实现未实现的两个方法，即 onCreate、onUpgrade，同时还要创建一个构造方法。代码如下：

```
public class MySQLiteOpenHelper extends SQLiteOpenHelper {
    /** *
     * @param context 上下文
     * @param name 数据库的名称
     * @param factory 游标工厂
     * @param version 数据库的版本从 1 开始
     */
    public MySQLiteOpenHelper(Context context, String name,
                        SQLiteDatabase.CursorFactory factory,int version) {
        super(context, name, factory, version);
    }
    //第一次创建数据库时会调用
    @Override
    public void onCreate(SQLiteDatabase db) {
    }
    //数据库版本更新时会自动调用
    @Override
```

```
    public void onUpgrade(SQLiteDatabase db, int oldVersion, int newVersion) {
    }
}
```

构造方法有 4 个参数，为了简化，通常会指定好要创建的数据库名，将游标工厂设置为空、版本号为 1，所以会用只传一个参数 Context 的构造方法代替，这里创建一个数据库名为 employee.db、版本号 1 的数据库。代码如下：

```
public class MySQLiteOpenHelper extends SQLiteOpenHelper {
private Context context;
public static final String DB_NAME="employee.db";//数据库名称
public static  int DB_VERSION=1;//数据库版本号
   public MySQLiteOpenHelper(Context context ) {
       super(context, DB_NAME, null, DB_VERSION);
       this.context = context;
   }
//第一次创建数据库时会调用
    @Override
    public void onCreate(SQLiteDatabase db) {
    }
    //数据库版本更新时会自动调用
    @Override
    public void onUpgrade(SQLiteDatabase db, int oldVersion, int newVersion) {
    }
  }
```

(2) 当构建出 MySQLiteOpenHelper 的实例后，调用它的两个非常重要的方法 getReadableDatabase()和 getWriteableDatabase()就能够创建数据库了，数据库文件会存在/data/data/<程序包名>/databases/目录下，当第一次执行时，如果检测系统中没有我们要创建的数据库时就会执行 onCreate()方法；如果检测出数据库已存在，onCreate()方法就不会执行。

(3) 在数据库 employee.db 中创建一个 emp 员工表，表中有_id(主键)、名字、年龄，代码如下：

```
@Override   //第一次创建数据库时会调用
   public void onCreate(SQLiteDatabase db) {
   // 构造要执行的 SQL 语句
   String sql="create table emp(_id  integer primary key autoincrement,name text,age text)";
   db.execSQL(sql);//执行建表语句
   Toast.makeText(context,"创建数据库 employee.db",Toast.LENGTH_LONG).show();
 }
```

可以看到 SQLite 的数据类型很简单，integer 表示整型，integer primary key autoincrement 表示主键是整型并且自动增长，其他都可以用 text 文本类型，当把不同类型的数据插入非主键的字段时会自动适应对应的类型，如 float 类型、字符类型等。

(4) 接下来创建一个 SQLiteActivity，对应的布局 activity_sqlite 中有一个按钮，单击时创建数据库。

activity_sqlite.xml 布局如下：

```
<?xml version="1.0" encoding="utf-8"?>
<LinearLayout xmlns:android="http://schemas.android.com/apk/res/android"
xmlns:app="http://schemas.android.com/apk/res-auto"
xmlns:tools="http://schemas.android.com/tools"
```

```
android:layout_width="match_parent"
android:layout_height="match_parent"
android:orientation="vertical"
>
    <Button
        android:onClick="create_db"
        android:text="创建数据库"
        android:textColor="#00f"
        android:textSize="20sp"
        android:layout_width="match_parent"
        android:layout_height="wrap_content" />
</LinearLayout>

```

修改 SQLiteActivity 中的代码如下：

```
public class SQLiteActivity extends AppCompatActivity {
    private MySQLiteOpenHelper helper;//数据库操作对象
@Override
    protected void onCreate(Bundle savedInstanceState) {
        super.onCreate(savedInstanceState);
        setContentView(R.layout.activity_sqlite);
        //初始化数据库操作对象 helper
        helper=new MySQLiteOpenHelper(SQLiteActivity.this);
    }
    //创建数据库
    public void create_db(View v)
    {
     SQLiteDatabase  writableDatabase = helper.getWritableDatabase();
    }
}
```

可先在 onCreate 的方法中实例化 MySQLiteOpenHelper 对象，把 SQLiteActivity.this 作为构造方法的参数传递进去。在创建数据库的按钮事件中调用 getWritableDatabase()方法，当程序执行后，单击了按钮，数据库就会创建出来。如果没有单击按钮，getWritableDatabase()方法没有被执行到，就不会创建数据库。

图 5-27　创建数据库

运行程序，当第一次单击按钮时，会检测到没有 employee.db，所以创建了名为 employee.db、版本号为 1 的数据库，当再次单击时发现数据库已被创建过，就不会再调用 onCreate (SQLiteDatabase db)方法了(见图 5-27)。

(5) 使用 SQLite_Expert_Professiona 工具查看数据库。

如果想看 employee.db 中的内容，可以把 employee.db 数据库导出保存到硬盘上，操作如图 5-28 所示，再通过 SQLite_Expert_Professiona 工具打开 employee.db。

安装好 SQLite_Expert_Professional 后，单击按钮打开 employee.db，可以查看数据库 employee.db 中的表 emp 和表中的字段，至此已经正确地创建了数据库和表，如图 5-29 所示。

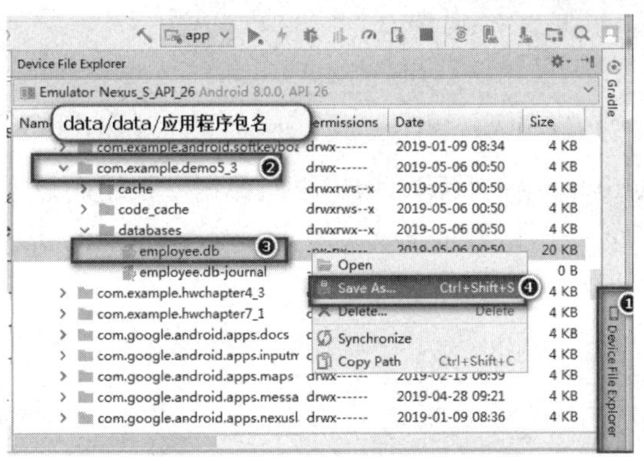

图 5-28 复制 employee.db

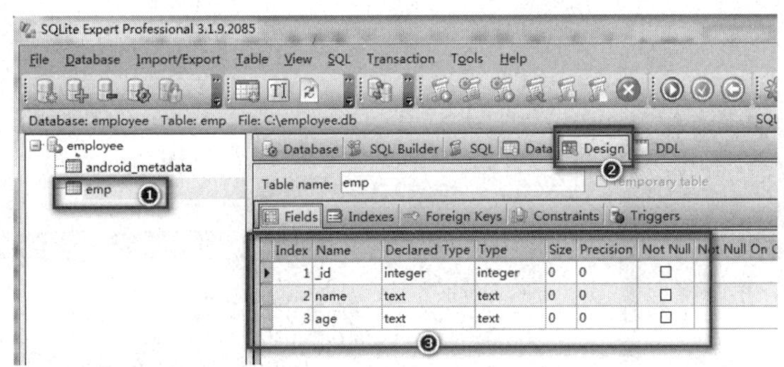

图 5-29 添加 employee.db 到数据库管理工具中并查看表结构

2. 更新数据库版本

(1) 数据库创建时版本从 1 开始，如果要更新数据库版本，可以给版本号一个公开的方法去操作它，当检测到版本号变化后，onUpgrade()会自动调用，所以可以在 onUpgrade()里做升级，以下是一个示例，调用 setDB_VERSION(int dB_VERSION)传入一个更高的版本号即可：

```
public static int getDB_VERSION() {
    return DB_VERSION;
}
public static void setDB_VERSION(int dB_VERSION) {
    DB_VERSION = dB_VERSION;
}
@Override  //数据库版本更新时会自动调用
public void onUpgrade(SQLiteDatabase db, int oldVersion, int newVersion) {
    //按照更新数据库的要求，新建表和插入数据
        //添加一张新表 salary
    String sql="create table salary(_id  integer primary key autoincrement,
emp_id integer,money text)";
    db.execSQL(sql);//执行建表语句
    Toast.makeText(context,"更新数据库 employee.db",Toast.LENGTH_LONG).show();
}
```

(2) 修改 activity_sqlite.xml 布局，多添加一个更新数据库的按钮(图 5-30)，代码如下：

```
<Button
    android:onClick="update_db"
    android:text="更新数据库"
```

```
    android:textColor="#00f"
    android:textSize="20sp"
    android:layout_width="match_parent"
    android:layout_height="wrap_content" />
```

(3) 修改 SQLiteActivity 中的代码，增加更新数据库按钮的点击事件，执行后导出数据库，可以看到数据库中多了一张表 salary，说明执行了 MySQLiteOpenHelper 中的 onUpgrade 方法，结果如图 5-31 所示。

```
//更新数据库
public void update_db(View v)
{
    helper=new MySQLiteOpenHelper(SQLiteActivity.this);
    SQLiteDatabase writableDatabase = helper.getWritableDatabase();
    helper.setDB_VERSION(2);

}
```

图 5-30 更新数据库

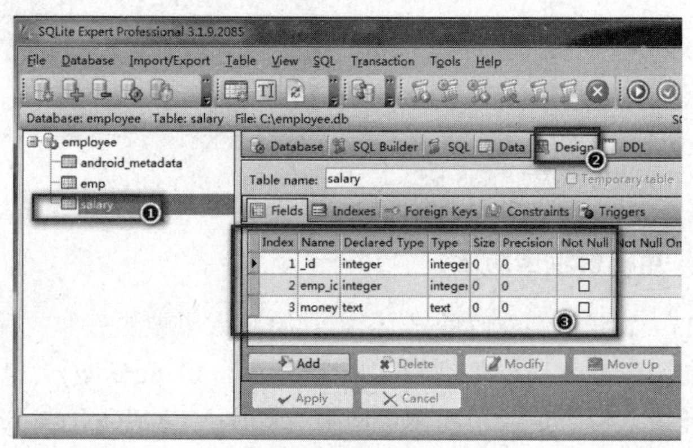

图 5-31 查看 salary 表

3. 添加数据

(1) 学会创建数据库和更新数据库后，接着要学习对表中的数据进行 CRUD 的操作了。CRUD 中的 C 代表创建(create)、R 代表检索(retrieve)、U 代表更新(update)、D 代表删除(delete)。SQLite 支持标准的 SQL 语法，为了操作简便，Android 提供了一些简单的方法用来进行 CRUD 操作：SQLiteDatabase writableDatabase = helper.getWritableDatabase();通过返回的对象 writableDatabase 可以对数据库进行添加、删除、修改操作。SQLiteDatabase readableDatabase = helper.getReadableDatabase();通过返回的对象 readableDatabase 可以对数据库进行查询操作。

(2) 要向表中添加数据，可以使用 SQLiteDatabase 提供的 insert()方法：long insert (String table, String nullColumnHack, ContentValues values)，其中，参数 table 代表想插入数据的表名，nullColumnHack 代表可以插入 null 值的数据列的列名，values 代表一行记录的数据。

insert 方法插入的一行记录使用 ContentValues 存放，ContentValues 类似于 Map，它提供了 put(String key,Xxx value)(其中 key 为数据列的列名)方法用于存入数据，getAsXxx (String key)方法用于取出数据。

insert 方法返回新添记录的行号，该行号是一个内部值，与主键 id 无关，发生错误返回-1。

插入一行数据，如果建表时_id 设置成自动增长，则该列不需要提供数据。

```
//插入数据
public void add(View v)
{
    SQLiteDatabase writableDatabase = helper.getWritableDatabase();
    ContentValues values=new ContentValues();
    //插入第一行数据
    values.put("name","陆伟");
    values.put("age",35);
    writableDatabase.insert("emp",null,values);
    values.clear();
    //插入第二行数据
    values.put("name","张新");
    values.put("age",22);
    writableDatabase.insert("emp",null,values);
}
```

4. 更新数据

要更新表中的数据，SQLiteDatabase 中提供了一个 update()方法：update(String table, ContentValues values, String whereClause, String[] whereArgs)，其中，参数 table 代表想要更新数据的表名，values 代表想要更新的数据，whereClause 代表更新条件，满足该 whereClause 子句的记录将会被更新，whereArgs 用于为 whereArgs 子句传递参数。

以下是更新 emp 表中 name 为“陆伟”的 age 为 28：

```
//更新数据
public void update(View v)
{
    SQLiteDatabase  writableDatabase = helper.getWritableDatabase();
    ContentValues values=new ContentValues();
    values.put("age",28);
    writableDatabase.update("emp",values,"name = ?",new String[]{"陆伟"});
}
```

5. 删除数据

要删除表中的数据，SQLiteDatabase 中提供了一个 delete()方法：delete(String table, String whereClause, String[] whereArgs)，其中，参数 table 代表想要删除数据的表名，whereClause 代表删除条件，满足该 whereClause 子句的记录将会被删除，whereArgs 用于为 whereArgs 子句传入参数。

以下是删除 emp 表中 age>25 的所有行：

```
//删除数据
public void del(View v)
{
    SQLiteDatabase writableDatabase = helper.getWritableDatabase();
```

```
    writableDatabase.delete("emp","age > ?",new String[]{"25"});
}
```

6. 查询数据

查询数据是 SQL 中最重要的操作，SQLiteDatabase 中提供了一个 query()方法来对数据进行查询：Cursor query(String table, String[] columns, String selection, String[] selectionArgs, String groupBy, String having, String orderBy)，这个 query 方法的参数 table 代表执行查询数据的表名，columns 代表要查询出来的列名，selection 代表查询条件子句，selectionArgs 用于为 selection 子句中占位符传入参数值，值在数组中的位置与占位符在语句中的位置必须一致，否则就会有异常，groupBy 用于控制分组，having 用于对分组进行过滤，orderBy 用于对记录进行排序。

通常只传需要的参数就行，调用 query()后会返回一个 Cursor 游标对象，查询到的数据通过游标对象取出来。游标对象会指向查询结果集第一行的前面，所以要通过 Cursor 游标对象的 moveToFirst()方向指向第一条记录，通过 moveToNext()指向下一条记录，如果没有记录了，moveToNext()返回值为空，所以通常取数据时都用循环来控制。

以下是取出 emp 表中所有数据的操作：

```
//查询数据
public void query(View v)//查询按钮的事件处理
{
    myquery();
}
Public void myquery()
{
SQLiteDatabase readableDatabase = helper.getReadableDatabase();
    //取出 emp 表中的所有数据
   Cursor cursor = readableDatabase.query("emp",null,null,null,null,null,null);
  //如果有数据
   while(cursor.moveToNext())
   {
       //遍历数据
       int _id=cursor.getInt(cursor.getColumnIndex("_id"));//取出_id
      String name= cursor.getString(cursor.getColumnIndex("name"));//取出 name
       int age=cursor.getInt(cursor.getColumnIndex("age"));//取出 age
       Log.i("TAG","查询数据: "+_id+" "+name+" "+age);
   }
    cursor.close();
}
```

如果要取出名字以“彭”开头的数据可以用：

```
Cursor cursor = readableDatabase.query("emp",new String[]{"_id,name,age"},"name like ?",
new String []{"彭%"},null,null,null);
```

以上对数据库的 CRUD 操作进行了简单的讲解，现在来看一下结果，修改 activity_sqlite.xml 布局，多添加 4 个按钮，即插入数据、删除数据、更新数据、查询数据：

```
<Button
    android:onClick="add"
    android:text="插入数据"
    android:textColor="#00f"
    android:textSize="20sp"
    android:layout_width="match_parent"
```

```
        android:layout_height="wrap_content" />
    <Button
        android:onClick="update"
        android:text="更新数据"
        android:textColor="#00f"
        android:textSize="20sp"
        android:layout_width="match_parent"
        android:layout_height="wrap_content" />
    <Button
        android:onClick="del"
        android:text="删除数据"
        android:textColor="#00f"
        android:textSize="20sp"
        android:layout_width="match_parent"
        android:layout_height="wrap_content" />
    <Button
        android:onClick="query"
        android:text="查询数据"
        android:textColor="#00f"
        android:textSize="20sp"
        android:layout_width="match_parent"
        android:layout_height="wrap_content"/>
```

在 SQLiteActivity 中按钮的对应事件处理方法 add(View v)、update(View v)、del(View v)、query(View v)的最后一行都调用 myquery()查询方法，程序执行起来，在日志文件中查看到结果如图 5-32 所示。

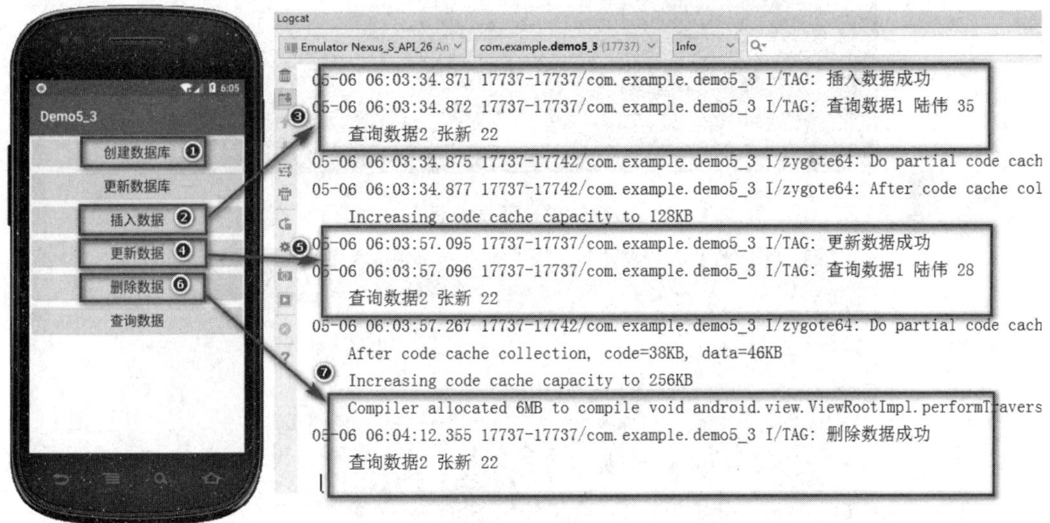

图 5-32　数据库操作结果

上面的操作也可以直接执行查询操作，在真正的项目中可能会碰到复杂得多的查询功能，更多高级的用法可自行慢慢摸索。

【任务实施】

1. 任务分析

本任务要完成记住登录功能。

(1) 程序界面使用任务 1 的登录界面，再新建一个注册界面。

(2) 创建一个 User 类用于封装用户数据。

(3) 创建 MySQLiteOpenHelper 类用于创建数据库和创建表。

(4) 创建一个 UserDao 类用于操作数据库的添加和查询数据。

(5) LoginActivity 用于登录，处理登录和注册按钮的点击事件。

(6) RegActivity 用于注册，处理注册按钮的点击事件。

2. 任务实现

(1) 使用任务 1 的工程 Project5_1，复制一份 Project5_1 工程，改名为 Project5_3 后用 Android Studio 打开。主 Activity 为登录 LoginActivity。在工程中新建一个 RegActivity，对应布局文件 activity_reg.xml，此处不展开代码，有需要可查看 Project5_3 工程源代码。界面效果如图 5-33 和图 5-34 所示。

图 5-33 登录界面

图 5-34 注册界面

注册界面启用左上角的返回功能 ← RegActivity，当单击时可以返回到 LoginActivity，这个功能可以在清单文件中实现：

```
<activity android:name=".RegActivity"
    android:label="RegActivity"
    android:parentActivityName=".LoginActivity" >
</activity>
```

(2) 新建一个 User 类用于封装用户数据：

```
public class User {
    private int _id;
    private String name;
    private String pass;
//无参构造   有参构造  get 和 set 方法   toString()方法此处略过
}
```

(3) 新建 MySQLiteOpenHelper 类用于创建数据库和创建表：

```
public class MySQLiteOpenHelper extends SQLiteOpenHelper {
   private Context context;

   public static final String DB_NAME = "user.db";//数据库名称
   public static int DB_VERSION = 1;//数据库版本号

   public static final String TABLE_NAME = "user";//表名
   public static final String FIELD_ID = "_id"; //表中字段名
   public static final String FIELD_NAME = "username";
   public static final String FIELD_PASS = "pass";

   public MySQLiteOpenHelper(Context context) {
      super(context, DB_NAME, null, DB_VERSION);
      this.context = context;
   }


   @Override   //第一次创建数据库时会调用
   public void onCreate(SQLiteDatabase db) {
      // 构造要执行的 SQL 语句
      String sql = "create table " + TABLE_NAME + "(" + FIELD_ID + "  integer
primary key autoincrement," + FIELD_NAME + "  text," + FIELD_PASS + "  text)";
      db.execSQL(sql);//执行建表语句
   }
   @Override
   public void onUpgrade(SQLiteDatabase db, int oldVersion, int newVersion) {
   }
}
```

数据库名、表名、列名都用常量描述，建表语句也用常量代替，注意建表时语句留有空格，以免与常量混在一起，不符合 SQL 语句格式。

(4) 新建一个 UserDao 类用于操作数据库的添加和查询数据：

```
public class UserDao {
    private MySQLiteOpenHelper helper;
    private Context context;
    public UserDao(Context context)
    {
        this.context = context;
        //初始化数据库操作对象 helper
        helper=new MySQLiteOpenHelper(context);
    }

     //插入数据:注册时调用
```

```
    public void addUser(String name,String pass)
    {
      SQLiteDatabase writableDatabase = helper.getWritableDatabase();
        ContentValues values=new ContentValues();
        //插入第一行数据
        values.put(MySQLiteOpenHelper.FIELD_NAME,name);
        values.put(MySQLiteOpenHelper.FIELD_PASS,pass);
        writableDatabase.insert(MySQLiteOpenHelper.TABLE_NAME,null,values);
        values.clear();
    }

    /**
     * 查询用户数据
     * @param inputname 输入的用户名
     * @return 用户对象
     */
    public User myquery(String inputname)
    {
        SQLiteDatabase  readableDatabase = helper.getReadableDatabase();
        //取出 user 表中的所有数据
           Cursor cursor = readableDatabase.query(MySQLiteOpenHelper.TABLE_NAME,
null,"username=?",new String[]{inputname},null,null,null);
        User user =null;
        //如果有数据
        if(cursor.moveToNext())
        {
            //遍历数据
   int _id=cursor.getInt(cursor.getColumnIndex(MySQLiteOpenHelper.FIELD_ID));//取出_id
            String name= cursor.getString(cursor.getColumnIndex
(MySQLiteOpenHelper.FIELD_NAME));//取出 username
            String pass=cursor.getString(cursor.getColumnIndex
(MySQLiteOpenHelper.FIELD_PASS));//取出 pass
            user =new User(_id,name,pass);
            Log.i("TAG","查询数据 user:"+user);
        }
        cursor.close();
         return user;
    }
}
```

(5) LoginActivity 用于登录，处理登录按钮的点击事件。

在 LoginActivity 类中添加 UserDao dao;成员变量，在登录按钮中添加以下代码，当单击“登录”按钮时，先凭用户名到数据库中查询，如果用户不存在，则提示先注册。

如果用户存在，则判断密码是否正确并给出相应的提示。当用户名存在并且密码正确时提示登录成功。代码如下：

```
dao = new UserDao(LoginActivity.this);
User user = dao.myquery(sname);

if (user == null) {
    Toast.makeText(LoginActivity.this, "用户不存在，请先注册", Toast.LENGTH_LONG).show();
} else {
    if (user.getPass().equals(spass)) {
        Toast.makeText(LoginActivity.this, "登录成功", Toast.LENGTH_LONG).show();
    } else {
        Toast.makeText(LoginActivity.this, "密码错误，登录失败
", Toast.LENGTH_LONG).show();
    }
}
```

处理 LoginActivity 中注册按钮的点击事件：

```
public void onReg(View v) {
    Intent intent = new Intent(LoginActivity.this, RegActivity.class);
    startActivity(intent);
}
```

(6) RegActivity 用于注册，处理注册事件。

先判断输入的用户名和密码是否为空，再判断两次输入的密码是否一致，当输入符合要求后，凭输入的用户名到数据库中查询用户是否存在，如果存在则不能注册；否则注册。这时才把用户名和密码的相关信息插入到数据库中。代码如下：

```
public class RegActivity extends AppCompatActivity {
    private EditText reg_et_name;//用户名
    private EditText reg_et_pwd1,reg_et_pwd2;//密码输入框
    UserDao dao;
    @Override
    protected void onCreate(Bundle savedInstanceState) {
        super.onCreate(savedInstanceState);
        setContentView(R.layout.activity_reg);
        reg_et_name = findViewById(R.id.reg_et_name);
        reg_et_pwd1 = findViewById(R.id.reg_et_pwd1);
        reg_et_pwd2 = findViewById(R.id.reg_et_pwd2);

        dao = new UserDao(RegActivity.this);
    }

    //注册按钮
    public void onReg(View v)
    {
        //获取输入的用户名和密码
        final  String sname=reg_et_name.getText().toString().trim();
        final  String spass1= reg_et_pwd1.getText().toString().trim();
        final  String spass2= reg_et_pwd2.getText().toString().trim();
        //如果用户名和密码输入为空
        if (TextUtils.isEmpty(sname) || TextUtils.isEmpty(spass1)||TextUtils.isEmpty(spass2)) {
            Toast.makeText(RegActivity.this, "用户名或密码不能为空", Toast.LENGTH_LONG).show();
            return;
        }
        if(!spass1.equals(spass2))
        {
            Toast.makeText(RegActivity.this, "两次输入密码不一致，请重输", Toast.LENGTH_LONG).show();
            return;
        }

        UserDao dao =new UserDao(RegActivity.this);
        //先查询用户名是否存在，不存在才能注册
        if(dao.myquery(sname)==null)
        {
            dao.addUser(sname,spass1);
            Toast.makeText(RegActivity.this, "注册成功",Toast.LENGTH_LONG).show();

        }
        else
        {
            Toast.makeText(RegActivity.this, "用户已存在",Toast.LENGTH_LONG).show();
```

```
            }
        }
}
```

3. 运行结果

运行程序，结果如下。

(1) 不输入用户名和密码直接单击“登录”按钮，结果如图 5-35(a)所示。

(2) 输入数据库中不存在的用户名，结果如图 5-35(b)所示。

(3) 两次密码不一致，结果如图 5-35(c)所示。

(4) 用户名和密码无误，成功注册，用户信息插入到数据库中，结果如图 5-35(d)所示。

(5) 用户名已存在，结果如图 5-35(e)所示。

(6) 从数据库中查询出无误的用户名和密码，登录成功，结果如图 5-35(f)所示。

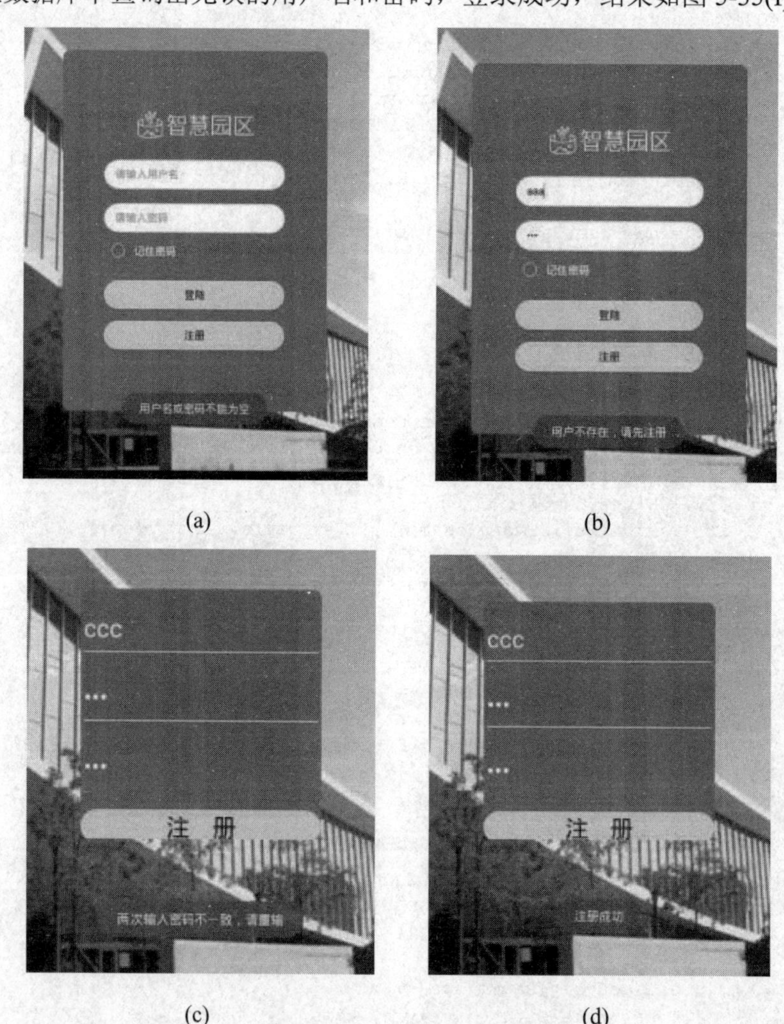

图 5-35　程序运行结果

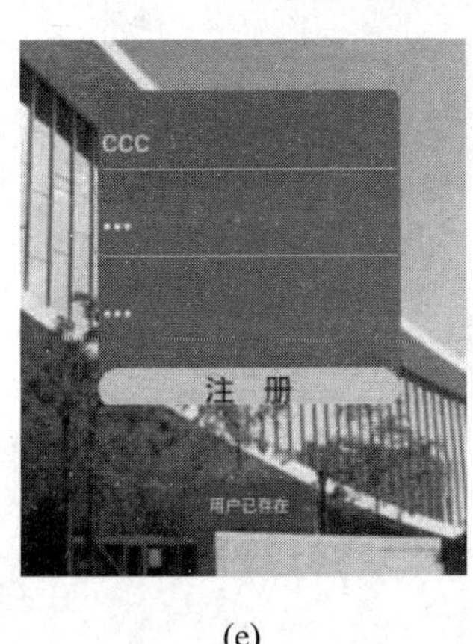

(e)

(f)

图 5-35 程序运行结果(续)

项 目 总 结

本项目主要讲解了 Android 中数据存储的 3 种方式，包括文件存储、SharedPreferences 存储和 SQLite 数据库存储。文件存储用了 openFileOutput 流和 openFileInput 流来操作，还实现了把图片存储到 SD 卡，同时处理了运行时权限申请。SharedPreferences 保存的是键值对的数据，SQLite 保存的是关系型数据，并讲解了数据库的创建和更新以及对数据的 CRUD 操作。

通过本项目的学习，读者应熟练掌握这 3 种数据存储的使用方法，并灵活应用到程序的开发中。

思考与练习

5-1 简述用 SharedPreferences 实现登录次数限定的思路。

5-2 简述如何使用 SQLiteOpenHelper 进行数据库创建和更新。

5-3 简述 SQLite 数据库分别用什么方法执行 CRUD 的操作。

项目 6 线程与消息处理

【项目描述】

在 Android 中，当需要执行耗时操作，如获取网络资源时，如果直接放在 main()方法所在的主线程中处理，会导致主线程阻塞，所以应该放在子线程中去处理。

本项目通过 4 个任务(见图 6-1)的学习，针对 Android 开发中必须掌握的线程和消息以及异步任务进行讲解，最后实现传感器数据的实时更新。

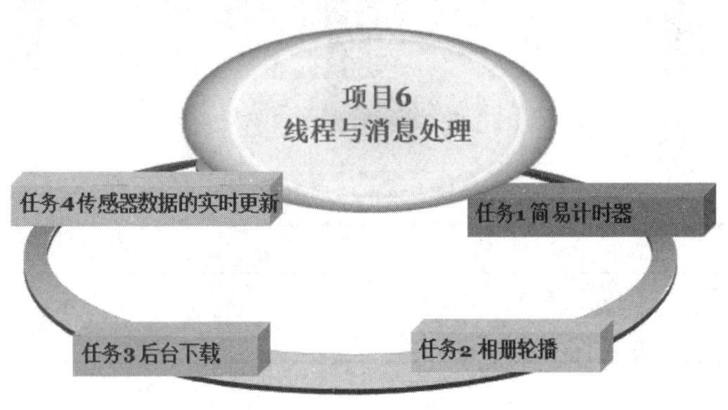

图 6-1　项目 6 任务图

【学习目标】

知识目标：

- 会线程的两种创建方式。
- 会使用 Handler 发送消息。
- 会使用 AsyncTask 执行异步任务。
- 学会 API 文档的基本使用方法。
- 知道 jar 包的基本使用方法。

技能目标：

- 能创建线程与启动线程。
- 能用消息机制发送和接收消息。
- 能使用异步任务进行网络数据的下载。
- 能利用 API 获取数字量传感器的数据。

任务 1　简易计时器

【任务描述】

扫码观看视频讲解

开发一个简易计时器，单击“开始计时”按钮开始计时，如图 6-2 所示。具体任务清单如表 6-1 所示。

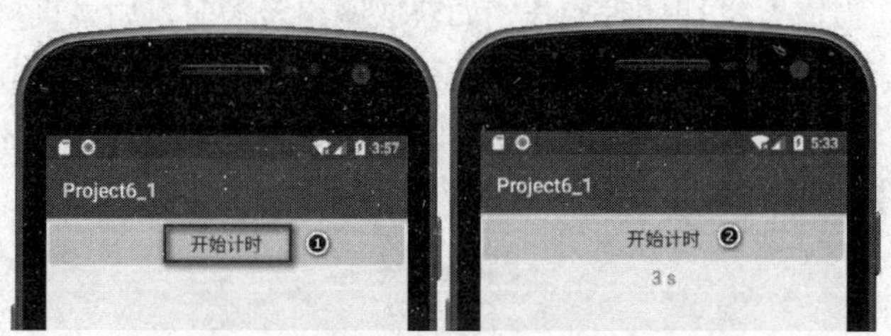

图 6-2　程序结果图

表 6-1　任务清单

任务课时	4 课时	任务组员数量	建议 1 人
任务组采用设备	PC 一台 模拟器或手机或工控平板		

【知识解析】

1. UI 线程

一般来说，Android 程序运行起来后，就会创建一个进程。也就是说，运行中的程序就是进程，而进程中的执行线路就是线程。进程负责竞争 CPU 资源，竞争到了进程就可以运行。Android 程序启动后会开启一个主线程，也叫 UI 线程，UI 线程中运行着许多重要的逻辑，如系统事件处理、用户输入事件处理、UI 绘制、Service、Alarm 等，如图 6-3 所示。

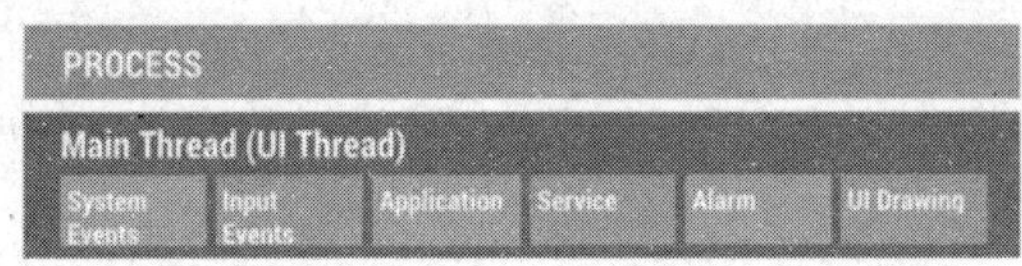

图 6-3　UI 线程包含的逻辑

2. 不能在主线程中执行耗时操作

在主线程中可以对用户触摸事件进行检测和响应、对用户输入进行处理、自定义 View 的绘制等。那么，在主线程中可以执行耗时操作吗？下面就来看一个例子。

(1) 布局 activity_main.xml：

```
<?xml version="1.0" encoding="utf-8"?>
<LinearLayout
    xmlns:android="http://schemas.android.com/apk/res/android"
    xmlns:app="http://schemas.android.com/apk/res-auto"
    xmlns:tools="http://schemas.android.com/tools"
    android:layout_width="match_parent"
    android:layout_height="match_parent"
    android:orientation="vertical"
    tools:context="com.example.thread.demo8_1.MainActivity">

    <Button
        android:onClick="btnClick"
        android:text="执行耗时操作"
        android:textSize="18sp"
        android:layout_width="match_parent"
        android:layout_height="wrap_content" />

    <TextView
        android:id="@+id/tvTime"
        android:gravity="center"
        android:layout_width="match_parent"
        android:layout_height="wrap_content"
        android:textSize="18sp"
        />

</LinearLayout>
```

布局中定义了两个控件：“开始计时”按钮中设置单击按钮后的响应方法为 click；单击按钮后时间显示在 TextView 上。

(2) 接下来修改 MainActivity 的代码，具体如下：

```
public class MainActivity extends AppCompatActivity {
    private TextView tvTime;

    @Override
    protected void onCreate(Bundle savedInstanceState) {
        super.onCreate(savedInstanceState);
        setContentView(R.layout.activity_main);
        tvTime = findViewById(R.id.tvTime);

    }
```

```
11.        //按钮的点击事件
12.     public void btnClick(View v) {
13.       try {
14.           Thread.sleep(10000);//主线程休眠 10s
15.       } catch (InterruptedException e) {
16.           e.printStackTrace();
17.       }
18.     }
19. }
```

第 14 行，在按钮的点击事件 btnClick()方法中，让主线程休眠 10s。

(3) 运行程序，单击“开始执行耗时操作”按钮，按返回键，发现界面卡死，过了一会儿，程序崩溃，报了应用程序没响应(isn’t responding)的错，如图 6-4 所示。

图 6-4　应用程序报 ANR 异常

(4) 查看 Logcat，发现报了一个 ANR，如图 6-5 所示。

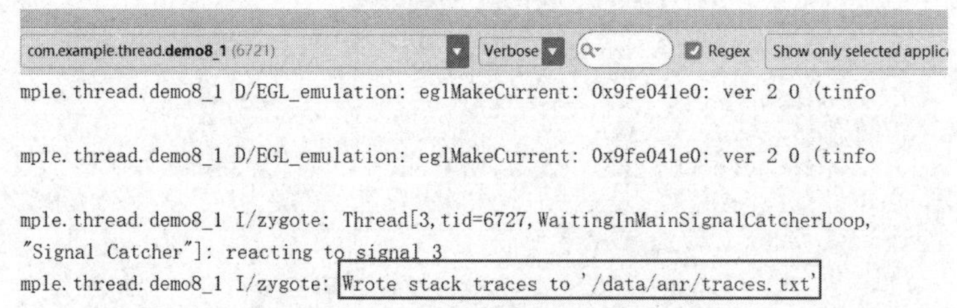

图 6-5　日志文件中的 ANR 异常提醒

因此，不能在主线程进行耗时操作。一般阻塞 UI 线程大于 5s 就会报 ANR(应用程序无响应的)异常，程序会终止并退出。

3. 线程的创建和启动

在主线程中执行大于 5s 的耗时操作可能会引起 ANR 异常，但在 App 中经常需要做一些耗时的操作，如网络请求、下载文件等，这些操作如果在主线程中执行，会导致主线程卡死，报 ANR 异常，所以通常在子线程中进行这些操作，那么如何创建子线程呢？

1) 继承 Thread 类

创建一个类 MyThraed 继承 Thread，重写 run()方法，把子线程要执行的代码写在 run()方法中，代码如下：

```
class MyThread  extends Thread{
    @Override
    public void run() {
        //在这里写要线程做的事情
    }
}
```

Thread 类的 start()方法可以启动线程，启动线程后会执行 Thread 类中的 run()方法。所以创建 MyThread 类的实例，调用 start()方法启动线程。

```
Thread thread= new MyThread();
thread.start();
```

2) 实现 Runnable 接口

另一种创建线程的方法是实现 Runnable 接口，这种方法适用于多个线程共享一个目标对象，非常适合多个线程处理同一份资源的情况，代码如下：

```
class MyRunnable implements Runnable{
    @Override
    public void run() {
        //在这里写要线程做的事情
    }
}
```

先创建一个 Runnable 的对象 myRunnable，把 myRunnable 作为线程构造方法的参数，调用线程的 start()方法启动线程，启动线程后会执行 MyRunnable 类中的 run()方法，代码如下：

```
MyRunnable myRunnable=new MyRunnable();
Thread thread = new Thread(myRunnable);
thread.start();
```

3) 匿名线程

很多时候我们不会专门创建一个类去完成子线程的操作。可以采用匿名内部类的方式创建匿名线程，这种写法在实际开发过程中更常用。匿名线程代码如下：

```
new Thread(new Runnable() {
    @Override
    public void run() {
        //在这里写要线程做的事情
    }
}).start();
```

4. 不能在子线程中更新 UI

接下来把计时的操作放到子线程中，把计时的秒数显示在文本框中。改写按钮点击事件的代码：

```
//计时按钮的点击事件
public void btnClick(View v) {
    new Thread(new Runnable() {
        @Override
        public void run() {
            while (true) {
                tvTime.setText(num++ + " s");
                try {
```

```
                    Thread.sleep(1000);
                } catch (InterruptedException e) {
                    e.printStackTrace();
                }
            }
        }
    }).start();
}
```

运行程序，过了一会儿程序崩溃了，如图 6-6 所示。查看 Logcat 日志，报错提示：只有在主线程中才能更新 View 组件，即不能在子线程中更新 UI，如图 6-7 所示。

图 6-6　程序崩溃了

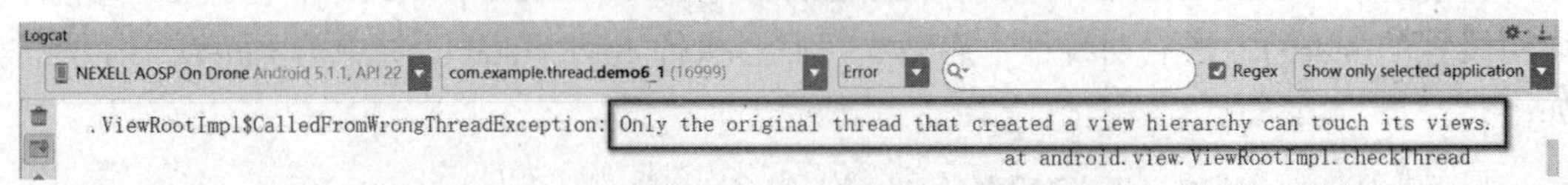

图 6-7　不能在非 UI 主线程更新 UI 的出错提示

5. 使用 runOnUiThread()方法更新 UI

Android 提供了 Activity.runOnUiThread(Runnable runnable)方法来更新 UI。如果当前线程是 UI 主线程，则直接执行 Runnable 的代码；否则将 Runnable 传递到 UI 线程的消息队列中，所以在子线程中可以调用 runOnUiThread()方法更新 UI。

```
runOnUiThread(new Runnable() {
    @Override
    public void run() {
      //在这里更新UI
    }
});
```

6. Timer 和 TimerTask

(1) Timer 是 JDK 中提供的一个定时器工具，它可以根据指定的时间、指定的执行周期在 UI 线程之外另起一个单独的线程来执行固定的任务 TimerTask，可以指定执行一次或者反复执行多次。

利用 Timer 执行计划任务的代码如下：

```
TimerStat task=new TimerStat(new Runnable(){
    @Override
    public void run() {
  //在这里执行任务
```

```
5.         }
6.     });
7.     Timer timer = new Timer();
8.     timer.schedule(task,1000,1000);//延时多久后执行，间隔多久重复执行
```

第 7 行 new 一个 Timer 子类，第 8 行调用的是 schedule()方法，在 schedule()方法中传入 TimerTask 对象，在 TimerTask 中重写 run()方法来指定具体的任务。

(2) schedule()方法：java.util.Timer.schedule(TimerTask task, long delay, long period)有 3 个参数：①TimerTask 类，使用时要继承该类，并实现 run()方法，因为 TimerTask 类实现了 Runnable 接口；②用户调用 schedule()方法后，要等待所设置的时间才可以第一次执行 run()方法；③第一次调用之后，从第二次开始每隔所设置的时间调用一次 run()方法。

(3) 终止 Timer 线程，调用 timer 的 cancel()方法即可。

```
timer.cancel();// 结束定时器
```

【任务实施】

1. 任务分析

本任务实现计时功能，计算出当前过去了多少秒。

(1) 界面设计一个 Button 和一个 TextView。当单击 Button 时在 TextView 上计时的时间。

(2) 用 Timer 启动一个定时任务 TimerTask，每次间隔 1s 执行一次。

(3) 在定时任务里启动一个线程，在线程的 Runnable 里用 Activity.runOnUiThread (Runnable)在 UI 线程中更新 TextView 的时间。

2. 任务实现

布局代码同上，这里列出 MainActivity 的代码：

```
public class MainActivity extends AppCompatActivity {
    private TextView tvTime;
    //初始时间为 0s
    private int num = 0;

    @Override
    protected void onCreate(Bundle savedInstanceState) {
        super.onCreate(savedInstanceState);
        setContentView(R.layout.activity_main);
        tvTime = findViewById(R.id.tvTime);
    }

    //计时按钮的点击事件
    public void btnClick(View v) {
        try {
            //1.定义时间定时器
            Timer timer = new Timer(true);//守护线程，守护线程会和主线程一起销毁
            //2.启动延时计时器。单击后延时 1 秒执行，每次间隔 1 秒后再次执行
            timer.schedule(task, 1000, 1000);
        } catch (Exception e) {
            e.printStackTrace();
        }
    }

```

```
    //3.定时器任务
    TimerTask task = new TimerTask() {
        @Override
        public void run() {
            tvTime.setText(num++ + "s");
            //4.实例化一个线程
            new Thread(new Runnable() {
                @Override
                public void run() {
                    runOnUiThread(new Runnable() {
                        @Override
                        public void run() {
                            // 5.更新时间
                            tvTime.setText(num++ + "s");
                        }
                    });
                }
            }).start();// 启动线程
        }
    };
}
```

至此，就完成了简易计时器。完整代码可查阅随书资料中的 Project6_1 工程源代码。

3. 运行结果

程序运行结果如图 6-8 所示。

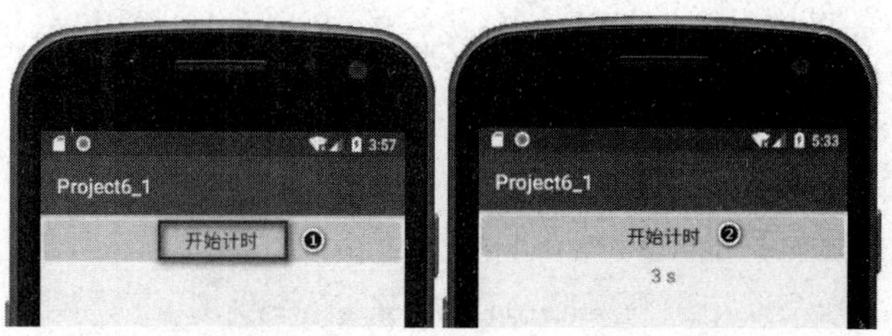

图 6-8　计时器运行结果

任务 2　相 册 轮 播

【任务描述】

本任务制作一个轮播相册，用消息机制完成。任务清单如表 6-2 所示。

扫码观看视频讲解

表 6-2　任务清单

任务课时	2 课时	任务组员数量	建议 1 人
任务组采用设备	PC 一台 模拟器或手机或工控平板		

【知识解析】

1. 异步消息处理机制

Handler 是 Android 提供的一套异步消息处理机制，用来解决在子线程中进行 UI 操作的问题。Handler 异步消息处理主要由 Message、Handler、MessageQueue 和 Looper 等 4 个部分组成。

1) Message

Message 用来在线程之间传递消息，传递消息时可以带一些简单数据，如图 6-9 所示。

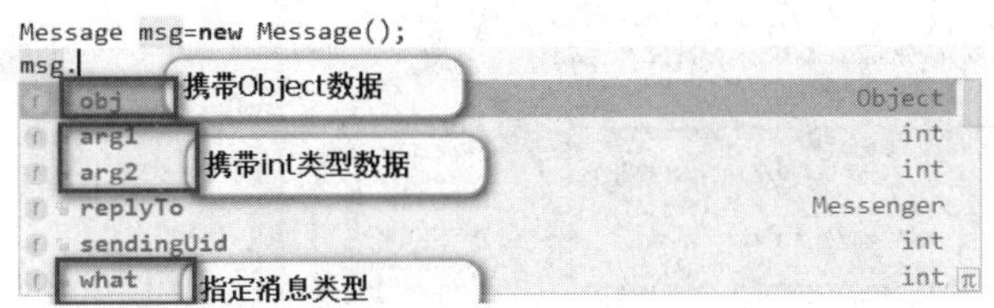

图 6-9 msg 可携带的参数类型

2) Handler

Handler 用来发送和处理消息。在主线程中创建 Handler 对象，重写 handlerMessage(Message msg)，在新线程中使用 Handler 的 sendMessage()方法发消息，而发出的消息经过处理后最终会传递到 handlerMessage(Message msg)中，再通过 msg 取出传过来的数据，如图 6-10 所示。

```
public static final int MSG_TYPE=1;
Handler handler=new Handler(){
    @Override
    public void handleMessage(Message msg) {
        switch (msg.what)
        {
            case MSG_TYPE:
                //在这里接收数据并处理业务
                Object obj = msg.obj;
                int val1 = msg.arg1;
                int val2 = msg.arg2;
        }
    }
};
```

在主UI线程中创建Handler
并接收数据和处理业务

```
public void click(View v)
{
    new Thread(new Runnable() {
        @Override
        public void run() {
            Message msg=new Message();
            msg.what=MSG_TYPE;
            msg.obj= new Object();
            msg.arg1=10;
            msg.arg2=20;
            handler.sendMessage();
        }
    }).start();
}
```

在新线程中发消息同时
可以带数据

图 6-10 Handler 发送消息和接收消息

3) MessageQueue

MessageQueue 是消息队列，每个线程中只会有一个 MessageQueue 对象，所有通过 Handler 发送的消息会一直存在于消息队列中等待被处理。

4) Looper

Looper 用来管理 MessageQueue，一个线程中只有一个 Looper 对象，调用 Looper 的 loop() 方法后，就会进入死循环中，一旦发现 MessageQueue 中有消息，就会将它取出并传递到 Handler 的 handlerMessage()方法中。

2. 消息机制的工作流程

消息机制的工作流程如下。

① 在主 UI 线程中创建 handler 对象。

② 在新线程中用 handler 对象发消息。

③ 消息会进入消息队列。

④ UI 线程从消息队列中取消息，如图 6-11 所示。

本项目任务 1 用到的 Activity.runOnUiThread 内部也是上述消息处理机制的封装。使用时注意导入的包是 android.os 下的，如 android.os.Handler、android.os.Message。

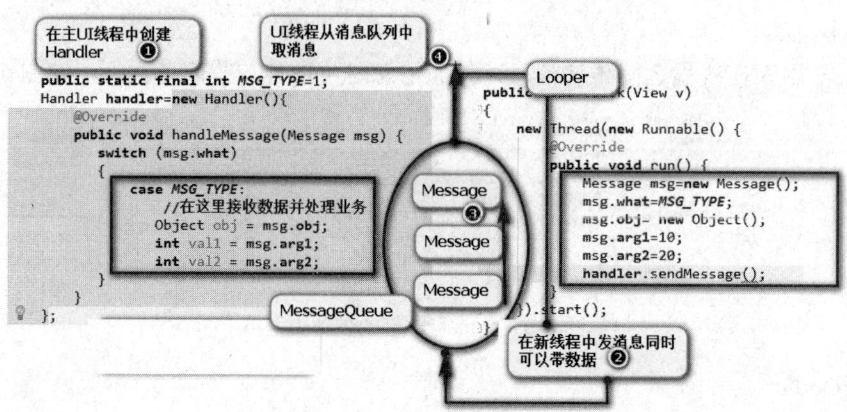

图 6-11　消息机制的工作流程

【任务实施】

1. 任务分析

自动轮播相册，可以用 Handler 机制来完成，在子线程中按时间间隔向 UI 线程发消息，UI 线程收到消息后切换图片。

(1) 把要轮播的图片放入 drawable 下，在程序中用数组存放图片的 ID。

(2) 布局文件中用 ImageView 控件展示图片。

(3) 主线程中声明一个 Handle 对象，重写 handleMessage()方法。

(4) 在子线程中创建 Message 对象，将它的 what 指定为要发的消息类型，调用 handler 的 sendMessage()将消息发出去，用定时任务每隔 1s 发一个消息。

(5) 主线程中在 handleMessage()方法里获取发过来的消息，判断消息类型无误后，从数组中取图片显示在 ImageView 控件上。

2. 任务实现

(1) 新建工程 Project6_2，创建的时候默认新建的 Activity 命名为 PhotoActivity，布局命名为 activity_photo。将随书资料中本项目任务 2 轮播的图片复制到 drawable 文件夹下，编写 activity_photo.xml 代码如下：

```
<?xml version="1.0" encoding="utf-8"?>
<LinearLayout xmlns:android="http://schemas.android.com/apk/res/android"
    xmlns:app="http://schemas.android.com/apk/res-auto"
    xmlns:tools="http://schemas.android.com/tools"
    android:layout_width="match_parent"
    android:layout_height="match_parent"
    tools:context=".PhotoActivity">
    <ImageView
```

```
        android:id="@+id/image"
        android:layout_width="match_parent"
        android:layout_height="match_parent" />

</LinearLayout>
```

(2) 修改 PhotoActivity 代码如下：

```
public class PhotoActivity extends AppCompatActivity {
    // 要轮播的图片，放在drawable文件夹下
    private int[] images = {R.drawable.pic_1, R.drawable.pic_2, R.drawable.pic_3, R.drawable. pic_4 };
    private ImageView image;
    //切换图片的消息标志
    public static final int JUMP = 1;
    public static int index = 0; //数组下标
    public static int len=0; //图片个数

    @Override
    protected void onCreate(Bundle savedInstanceState) {
        super.onCreate(savedInstanceState);
        setContentView(R.layout.activity_photo);
        image = findViewById(R.id.image);
        len=images.length;
        final Handler handler = new Handler() {
            @Override
            public void handleMessage(Message msg) {
                //接收消息
                super.handleMessage(msg);
                //判断是不是要接收的消息类型
                if (msg.what == JUMP) {

                    image.setImageResource(images[index++]);
                    //判断是不是最后一张，如果是，从头开始
                    if(index>len-1)
                        index=0;
                }
            }
        };
        new Timer().schedule(new TimerTask() {
            @Override
            public void run() {
                //声明消息对象
                Message msg = new Message();
                //声明消息类型
                msg.what = JUMP;
                //发送消息
                handler.sendMessage(msg);
            }
        }, 1000, 2000);
    }
}
```

至此，就完成了相册轮播。完整代码可查阅随书资料中的 Project6_2 工程源代码。

3. 运行结果

运行程序，结果如图 6-12 所示。

图 6-12　程序运行结果

任务 3　后 台 下 载

【任务描述】

本案例用 AsyncTask 异步任务完成百度 Logo 图片的下载。任务清单如表 6-3 所示。

扫码观看视频讲解

表 6-3　任务清单

任务课时	4 课时	任务组员数量	建议 1 人
任务组采用设备	PC 一台 模拟器或手机或工控平板		

【知识解析】

1. AsyncTask 异步任务的 3 个参数

AsyncTask 是 Android 提供的一个工具，内部封装了异步消息处理机制。使用 AsyncTask 可以很简单地从子线程切换到 UI 主线程。

AsyncTask 是一个抽象类，要使用它必须创建一个子类去继承它，继承时需要输入 3 个泛型参数，如图 6-13 所示。

图 6-13　AsyncTask 的 3 个参数

参数①Params：在执行 AsyncTask 时需要传入的参数，可用于在后台任务中使用，比如若要下载，就把要下载资源的地址传进来，则它是 String 类型。

参数②Progress：后台任务执行时，如果需要在界面上显示当前的进度，则使用这里指定的泛型参数为进度单位。

参数③Result：当任务执行完毕后，如果需要对结果进行返回，则使用这里指定的泛型参数作为返回值类型。

比如要下载网络图片，需要把图片所在的地址传进去让后台执行，所以参数①是 String，下载的时候需要显示进度条，所以参数②是 Integer，下载的结果存放在字节数组中，所以参数③是 byte[]：

```
public class MyDownTask extends AsyncTask<String, Integer, byte[]> {
    ......
}
```

2. AsyncTask 异步任务的 4 个方法

上面只是定义了一个异步任务，代码还在报错，提示我们必须实现方法，如图 6-14 所示。

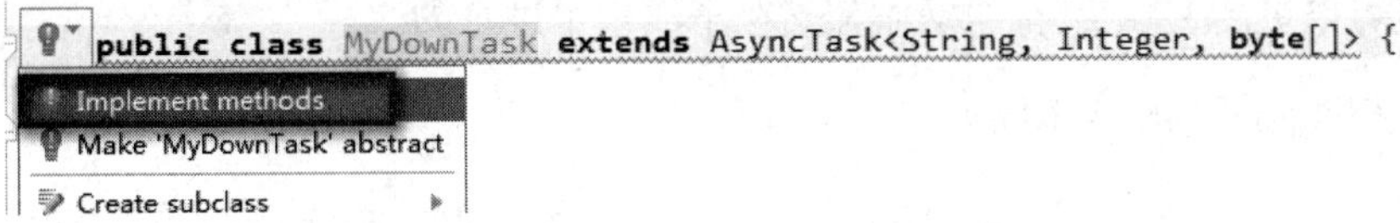

```
public class MyDownTask extends AsyncTask<String, Integer, byte[]> {
  Implement methods
  Make 'MyDownTask' abstract
  Create subclass
```

图 6-14 AsyncTask 必须实现后台下载的方法

按操作提示，发现必须实现一个 doInBackground()的方法，其中该方法的参数由①决定，方法的返回值类型由图 6-15 中的②决定，也就是说定义 AsyncTask 时的 3 个参数会影响 doInBackground()的写法，如图 6-15 所示。

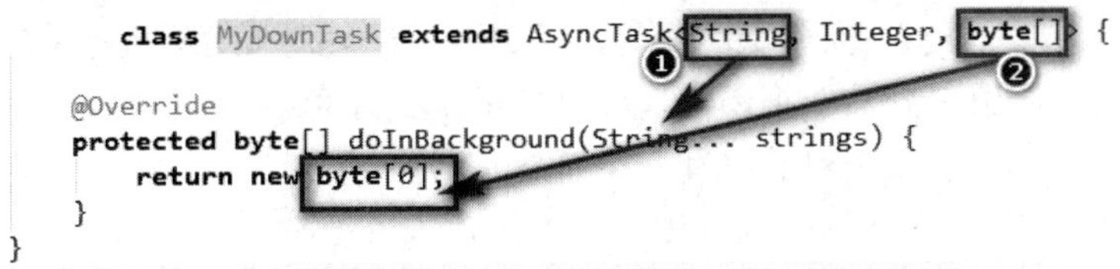

```
class MyDownTask extends AsyncTask<String, Integer, byte[]> {

    @Override
    protected byte[] doInBackground(String... strings) {
        return new byte[0];
    }
}
```

图 6-15 AsyncTask 的参数与返回值的关系 1

如果把要传送的用户名和密码放在集合中，可以用 List 参数，Void 代表没有返回值，如图 6-16 所示。

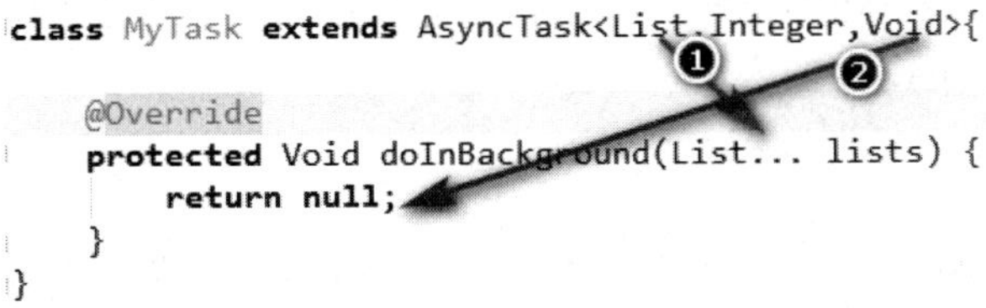

```
class MyTask extends AsyncTask<List,Integer,Void>{

    @Override
    protected Void doInBackground(List... lists) {
        return null;
    }
}
```

图 6-16 AsyncTask 的参数与返回值的关系 2

要完成异步任务，通常会重写 4 个方法，如图 6-17 所示。

图 6-17　AsyncTask 的 4 个方法

AsyncTask 的工作过程：①做初始化工作，它会在②之前执行。②完成最重要的后台任务。②执行的过程中如果有更新进度条则在③里进行。②执行结束会把结果传递给④的参数，在这里可以更新 UI。所有的操作都是在子线程中进行的。

3. 执行 AsyncTask 异步任务

```
new MyDownTask().execute(参数);
```

注意 这 4 个方法只有 doInBackground()是必需的，其他 3 个方法按需要而定。

【任务实施】

1. 任务分析

下载百度新闻的 Logo，需要联网操作，并且是一个耗时操作，AsyncTask 使用封装好的异步消息机制，使用它只需要把内容写在规定的地方即可。

(1) 布局设计一个进度条对话框和一个开始下载按钮。

(2) 编写异步任务，把百度 Logo 图片的地址作为参数传进 execute(IMAGE_PATH)中。

(3) 在异步任务的 doInBackground()里用 HttpURLConnection 进行联网操作，建立连接通道后获取 InputStream 流，从流里取数据并进行进度条的更新。

(4) 在异步任务的 onPreExecute()方法中显示进度条对话框。

(5) 在异步任务的 onProgressUpdate()方法中更新进度。

(6) 在异步任务的 onPostExecute()方法中处理结果，把返回的字节数组封装到 Bitmap 中，然后显示在 ImageView 控件中。

2. 任务实现

(1) 新建工程 Project6_3，编写布局 activity_main.xml，代码如下：

```
<LinearLayout xmlns:android="http://schemas.android.com/apk/res/android"
    xmlns:tools="http://schemas.android.com/tools"
    android:layout_width="match_parent"
    android:layout_height="match_parent"
    android:orientation="vertical"
```

```
    tools:context=".MainActivity" >

    <ImageView
        android:id="@+id/iv"
        android:layout_gravity="center_horizontal"
        android:layout_width="wrap_content"
        android:layout_height="wrap_content"
        android:src="@mipmap/ic_launcher"
        android:layout_weight="1"

        />

    <Button
        android:onClick="click"
        android:layout_width="match_parent"
        android:layout_height="wrap_content"
        android:text="单击获取图片"

        />

</LinearLayout>
```

(2) 修改 MainActivity 代码如下：

```
public class MainActivity extends AppCompatActivity {
    private ImageView iv;
    // 进度条对话框
    private ProgressDialog dialog;
    //图片地址
    private final String IMAGE_PATH="https://box.bdimg.com/static/fisp_static/common/img/searchbox/logo_news_276_88_1f9876a.png";

    @Override
    protected void onCreate(Bundle savedInstanceState) {
        super.onCreate(savedInstanceState);
        setContentView(R.layout.activity_down_pic);

        iv = (ImageView) findViewById(R.id.iv);

        dialog = new ProgressDialog(DownPicActivity.this);
        dialog.setCancelable(false);// 不能移除对话框
        dialog.setTitle("提示");
        dialog.setMessage("下载中，请等待......");
        // 对话框显示
        dialog.setProgressStyle(ProgressDialog.STYLE_HORIZONTAL);
    }

    /*
     * Params 在执行异步任务时需要传入的参数，可用于在后台任务中使用
     * Progress：后台执行时，如果需要在界面上显示当前的进度，这里指定的是进度的单位
     * Result:用于当任务执行完毕后，如果需要对结果进行处理，这里指定的是结果的返回值的类型
     */
    public class MyTask extends AsyncTask<String, Integer, byte[]> {
      //2.显示进度条对话框
        @Override
        protected void onPreExecute() {
            // TODO Auto-generated method stub
            super.onPreExecute();
            dialog.show();
        }

```

```
        // 3.执行后台下载图片的任务
        @Override
        protected byte[] doInBackground(String... params) {
            ByteArrayOutputStream out=null;
            HttpURLConnection conn=null;
            byte[] result = null;
            InputStream inputStream = null;
            //字节数组输出流
            ByteArrayOutputStream outputStream = new ByteArrayOutputStream();
            try {
                URL url = new URL(params[0]);
                conn=(HttpURLConnection) url.openConnection();
                conn.setReadTimeout(5000);
                conn.setConnectTimeout(5000);//设置连接超时
                conn.setRequestMethod("GET");//用 GET 的方式发请求
                conn.setRequestProperty("accept", "*/*");
                conn.setRequestProperty("connection", "Keep-Alive");
                //文件长度
                long file_length = conn.getContentLength();
            if(conn.getResponseCode()==200) {
                //获取字节流
                inputStream = conn.getInputStream();
                int len = 0;
                int total_length = 0;
                byte[] temp = new byte[1024];
                //从字节流中读取数据
                while ((len = inputStream.read(temp)) != -1) {
                    total_length += len;
                    //计算进度条
                    int progress_value = (int) ((total_length / (float) file_length) * 100);
                    //4.更新进度值
                    publishProgress(progress_value);
                    //输出到字节输出流中
                    outputStream.write(temp, 0, len);
                }
                //5.返回存放下载好的图片字节数组
                result = outputStream.toByteArray();
            }
        } catch (Exception e) {
            e.printStackTrace();
        }finally{
            if(conn!=null)
                conn.disconnect();
        }
            return result;
        }

       //6.更新进度条
        @Override
        protected void onProgressUpdate(Integer... values) {
            super.onProgressUpdate(values);
            dialog.setProgress(values[0]);
        }

        // 7.任务结束后的处理
        @Override
        protected void onPostExecute(byte[] result) {
            super.onPostExecute(result);

```

```
            // 把返回的结果从字节数组转换成Bitmap
            Bitmap bm = BitmapFactory.decodeByteArray(result, 0, result.length);
            //8.显示在控件上
            iv.setImageBitmap(bm);
            //9.关闭进度条
            dialog.dismiss();
            if (result != null) {
                Toast.makeText(DownPicActivity.this, "下载成功", Toast.LENGTH_SHORT)
                        .show();
            } else {
                Toast.makeText(DownPicActivity.this, "下载失败", Toast.LENGTH_SHORT)
                        .show();

            }
        }
    }
   //按钮的点击事件
    public void click(View view) {
        //1.执行异步任务
        new MyTask().execute(IMAGE_PATH);

    }

}
```

(3) 在清单文件中添加访问网络的权限，代码如下：

```
<uses-permission android:name="android.permission.INTERNET"/>
```

至此就完成了后台下载的功能。完整代码可查阅随书资料中的 Project6_3 工程源代码。

3. 运行结果

运行程序，结果如图 6-18 所示。

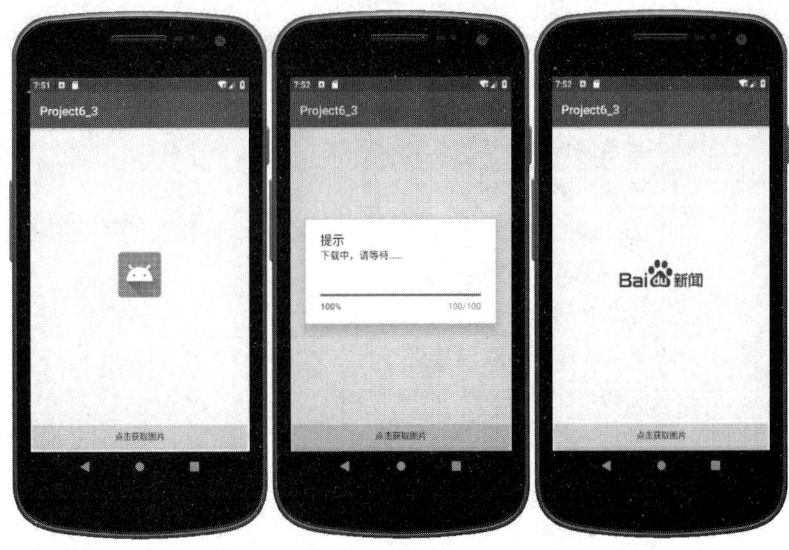

图 6-18 下载百度新闻 Logo

上面的代码用到了 HttpUrLConnec 访问网络，网络的相关内容在项目 10 中介绍。

任务 4　传感器数据的实时更新

【任务描述】

初始项目 Project6_4Init 已经实现了引导页、用户注册、登录、室内环境采集系统界面和园区监控系统界面的界面效果。本任务在初始项目 Project6_4Init 的基础上实现温湿度、光照、烟雾数据的实时更新，效果如图 6-19 所示。任务清单如表 6-4 所示。

图 6-19　程序运行效果

表 6-4　任务清单

任务课时	4 课时	任务组员数量	建议 3 人
任务组采用设备	四输入模块一个 温湿度传感器一个 光照传感器一个 ZigBee 协调器一个 ADAM-4150 模块一个 烟雾传感器一个 RS485 转 RS232 转接头一个 公母串口线两条 PC 一台 工控平板		

【拓扑图】

本任务的拓扑图如图 6-20 所示。

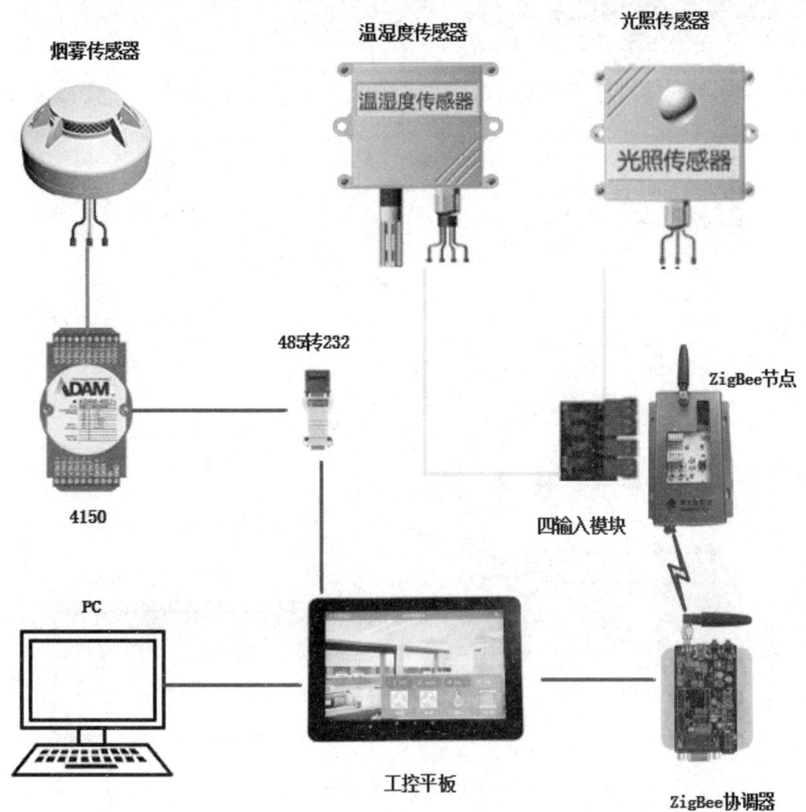

图 6-20　拓扑图

注意：本任务的烟雾传感器接 4150 模块的 DI4 口，温湿度传感器、光照传感器分别接四输入模块的 IN2、IN3、IN1 口。

【知识解析】

(1) 针对连接在数字量采集器 ADAM-4150 上的设备、通过四输入连接的设备、ZigBee 方式组网的设备，提供了\armeabi-v7a\libserial_port.so、hardware.jar 和 hardware-sources.jar 这 3 个接口文件。\armeabi-v7a\libserial_port.so 是底层操作串口的动态连接库，在此不做讲解。本任务主要用 hardware.jar 和 hardware-sources.jar 包类中的方法获取传感器的数据。

① 把 armeabi-v7a 文件夹、hardware.jar 和 hardware-sources.jar 放入 libs 目录下，并在 app 的 build.gradle 文件中的 android 闭包中添加代码：

```
sourceSets {
    main {
        jniLibs.srcDirs = ['libs']
    }
}
```

② hardware.jar 运行环境是 Java 8，需要在 app 的 build.gradle 文件中的 android 闭包中添加代码，以支持 Java 8 中的新特性使用：

```
compileOptions {
    targetCompatibility 1.8
```

```
    sourceCompatibility 1.8
}
```

③ 单击同步，这样程序就可以使用这 3 个类库文件中的方法了，如图 6-21 所示。

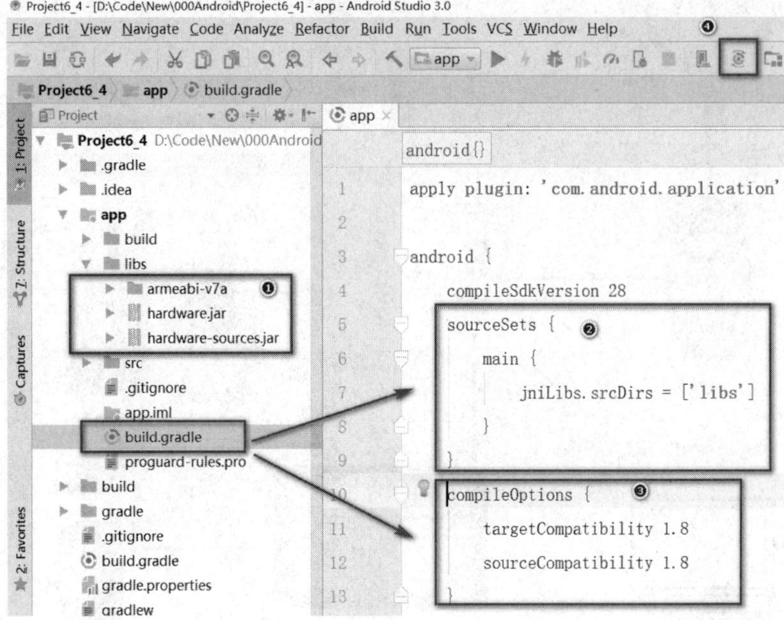

图 6-21　添加库文件到工程中

(2) ADAM-4150 设备控制说明。

以下是串口直连情况下的方法。

① 初始化，代码如下：

```
Modbus4150 modbus4150 = new Modbus4150(DataBusFactory.newSerialDataBus(2, 9600));
```

代码中，2 表示连接的串口 2(根据实际连接情况修改)，9600 表示比特率。

② 获取数据，代码如下：

```
modbus4150.getVal(0, new MdBus4150SensorListener() {
    @Override
    public void onVal(int i) {
    }
});
```

代码中，0 表示传感器连接 DI0(根据实际连接情况修改)，MdBus4150SensorListener 是 4150 传感器数据回调的接口，数据会返回到 onVal()方法的参数 i 中。连接的如果是烟雾传感器，1 表示有烟，0 表示无烟。红外对射 1 表示无人，0 表示有人。微动开关 1 表示被触发。

③ 打开继电器，代码如下：

```
modbus4150Serial.openRelay(1, new MdBus4150RelayListener() {
    @Override
    public void onCtrl(boolean isSuccess) {
        System.out.println(isSuccess);
    }
});
```

代码中，1 表示继电器连接 DO1(根据实际连接情况修改)，isSuccess 表示是否操作成功。

④ 关闭继电器，代码如下：

```
modbus4150Serial.closeRelay(1, new MdBus4150RelayListener() {
    @Override
    public void onCtrl(boolean isSuccess) {
        System.out.println(isSuccess);
    }
});
```

代码中，1 表示继电器连接 DO1(根据实际连接情况修改)，isSuccess 表示是否操作成功。

(3) ZigBee 设备数据获取说明。

以下是串口直连情况下的方法。

① 初始化，代码如下：

```
Zigbee zigBee = new Zigbee(DataBusFactory.newSerialDataBus(2, 38400));
```

代码中，2 表示连接的串口 2(根据实际连接情况修改)，38400 表示比特率。

② 获取四输入数据，代码如下：

```
double[] vals = zigBee .getFourEnter();
```

vals 数组分别表示四输入从 1 到 4 的端口的电流值，需要根据实际接入情况转化为实际传感值，使用 FourChannelValConvert 类的静态方法来转换。各种转换代码如下：

```
//温度转换
public static double getTemperature(double value)
说明：如温度传感器接的是 in1，则对应的电流值为 vals[0]，温度值为 FourChannelValConvert.getTemp(vals[0])

//湿度转换
public static double getHumidity(double value)

//光照转换
public static double getLight(double value)

//噪声转换
public static double getNoice(double value)

//二氧化碳转换
public static double getCO2(double value)
```

【任务实施】

1. 任务分析

根据【知识解析】中的 jar 包说明，分别封装 ZigBee 设备和 ADAM-4150 模块的操作类，然后通过 Handler 每隔 1s 调用 Runnable 的方式实现数据的实时更新。

(1) 引入 ZigBee 设备和 ADAM-4150 模块的 jar 包以及串口操作的 so 文件并进行配置。

(2) 配置 Java 8 的运行环境。

(3) 封装(串口直连四输入)ZigBee 设备的操作类。

(4) 封装 ADAM-4150 模块的操作类。

(5) 编写室内环境采集系统对应的 Fragment 类 EnvironmentFragment.java 代码，实现传

感器数据的实时更新。

2. 任务实现

(1) 把随书资料下的 armeabi-v7a 文件夹、hardware.jar 和 hardware-sources.jar 放入 Project6_4Init 工程的 libs 目录下，并在 app 的 build.gradle 文件的 android 闭包中添加如下代码：

```
sourceSets {
    main {
        jniLibs.srcDirs = ['libs']
    }
}
```

(2) 在 app 的 build.gradle 文件的 android 闭包中添加配置 Java 8 运行环境的代码，然后点击同步：

```
compileOptions {
    targetCompatibility 1.8
    sourceCompatibility 1.8
}
```

(3) 在 com.project.sensor.project6_4.base 包下新建 ZigBee 设备的操作类 ZigBeeOperator.java。代码如下：

```
public class ZigbeeOperator {
    private Zigbee zigBee;
    private double[] zigBeeVals;
    private DecimalFormat df;

    //打开串口并获得四输入数据
    public void openPort(int com, int baudRate) {
        zigBee = new Zigbee(DataBusFactory.newSerialDataBus(com, baudRate));
        df = new DecimalFormat("0.00");
    }

    public Double getTemp() {
        try {
            zigBeeVals = zigBee.getFourEnter();
            if(zigBeeVals==null){
                //如果获取到四输入的数组为空，返回温度值为0.00
                return 0.00;
            }
            //温度接in2口
            Double temp = FourChannelValConvert.getTemperature(zigBeeVals[1]);
            //保留两位小数
            Double tempValue = Double.valueOf(df.format(temp));
            return tempValue;
        } catch (Exception e) {
            e.printStackTrace();
        }
        return 0.00;
    }

    public double getHumi() {
        try {
            zigBeeVals = zigBee.getFourEnter();
            if(zigBeeVals==null){
                //如果获取到四输入的数组为空，返回湿度值为0.00
```

```
                return 0.00;
            }
            //湿度接 in3 口
            Double humi = FourChannelValConvert.getHumidity(zigBeeVals[2]);
            //保留两位小数
            Double humiValue = Double.valueOf(df.format(humi));
            return humiValue;
        } catch (Exception e) {
            e.printStackTrace();
        }
        return 0.00;
    }

    public double getLight() {
        try {
            zigBeeVals = zigBee.getFourEnter();
            if(zigBeeVals==null){
                //如果获取到四输入的数组为空，返回光照值为 0.00
                return 0.00;
            }
            //光照接 in1 口
            double light = FourChannelValConvert.getLight(zigBeeVals[0]);
            //保留两位小数
            Double lightValue = Double.valueOf(df.format(light));
            return lightValue;
        } catch (Exception e) {
            e.printStackTrace();
        }

        return 0.00;
    }

    //关闭串口，释放资源
    public void closePort() {
        if (zigBee != null) {
            zigBee.stopConnect();
        }
    }

}
```

(4) 在 com.project.sensor.project6_4.base 包下新建 ADAM-4150 模块的操作类 Adam4150.java。代码如下：

```
public class Adam4150 {
    private Modbus4150 modbus4150;
    private int hasPerson=1;//默认无人
    private int switchState;
    private int hasSmoke;

    public void openPort(int com, int baudRate) {
        modbus4150 = new Modbus4150(DataBusFactory.newSerialDataBus(com, baudRate));
    }

    //获得烟雾的数据
    public String getSmoke(int port) {
        try {
            modbus4150.getVal(port, new MdBus4150SensorListener() {
                @Override
                public void onVal(int i) {
```

```
                    hasSmoke = i;
                }
            });
        } catch (Exception e) {
            e.printStackTrace();
        }
        if (hasSmoke == 1) {
            return "有烟";
        }
        return "无烟";
    }
    //获得红外对射的数据
    public String getInfrared(int port) {
        try {
            modbus4150.getVal(port, new MdBus4150SensorListener() {
                @Override
                public void onVal(int i) {
                        hasPerson = i;
                }
            });
        } catch (Exception e) {
            e.printStackTrace();
        }
        if (hasPerson == 0) {
            return "有人";
        }
        return "无人";
    }
    //获得微动开关状态
    public boolean getMicroSwitch(int port) {
        try {
            modbus4150.getVal(port, new MdBus4150SensorListener() {
                @Override
                public void onVal(int i) {
                        switchState=i;
                }
            });
        } catch (Exception e) {
            e.printStackTrace();
        }
        if (switchState == 0) {
            return false;
        }
        return true;
    }

    //开继电器
    public void openRelay(final int port) {
        try {
            modbus4150.openRelay(port, isSuccess -> System.out.println("开 DO" + port + "对应的继电器"+isSuccess));
        } catch (Exception e) {
            e.printStackTrace();
        }
    }


    //关继电器
    public void closeRelay(final int port) {
        try {
```

```
                modbus4150.closeRelay(port, isSuccess -> System.out.println("关 DO" + port + "对应的继电器"+isSuccess));
            } catch (Exception e) {
                e.printStackTrace();
            }
        }

        //关闭串口，释放资源
        public void closePort() {
            modbus4150.stopConnect();
        }
}
```

(5) 修改自定义的 Fragment 基类的代码，在其中添加 SharePreferenceUtil 和 Adam50 的实例，BaseFragment 的代码如下：

```
public abstract class BaseFragment extends Fragment {
    public static SharePreferenceUtil sharePreferenceUtil;
    public Adam4150 adam4150;

    @Override
    public void onActivityCreated(@Nullable Bundle savedInstanceState) {
        super.onActivityCreated(savedInstanceState);
        if (adam4150 == null) {
            adam4150 = new Adam4150();
            adam4150.openPort(Constants.COMS[0], 9600);
        }
        if (sharePreferenceUtil == null) {
            sharePreferenceUtil = SharePreferenceUtil.getInstant(getActivity());
        }
    }

}
```

(6) 在室内环境采集系统对应的 Fragment 类 EnvironmentFragment.java 初始化的方法 initView()中添加初始化组件和打开 ZigBee 协调器连接的串口并初始化，执行获取数据的 Runnable，代码如下：

```
      //定义全局变量 ZigBeeOperator
       private ZigbeeOperator zigBeeOperator;
       //private void initView() {
       tvTemp = getActivity().findViewById(R.id.tv_temp);
       tvHumi = getActivity().findViewById(R.id.tv_humi);
       tvLight = getActivity().findViewById(R.id.tv_light);
       tvSmoke = getActivity().findViewById(R.id.tv_smoke);
       //实例化 ZigBeeOperator 对象
       zigBeeOperator = new ZigbeeOperator();
       //打开 ZigBee 协调器连接的串口并初始化
       zigBeeOperator.openPort(Constants.COMS[1], 38400);
       // 执行更新数据的 Runnable
           handler.post(runnable);
       //}
```

(7) 在 initView()方法下方添加更新数据的 Runnable，实现温湿度、光照、烟雾、文本的实时更新，代码如下：

```
//更新数据的 Runnable
private Runnable runnable=new Runnable() {
     @Override
```

```
        public void run() {
            tempData = zigBeeOperator.getTemp();
            humiData = zigBeeOperator.getHumi();
            lightData = zigBeeOperator.getLight();
            smokeData = adam4150.getSmoke(1);//烟雾接 DI1
            tvTemp.setText(tempData + "℃");
            tvHumi.setText(humiData + "%RH");
            tvLight.setText(lightData + "Lx");
            tvSmoke.setText(smokeData);
            handler.postDelayed(runnable,1000);//1s 后调用 Runnable
        }
    };
```

至此，便完成了传感器数据的实时更新。完整代码可查阅随书资料中的 Project6_4 工程源代码。

3. 运行结果

运行程序，结果如图 6-22 所示。

图 6-22　运行程序结果

项 目 总 结

本项目主要讲解了 Android 的多线程开发、Handler 消息机制以及 AsyncTask 异步任务，并通过 Handler 和 Runnable 实现了传感器数据的实时更新。

思考与练习

6-1　简述消息机制。

6-2　简述 Handler 是如何发送消息和接收消息的。

项目7 服务与广播

【项目描述】

在前面的项目中，读者基本掌握了 Activity 的使用方法。Activity 是 Android 四大核心组件之一，此外，还有 ContentProvider(内容提供者)、BroadcastReceiver(广播)和 Service(服务)。

BroadcastReceiver 用于在不同的组件之间传递消息，只要将消息内容封装到 Intent 对象中，然后用广播传递它即可。

Service 不提供界面，用于在后台完成各种耗时的逻辑，或者去执行某些长期运行的任务，当退出程序的时候，程序还能在后台继续运行。

本项目通过对两个任务(见图 7-1)的学习，针对 Android 开发中必须掌握的广播和服务进行讲解，最后实现室内环境采集系统的功能，检测到有烟的时候点亮三色灯中的橙灯，没烟的时候橙灯灭；在登录系统成功的时候启动服务；在服务中实时判断微动开关是否被触发，如果微动开关被触发，灭设备上的红灯和橙灯，亮绿灯，同时发送广播，关闭界面上的红灯和橙灯。

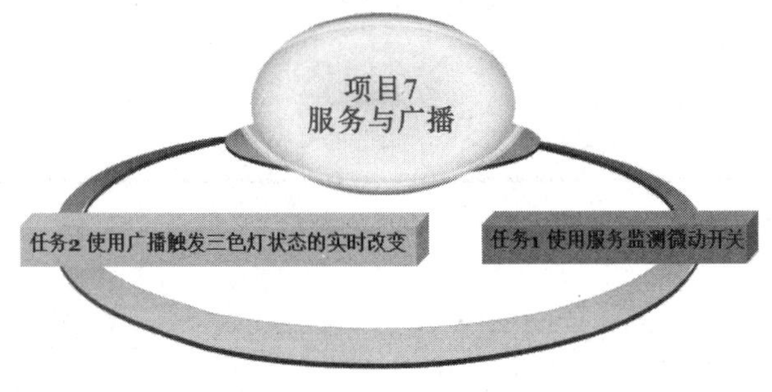

图 7-1　项目 7 任务图

【学习目标】

知识目标：

- 会以静态和动态方式注册广播。
- 会发送和接收自定义的广播。
- 会两种方式启动和停止服务。

技能目标：

- 能在项目中灵活运用广播。
- 能实现后台运行。

任务 1　使用服务监测微动开关

【任务描述】

本任务通过实现室内环境采集系统，实时检测到有烟的时候点亮三色灯中的橙灯，没烟的时候橙灯灭；在登录系统成功的时候启动服务；在服务中实时判断微动开关是否被触发，如果微动开关被触发，灭设备上的红灯和橙灯，亮绿灯。任务清单如表 7-1 所示。

表 7-1　任务清单

任务课时	4 课时	任务组员数量	建议 3 人
任务组采用设备	四输入模块一个 温湿度传感器一个 光照传感器一个 ZigBee 协调器一个 ZigBee 智能节点盒一个 ADAM-4150 一个 烟雾传感器一个 微动开关一个 三色灯一个 RS485 转 RS232 转换头一个 公母串口线两条 PC 一台 工控平板		

【拓扑图】

本任务的拓扑结构图如图 7-2 所示。

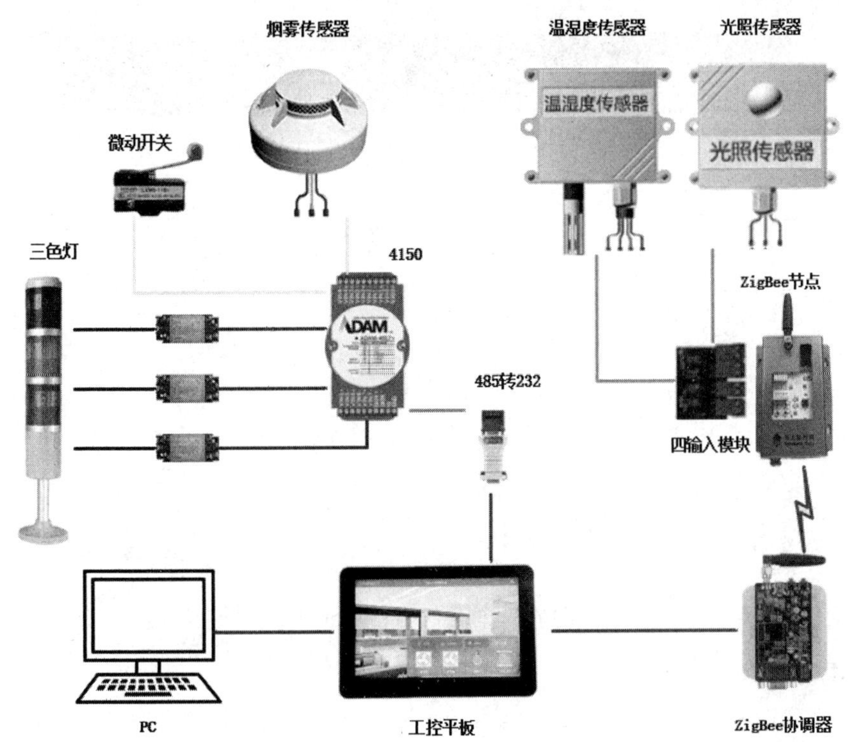

图 7-2 拓扑图

注意：本任务的三色灯-红灯、三色灯-绿灯、三色灯-橙灯所接继电器分别接 ADAM-4150 模块的 DO3、DO4、DO5 口，烟雾传感器接 ADAM-4150 模块的 DI4 口，微动开关接 ADAM-4150 模块的 DI5 口，温湿度传感器、光照传感器分别接四输入模块的 IN2、IN3、IN1 口。

【知识解析】

1. 服务的概念

服务即 Service，Service 是 Android 中自带的一个类，是 Android 四大核心组件之一。服务的运行不依赖于任何用户界面，不与用户产生 UI 交互，它主要用于在后台处理一些耗时的逻辑，或者去执行某些需要长期运行的任务。即使程序被切换到后台，或者用户开启另一个应用程序，服务仍然能够正常运行。其他的应用组件可以启动 Service，可以与 Service 绑定并与之交互，甚至是跨进程通信，如 Service 可以在后台执行网络请求、播放音乐、执行文件读写等操作。

2. 定义一个服务

至此已经知道了什么是服务，那么如何去定义一个服务呢？

(1) 新建工程 Demo7_1，在 Demo7_1 中右击 com.example.demo.demo7_1，选择快捷菜单中的 New 命令，再选择 Service→Service，弹出图 7-3 所示的窗口。

(2) 可以看到，服务默认被命名为 MyService，Exported 属性表示是否允许除当前程序以外的其他程序访问本服务，Enabled 属性表示是否启用这个服务。勾选这两个属性，单击 Finish 按钮完成创建。

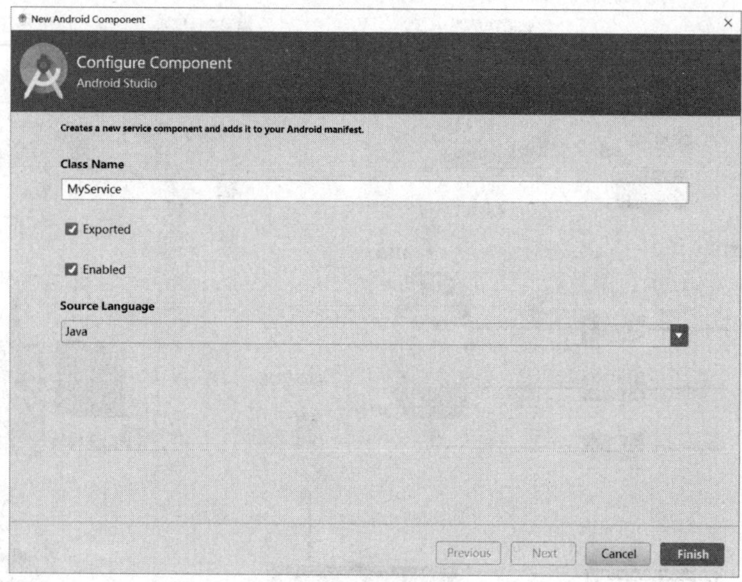

图 7-3　创建服务 MyService

(3) MyService 中的代码如下：

```
public class MyService extends Service {
    public MyService() {
    }
    @Override
    public IBinder onBind(Intent intent) {
        // TODO: Return the communication channel to the service.
        throw new UnsupportedOperationException("Not yet implemented");
    }
}
```

从中可以看出 MyService 继承自 Service 类，表明 MyService 是一个服务。MyService 类中除了一个构造方法外还有一个 onBind()方法，这个方法是 Service 中唯一一个抽象方法，所以必须在子类中实现。

(4) 服务定义好了，接下来需要在服务中处理一些事情，处理事情的逻辑需要重写 Service 的其他方法，代码如下：

```
public class MyService extends Service {
    ...
    @Override
    public void onCreate() {
        super.onCreate();
    }

    @Override
    public int onStartCommand(Intent intent, int flags, int startId) {
        return super.onStartCommand(intent, flags, startId);
    }

    @Override
    public void onDestroy() {
        super.onDestroy();
    }
}
```

上面的代码中，重写了 onCreate()、onStartCommand()、onDestroy()这 3 个方法，它们是服务中最常用的 3 个方法。onCreate()会在服务创建的时候被调用，onStartCommand()会在服务启动的时候被调用，onDestroy()方法会在服务销毁的时候被调用。

通常，如果希望服务已启动马上去执行某个动作，可以将代码写在 onStartCommand()方法中。当服务销毁的时候，需要在 onDestroy()方法中回收那些不使用的资源。值得注意的是，每个 Service 都必须在 AndroidManifest.xml 清单文件中进行注册才能生效。其实，这是四大组件共同的特点。通过前面的方法去创建 Service，Android Studio 已经帮我们将这一步完成了，代码如下：

```
<?xml version="1.0" encoding="utf-8"?>
<manifest xmlns:android="http://schemas.android.com/apk/res/android"
        package="com.example.demo.demo7_2">
    <application
        android:allowBackup="true"
        android:icon="@mipmap/ic_launcher"
        android:label="@string/app_name"
        android:roundIcon="@mipmap/ic_launcher_round"
        android:supportsRtl="true"
        android:theme="@style/AppTheme">
       ...
        <service
            android:name=".MyService"
            android:enabled="true"
            android:exported="true">
        </service>
    </application>
</manifest>
```

这样，一个服务就定义好了。

3. Service 的两种启动方式

Service 有两种启动方式：①通过 startService()来启动；②通过 bindService()来启动。但是它们的使用场景有所不同。

1) startService 方式

其他组件可以调用 startService()方法启动一个 Service。当应用组件(如 Activity)通过调用 startService()启动服务时，服务即处于“启动”状态。一旦启动，服务将一直在后台运行，即使启动服务的组件销毁也不受影响，除非手动调用才能停止服务。

这种方式启动和停止服务代码如下：

```
//启动服务参数一 ：上下文  参数二：要启动的服务类对应的字节码
    Intent start = new Intent(MainActivity.this,MyService.class);
    startService(start);
//停止服务参数一 ：上下文  参数二：要启动的服务字节码
    Intent stop = new Intent(MainActivity.this,MyService.class);
    stopService(stop);
```

startService 启动方式对应的 Service 生命周期如图 7-4 所示。

说明：通过 startService 启动服务，若服务未启动，会先执行 onCreate 函数(若服务已启动则不执行此函数)，再执行 onStartCommand 函数。由此可知，多次调用 startService 传入相同参数不会启动多个服务(onStartCommand 函数会执行多次)，所以最终只需要调用一次

stopService 或 stopSelf 函数停止服务；可以将 Service 的处理逻辑放入 onStartCommand 函数中。服务一直运行，在程序退出后服务也不会停止，直到 stopService 或 stopSelf 函数被调用为止，当然可能被系统回收。

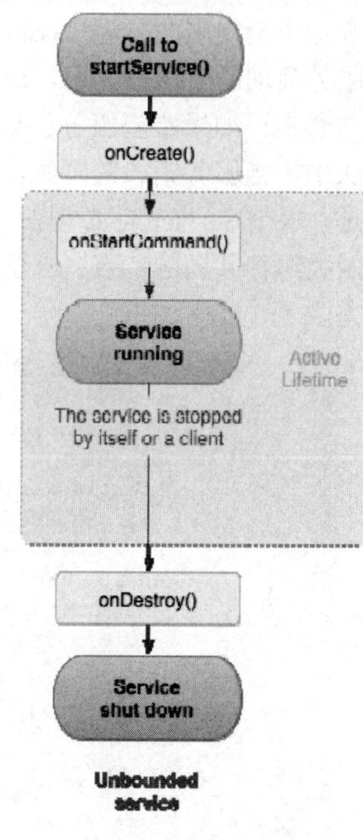

图 7-4　服务的生命周期

接下来通过 startService 方式启动服务播放音乐。

(1) 在工程 Demo7_1 的 activity_main.xml 文件中添加两个按钮，用于启动服务和停止服务，代码如下：

```
<?xml version="1.0" encoding="utf-8"?>
<LinearLayout xmlns:android="http://schemas.android.com/apk/res/android"
              android:orientation="vertical"
              android:layout_width="fill_parent"
              android:layout_height="fill_parent"
    >
    <Button
        android:id="@+id/startService"
        android:layout_width="match_parent"
        android:layout_height="wrap_content"
        android:text="start 方式启动服务"
        android:onClick="click"
        />
    <Button
        android:text="start 方式停止服务"
        android:id="@+id/stopService"
        android:layout_width="match_parent"
        android:layout_height="wrap_content"
        android:onClick="click"
        />
</LinearLayout>
```

(2) 将随书资料中歌曲文件的 yypl.mp3 文件复制到工程的 res→raw 文件夹下。

(3) 创建 MyService 类，在 onCreate()方法中初始化 MediaPlayer，在 onStartCommand()方法中播放音乐，在 onDestroy()方法中停止播放。代码如下：

```
public class MyService extends Service {
    private static final String TAG = "MyService";
    private MediaPlayer mp; // 媒体播放器对象
    //Service 创建时调用
    @Override
    public void onCreate() {
        Log.e(TAG, "Service onCreate--->");
        initMediaPlayer(); // 初始化音乐
    }
    @Override
    public int onStartCommand(Intent intent, int flags, int startId) {
        Log.e(TAG, "Service onStartCommand--->");
        play();
        return super.onStartCommand(intent, flags, startId);
    }

    //当 Service 不再使用时调用
    public void onDestroy() {
        Log.e(TAG, "Service onDestroy--->");
        stop();
    }
```

```

    @Nullable
    @Override
    public IBinder onBind(Intent intent) {
        return null;
    }

    /**
     * 初始化音乐播放器
     */
    private void initMediaPlayer() {
        try {
            //通过音频资源的 id 来创建一个 MediaPlayer 实例
            mp = MediaPlayer.create(MyService.this, R.raw.yypl);
            //设置可以重复播放
            mp.setLooping(true);
        } catch (Exception e) {
            e.printStackTrace();
        }
    }
    /**
     * 播放音乐
     */
    private void play() {
        //如果 MediaPlayer 不为空，播放音乐
        if (mp != null) {
            mp.start();
            //播放音乐时发生错误的事件处理
            mp.setOnErrorListener(new MediaPlayer.OnErrorListener() {
                public boolean onError(MediaPlayer mp, int what, int extra) {
                    try {
                        //释放资源
                        mp.release();
                    } catch (Exception e) {
                        e.printStackTrace();
                    }
                    return false;
                }
            });
        }
    }
    private void stop() {
        //如果 MediaPlayer 不为空，停止播放并释放资源
        if (mp != null) {
            mp.stop();
            //释放资源
            mp.release();
        }
    }
}
```

(4) 在 MainActivity 实现在“start 方式启动服务”按钮的点击事件中启动服务，在“start 方式停止服务”按钮的点击事件中停止服务。代码如下：

```
public class MainActivity extends Activity {
    public Intent intent = new Intent();

    @Override
    protected void onCreate(Bundle savedInstanceState) {
        super.onCreate(savedInstanceState);
```

```
        setContentView(R.layout.activity_main);
        // 设置 Class 属性
        intent.setClass(MainActivity.this, MyService.class);
    }


    public void click(View v) {
        switch (v.getId()) {
            case R.id.startService: //启动服务
                startService(intent);
                break;
            case R.id.stopService: //停止服务
                Log.i("TAG", "停止服务......");
                stopService(intent);
                break;
                }
    }

    @Override
    protected void onDestroy() {
        super.onDestroy();
        Log.i("TAG", "停止绑定方式的服务......");
    }

}
```

(5) 运行程序，结果如图 7-5 所示。单击“绑定方式启动服务”按钮，再单击“播放音乐”按钮开始播放音乐，单击“接触绑定”按钮音乐停止。

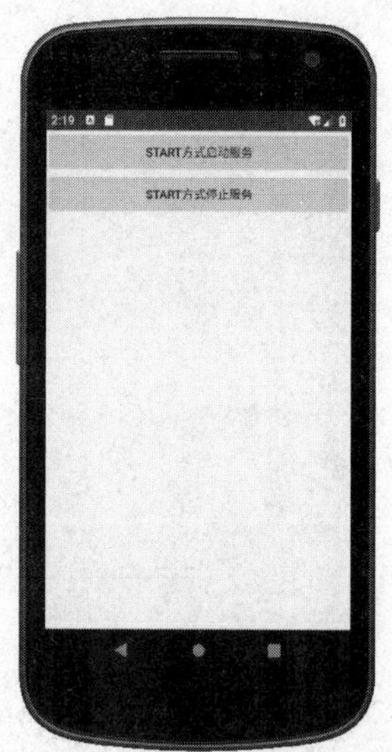

图 7-5 程序结果图

2) bindService 方式

在前一个案例中学习了以 startService 方式启动服务，虽然服务是启动了，但好像跟 Activity 的关系不大，只是进行了服务的启动，然后 Service 就忙自己的事去了。而通常希望在 Activity 上可以指定让 Service 去执行什么任务，如下载电视剧，用户要下载哪一集呢？在 Activity 上选择好后单击“下载”按钮就可以启动一个 Service 去后台下载，要实现这样的效果，就得想办法让 Activity 和 Service 建立关联。

前面的案例中有一个 onBind()方法没有用到，这个方法就是用于和 Activity 建立关联的。要想实现 Activity 指挥 Service 做具体的事情，必须让这个方法绑定 Activity 需要让 Service 做事情的代理者，然后在 Activity 中与 Service 通过 ServiceConnection 连接器连接起来，下面就来实现这个功能。

(1) 在 Service 类中返回服务要做的代理者。

前面案例中的 MyService 类的 onBind()方法返回了 null：

```
public IBinder onBind(Intent intent) {
        Log.i("TAG","MyService onBind......");
        return null;
}
```

要想服务与 Activity 建立关联，需要在 onBind()方法中返回一个 IBinder 对象，这个 IBinder 对象就是一个代理者，想让服务做什么事情，就交给这个代理者去处理。

IBinder 是一个接口，它有一个实现类 Binder，可以写一个类继承自 Binder，在类中写要让服务执行的方法即可。

我们可以在 Binder 的实现类中写一个播放音乐的方法，用于播放音乐，然后重写 onUnbind()方法用于解除与 Activity 的绑定。在 Demo7_1 的 MyService 中进行改写，代码如下：

```
public class MyService extends Service {
    private static final String TAG = "MyService";
    private MediaPlayer mp; //媒体播放器对象
    //绑定服务时调用
    @Override
    public IBinder onBind(Intent intent) {
        Log.e(TAG, "Service onBind--->");
        return new MyBinder();
    }
    //新增的服务提供的方法可以有多个
    public class MyBinder extends Binder {
        public void startMusic(){
            play();
        }
    }
    //当解除绑定时调用
    public boolean onUnbind(Intent intent) {
        Log.e(TAG, "Service onUnbind--->");
        return super.onUnbind(intent);
    }

  其他略......
}
```

(2) 绑定方式启动和停止 Service。

绑定方式启动服务，要是用 bindService(Intent service, ServiceConnection conn, int flags)方法，该方法接收 3 个参数：第一个参数是 Intent 对象；第二个参数是一个 ServiceConnection 的实例；第三个参数是一个标志位，一般传入 BIND_AUTO_CREATE，表示在 Activity 和 Service 建立关联后自动创建 Service，这会使得 MyService 中的 onCreate()方法得到执行，但 onStartCommand()方法不会执行。

参数二 ServiceConnection 是一个接口，接口中有两个方法，分别是 onServiceConnected()和 onServiceDisconnected()，这两个方法分别会在 Activity 与 Service 建立关联和解除关联的时候调用。需要写一个类实现 ServiceConnection 接口，重写的在 onServiceConnected()方法中，将参数 IBinder 向下转型得到 MyService.MyBinder 实例。然后就可以在 Activity 中根据具体场景来调用 MyBinder 中公有方法。

在 MainActivity 中添加代码如下：

```
public class MainActivity extends Activity {
    public Intent intent = new Intent();
    private MyService.MyBinder mybinder;
    private MyServiceConnection connection;
    其他代码略......
    public void click(View v) {
       其他代码略......
        switch (v.getId()) {
                case R.id.bindService://绑定方式启动服务
                 Log.i("TAG", "绑定方式启动服务......");
```

```
                connection = new MyServiceConnection();
                // 绑定 Service
                bindService(intent, connection, Service.BIND_AUTO_CREATE);
                break;
            case R.id.unbindService://解除绑定的服务
                Log.i("TAG", "停止绑定方式的服务......");
                if (connection != null) {
                    unbindService(connection);
                    connection = null;
                }
                break;
            case R.id.playMusic://单击可以启动后台的听音乐功能
                if(mybinder!=null){
                mybinder.startMusic();
                }
        }
    }
    private class MyServiceConnection implements ServiceConnection {
        @Override
        public void onServiceConnected(ComponentName name, IBinder service) {
        //通过 IBinder 的内部类返回 Service 的实例，Activity 通过这个实例调用服务的方法
            mybinder = (MyService.MyBinder) service;
        }

        @Override
        public void onServiceDisconnected(ComponentName name) {
            mybinder=null;
        }
    }

}
```

运行程序，结果如图 7-6 所示。单击“绑定方式启动服务”按钮，再单击“播放音乐”按钮开始播放音乐，单击“解除绑定”按钮音乐停止。

(3) 绑定方式启动服务的生命周期。

bindService 启动方式对应的 Service 生命周期如图 7-7 所示。

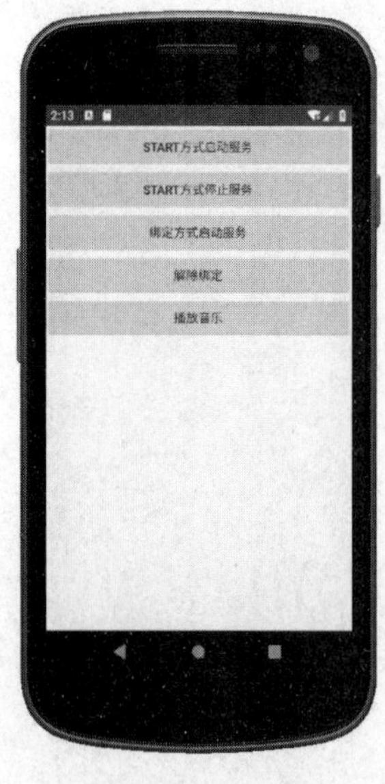

图 7-6 程序结果图

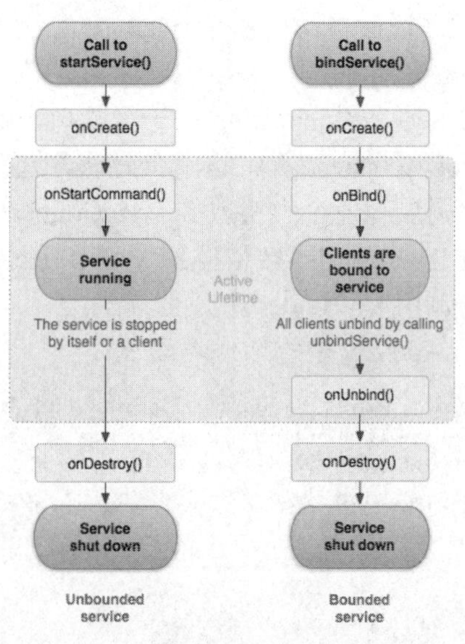

图 7-7 两种服务的生命周期对比

通过 bindService 绑定服务，若服务未启动，会先执行 Service 的 onCreate()函数，再执行 onBind()函数，最后执行 ServiceConnection 对象的 onServiceConnected()函数。若服务已启动但尚未绑定，则先执行 onBind()函数，再执行 ServiceConnection()对象的 onServiceConnected()函数。若服务已绑定成功，则直接返回。这里不会自动调用 onStartCommand()函数。

通过 unbindService 解除绑定服务，若已绑定成功，会先执行 Service 的 onUnbind()函数，再执行 onDestroy()函数，注意这里不会执行 ServiceConnection 对象的 onServiceDisconnected()函数，因为该函数是在服务所在进程被杀死或是死机时被调用。

【任务实施】

1. 任务分析

本任务在 Project7_1Init 基础上实现。初始项目 Project7_1Init 是项目 6 任务 4 完成后的代码。

(1) 在室内环境采集对应的 Fragment 类 EnvironmentFragment 中实时判断，有烟雾的时候橙灯亮，没烟雾的时候橙灯灭。

(2) 在 Service 中实时判断，如果微动开关被触发，灭红灯和橙灯，亮绿灯。

(3) 在 AndroidManifest.xml 清单文件中注册 Service。

(4) 登录成功的时候启动 Service。

2. 任务实现

(1) 在 EnvironmentFragment.java 更新传感器数据 Runnable 的 run()方法中，添加以下代码，如果有烟雾并且 SharePreferences 中橙灯的标记为 false，亮橙灯，同时改变标记状态；如果没烟雾并且 SharePreferences 中橙灯的标记为 true，灭橙灯，同时改变标记状态。

```
if(smokeData.equals("有烟")&&!sharePreferenceUtil.getIsOrangeOn()){
                //亮三色灯的橙灯
                adam4150.openRelay(Constants.RELAYPORTS[5]);
                //设置 SharePreferences 中橙灯标记状态为 true
                sharePreferenceUtil.setIsOrangeOn(true);
                //切换图片
                cbSmoke.setChecked(true);

            }else if(smokeData.equals("无烟")&&sharePreferenceUtil.getIsOrangeOn()){
                //灭三色灯中的橙灯
                adam4150.closeRelay(Constants.RELAYPORTS[5]);
                //设置 SharePreferences 中橙灯标记状态为 false
                sharePreferenceUtil.setIsOrangeOn(false);
                //切换图片
                cbSmoke.setChecked(false);
            }
```

(2) 在 com.newland.smarkpark.service 包下创建自定义的 Service 类 MyService.java，在 onCreate()方法中启动一个线程，在线程中每隔 0.5 秒判断一次，如果微动开关被触发，灭红灯和橙灯，亮绿灯，代码如下：

```
/**
 * 后台服务
 */
public class MyService extends Service {
```

```
    private boolean isRun = true;
    //根据对应的警报，创建对应的提示框
    private Adam4150 adam4150;
    private Intent intent;
    private SharePreferenceUtil sharePreferenceUtil;
    private Boolean isRedOn, isOrangeOn;//标记红灯和橙灯的当前亮灭状态

    @Override
    public void onCreate() {
        super.onCreate();
        sharePreferenceUtil = SharePreferenceUtil.getInstant(this);
        adam4150 = new Adam4150();
        adam4150.openPort(Constants.COMS[0], 9600);
        new Thread(new Runnable() {
            @Override
            public void run() {
                while (isRun) {
                    try {
                        isRedOn = sharePreferenceUtil.getIsRedOn();
                        isOrangeOn = sharePreferenceUtil.getIsOrangeOn();
                        //如果微动开关被触发
                        if (adam4150.getMicroSwitch(Constants.DIPORTS[2])) {
                            //如果红灯亮，灭红灯
                            if (isRedOn) {
                                adam4150.closeRelay(Constants.RELAYPORTS[3]);
                                sharePreferenceUtil.setIsRedOn(false);
                            }
                            //如果橙灯亮，灭橙灯
                            if (isOrangeOn) {
                                adam4150.closeRelay(Constants.RELAYPORTS[5]);
                                sharePreferenceUtil.setIsOrangeOn(false);
                            }
                            //亮绿灯
                            adam4150.openRelay(Constants.RELAYPORTS[4]);
                        }
                        Thread.sleep(500);
                    } catch (InterruptedException e) {
                        e.printStackTrace();
                    }
                }
            }
        }).start();
    }


    @Nullable
    @Override
    public IBinder onBind(Intent intent) {
        return null;
    }

    /**
     * 服务退出时，停止线程，关闭 4150 串口
     */
    @Override
    public void onDestroy() {
        super.onDestroy();
        isRun = false;
        //关闭串口
        adam4150.closePort();
    }
}
```

(3) 在 AndroidManifest.xml 清单文件中注册 MyService，在 application 标签对中间添加如下代码：

```
<!--注册服务-->
<service android:name=".service.MyService"></service>
```

(4) 登录成功的时候启动 MyService，在 LoginActivity 中跳转到主页面 HomeActivity 之前添加启动 MyService 的代码，具体如下：

```
//启动服务
startService(new Intent(LoginActivity.this, MyService.class));
```

至此，便完成了使用服务监测微动开关状态的功能。完整代码可查阅随书资料中的 Project7_1 工程源代码。

3. 运行结果

运行程序，结果如图 7-8 和图 7-9 所示，发现有烟雾或有人的时候，触发微动开关，设备上的红灯和橙灯灭了，但是界面上的灯没有变化，因为并没有改变界面上灯的显示状态。

图 7-8　微动开关触发时室内环境采集系统

图 7-9　微动开关触发时园区监控系统

任务2 使用广播触发三色灯状态的实时改变

【任务描述】

本项目任务 1 中已经实现了微动开关被触发，关闭设备上的橙灯和红灯，但是室内环境采集系统界面上的橙灯和园区监控系统界面上的红灯并未改变。本任务将实现微动开关被触发关闭设备上红灯和橙灯的同时，发送广播，将界面上的红灯和橙灯也关闭。任务清单如表 7-2 所示。

表 7-2 任务清单

任务课时	6 课时	任务组员数量	建议 3 人
任务组采用设备	四输入模块一个 温湿度传感器一个 光照传感器一个 ZigBee 协调器一个 ZigBee 智能节点盒一个 ADAM-4150 一个 烟雾传感器一个 红外对射传感器一个 微动开关一个 三色灯一个 RS485 转 RS232 转接头一个 公母串口线两条 PC 一台 工控平板		

【拓扑图】

本任务拓扑结构如图 7-10 所示。

注意：本任务的三色灯-红灯、三色灯-绿灯、三色灯-橙灯所接继电器分别为 ADAM-4150 的 DO3、DO4 和 DO5 口，烟雾传感器接 ADAM-4150 模块的 DI4 口，微动开关接 ADAM-4150 的 DI5 口，温湿度传感器、光照传感器分别接四输入模块的 IN2、IN3、IN1 口。

【知识解析】

1. 广播的类型

Android 中的广播主要分为两种类型，即标准广播和有序广播。

标准广播也叫无序广播，是一种完全异步执行的广播，广播发出后，所有的广播接收者几乎同时接收到这条广播消息，这种广播的效率比较高，并且无法被截断，如图 7-11 所示。

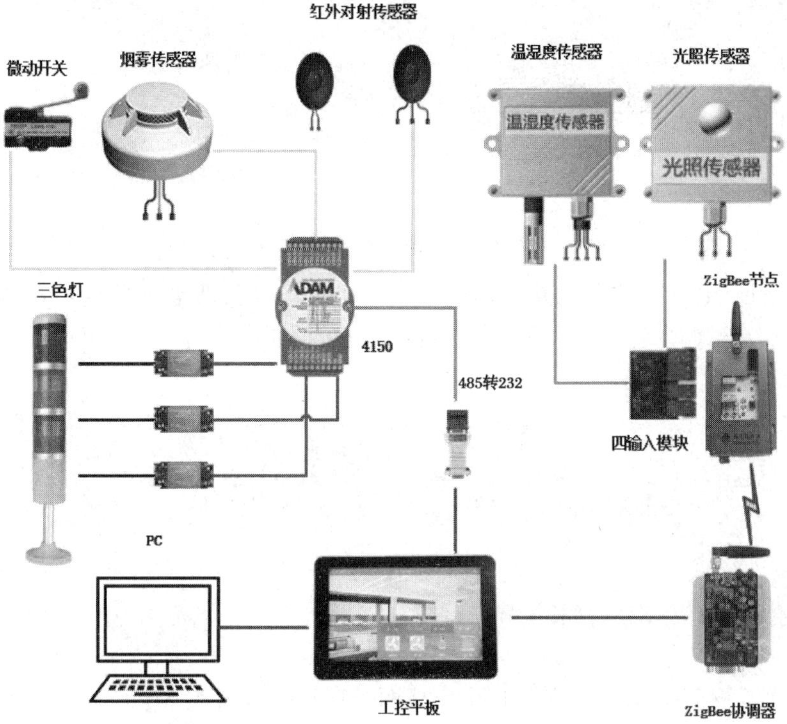

图 7-10　拓扑图

有序广播是一种同步执行的广播，广播发出后，同一时间只会有一个广播接收者能够收到这条广播消息，当这个广播接收者中的逻辑执行完毕后，广播才会继续传递。显然，接收的广播接收者是有先后顺序的，优先级高的广播接收者就可以先收到广播消息，并且前面的广播接收者可以截断正在传递的广播，这样后面的广播接收者就无法收到广播消息了，如图 7-12 所示。

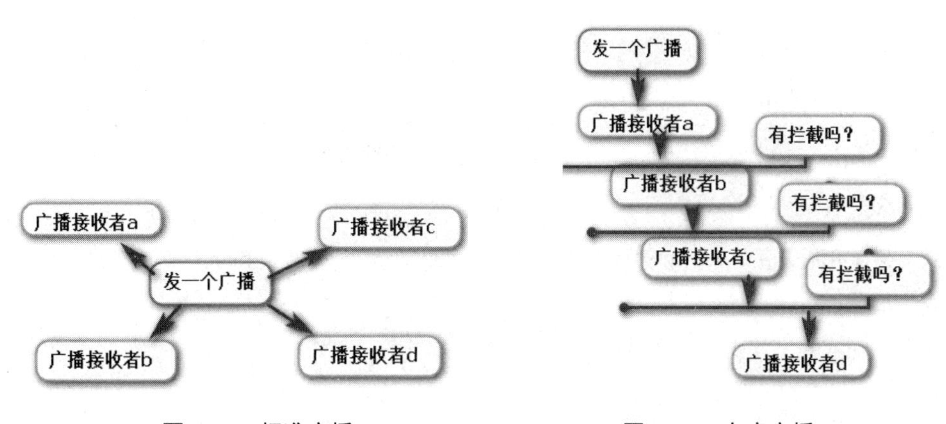

图 7-11　标准广播　　　　图 7-12　有序广播

2. 发送标准广播和有序广播

要想接收广播就必须成为一个广播接收者，广播接收者需要继承 BroadcastReceiver，如图 7-13 所示。

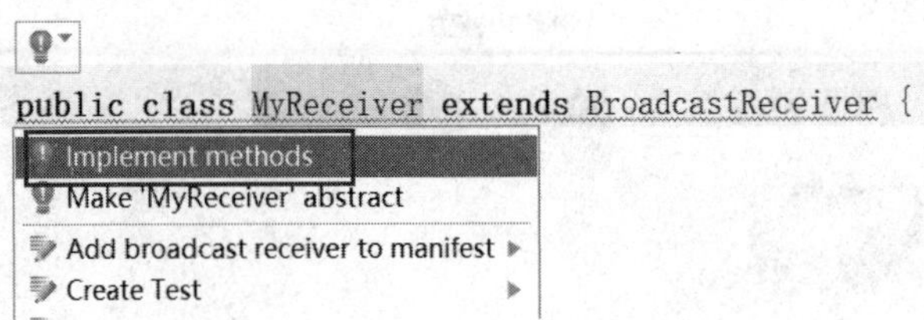

图 7-13　继承自广播接收者

按上面出现的提示重写 onReceive()方法，当该广播接收者接收到广播消息的时候，onReceive()方法自动被调用，广播消息被封装进 Intent 的参数中，所以可以通过该参数获取广播发来的消息。代码如下：

```
public class MyReceiver extends BroadcastReceiver {
    @Override
    public void onReceive(Context context, Intent intent) {
        //在这里处理业务
    }
}
```

接下来分别发送标准广播和无序广播。

1)　发送标准广播

(1)　创建工程 Demo7_2，在工程中创建两个广播接收者类 MyReceiver 和 MyReceiver1 来接收广播，都继承自 BroadcastReceiver，实现其方法 onReceive()，打印“接收到消息”来确定是否收到消息。代码如下：

```
    //自定义广播接收者类
public class MyReceiver extends BroadcastReceiver {
    @Override
    public void onReceive(Context context, Intent intent) {
        System.out.println("MyReceiver 接收到消息");
    }
}
```

```
public class MyReceiver1 extends BroadcastReceiver {

    @Override
    public void onReceive(Context context, Intent intent) {
        System.out.println("MyReceiver1 接收到消息");
    }
}
```

(2)　在 activity_main.xml 文件中分别创建“发送标准广播”和“发送有序广播”两个按钮，代码如下：

```
<?xml version="1.0" encoding="utf-8"?>
<LinearLayout xmlns:android="http://schemas.android.com/apk/res/android"
              xmlns:tools="http://schemas.android.com/tools"
              android:layout_width="match_parent"
              android:layout_height="match_parent"
              android:orientation="vertical">

    <Button
        android:id="@+id/btnBroadcastReceiver"
        android:layout_width="wrap_content"
        android:layout_height="wrap_content"
```

```
            android:text="发送标准广播"/>
        <Button
            android:id="@+id/btnOrderBroadcastReceiver"
            android:layout_width="wrap_content"
            android:layout_height="wrap_content"
            android:text="发送有序广播"/>
    </LinearLayout>
```

(3) 在 MainActivity“发送标准广播”按钮的点击事件中添加发送无序广播的代码，代码如下：

```
        Intent intent = new Intent();
        intent.setAction("com.example.demo.action.MyReceiver");
        sendBroadcast(intent, null);//有序广播需要用 sendOrderedBroadcast()方法发送
```

(4) 在清单文件 AndroidManiferst.xml 中配置广播，代码如下：

```
        <receiver android:name=".MyReceiver">
    <intent-filter  >
        <action android:name="com.example.demo.action.MyReceiver" />
    </intent-filter>
</receiver>
<receiver android:name=".MyReceiver1"  >
    <intent-filter >
        <action android:name="com.example.demo.action.MyReceiver" />
    </intent-filter>
</receiver>
```

(5) 双击 app 对应的 build.gradle 文件，将目标 SDK 版本改成 25：

```
targetSdkVersion 25
```

(6) 单击“发送广播”按钮，查看控制台，发现两个接收者都接收到了，日志信息如图 7-14 所示。

```
I/System.out: MyReceiver接收到消息
I/System.out: MyReceiver1接收到消息
```

图 7-14　Log 日志信息

多学一招：

在 Android 8.0(SDK 26)及以上的平台，应用不能对大部分的广播进行静态注册，也就是说，大部分的广播都不能在 AndroidManifest.xml 文件中进行静态注册。如果仍然想用静态注册，可以将 gradle 中的 targetSdkVersion 设置为小于 26。

2) 发送有序广播并在 MyReceiver 中进行截断

(1) 在 MyReceiver 的 onReceive 方法中添加 abortBroadcast();终止广播。

(2) 同时在清单文件 AndroidManifest.xml 中设置了 priority 优先级，设置 MyReceiver 的优先级为 10，设置 MyReceiver1 的优先级为 1。

```
<receiver android:name=".MyReceiver" >
            <intent-filter  android:priority="10">
                <action android:name="com.example.demo.action.MyReceiver" /
            </intent-filter>
        </receiver>
        <receiver android:name=".MyReceiver1">
```

```
            <intent-filter android:priority="1">
                <action android:name="com.example.demo.action.MyReceiver" />
            </intent-filter>
        </receiver>
```

(3) 在 MainActivity“发送标准广播”按钮的点击事件中添加发送无序广播的代码，具体如下：

```
case R.id.btnOrderBroadcastReceiver://发送有序广播
            Intent intent0 = new Intent();
            intent0.setAction("com.example.demo.action.MyReceiver");
            //该方法第 2 个参数决定该广播的级别，级别数值是在 -1000 ～ 1000 之间，值越大，发送的优先级越高
           sendOrderedBroadcast(intent0, null);//有序广播需要用 sendOrderedBroadcast()方法发送
            break;
```

(4) 执行程序，可在控制台看到，只有 MyReceiver 接收到消息，如图 7-15 所示，证明广播被终止了。

```
I/System.out: MyReceiver接收到消息
```

图 7-15　Log 日志信息

3. 广播的静态注册和动态注册及其优先级

在 Android 中注册广播有两种方式，即代码中注册(动态注册)以及在配置文件中注册(静态注册)。在 Android 8.0(SDK 26)及以上的平台上，应用不能对大部分的广播进行静态注册，也就是说，大部分的广播都不能在 AndroidManifest.xml 文件中进行静态注册。

两者及其接收广播的区别如下。

(1) 动态注册广播不是常驻型广播，也就是说广播跟随 Activity 的生命周期。注意在 Activity 结束前，移除广播接收者；静态注册是常驻型，也就是说当应用程序关闭后，如果有信息广播，程序也会被系统调用自动运行。

(2) 当广播为有序广播时：优先级高的先接收(不分静态和动态)；同优先级的广播接收者，动态优先于静态。

(3) 同优先级的同类广播接收者，对于静态，先扫描的优先于后扫描的；对于动态，先注册的优先于后注册的。

(4) 当广播为默认广播时：无视优先级，动态广播接收者优先于静态广播接收者。同优先级的同类广播接收者，对于静态，先扫描的优先于后扫描的；对于动态，先注册的优先于后注册的。

接下来分别进行动态注册和静态注册，并同时查看在无序和有序两种情况下发送分别是什么样的情况。

(1) 在 Demo7_2 中创建两个广播接收者，一个是接收静态注册，一个是接收动态注册，分别为 StaticReceiver 和 DynamicReceiver，并在其 onReceive()方法中弹出不同的提示信息，代码如下：

```
      Toast.makeText(context, "静态注册", Toast.LENGTH_SHORT).show();
```

```
       Toast.makeText(context, "动态注册", Toast.LENGTH_SHORT).show();
```

(2) 以静态方式注册广播接收者 StaticReceiver，在清单文件 AndroidManifest.xml 中添加以下代码：

```
      <receiver android:name="com.example.demo.demo7_2.StaticReceiver">
              <intent-filter>
                  <action android:name="com.example.demo.action.Receiver" />
              </intent-filter>
 </receiver>
```

(3) 在 MainActivity 中动态注册 DynamicReceiver，代码如下：

```
//获取一个过滤器
IntentFilter filter = new IntentFilter();
filter.addAction("com.example.demo.action.Receiver");
registerReceiver(dynamic, filter);//注册
```

(4) 单击“无序下同时发送静态和动态广播”按钮后发送无序广播，在 MainActivity 中添加如下代码：

```
     case R.id.btnDisorderedLaunch://无序下同时发送静态和动态广播
                Intent intent1 = new Intent();
                intent1.setAction("com.example.demo.action.Receiver");
                sendBroadcast(intent1);//发送无序广播
                break;
```

(5) 运行程序，发现动态注册优先于静态注册，如图 7-16 所示。

接下来调整广播接收者的优先级。

① 将静态注册的广播接收者 StaticReceiver 优先级设置为 100，代码如下：

```
<receiver android:name="com.example.demo.demo7_2.StaticReceiver">
            <intent-filter android:priority="100">
                <action android:name="com.example.demo.action.Receiver" />
            </intent-filter>
        </receiver>
```

② 将动态注册的广播接收者 DynamicReceiver 优先级设置为 1，代码如下：

```
DynamicReceiver dynamic = new DynamicReceiver();
//获取一个过滤器
IntentFilter filter = new IntentFilter();
filter.addAction("com.example.demo.action.Receiver");
filter.setPriority(1);//设置优先级
registerReceiver(dynamic, filter);//注册
```

设置静态注册的广播接收者优于动态注册的广播接收者，发现还是动态注册的广播接收者先接收到广播消息，运行效果如图 7-17 所示。

接下来发送有序广播。

(1) 单击“有序下同时发送静态和动态广播”按钮后发送有序广播，在 MainActivity 中添加如下代码：

```
case R.id.btnOrderlyLaunch://有序下同时发送静态和动态广播
            Intent intent2 = new Intent();
            intent2.setAction("com.example.demo.action.Receiver");
            sendOrderedBroadcast(intent2, null);//发送有序广播
            break;
```

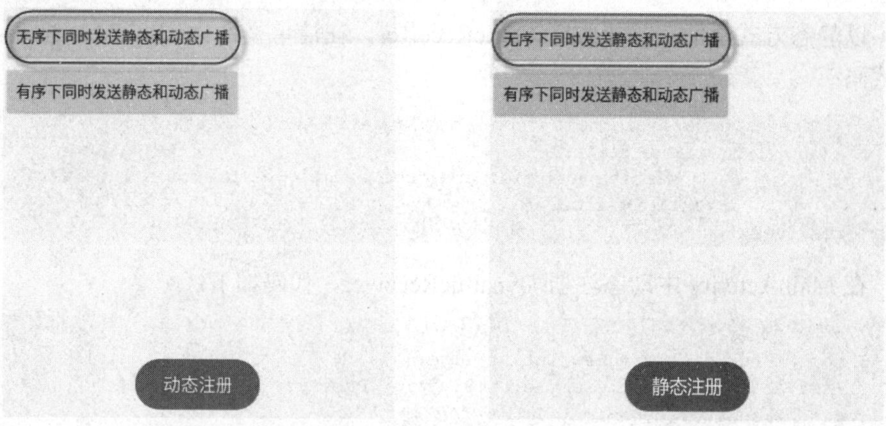

图 7-16 广播案例效果图 1　　　　图 7-17 广播案例效果图 2

(2) 运行程序单击“有序下同时发送静态和动态广播”按钮，发现两种注册方式中，静态方式的广播先进行了注册(此时静态优先级是 100，动态优先级为 1)，如图 7-18 所示。

如果是同优先级情况下又是谁先执行呢？我们将优先级都去掉。执行程序，发现动态注册的广播接收者先收到广播消息，如图 7-19 所示。

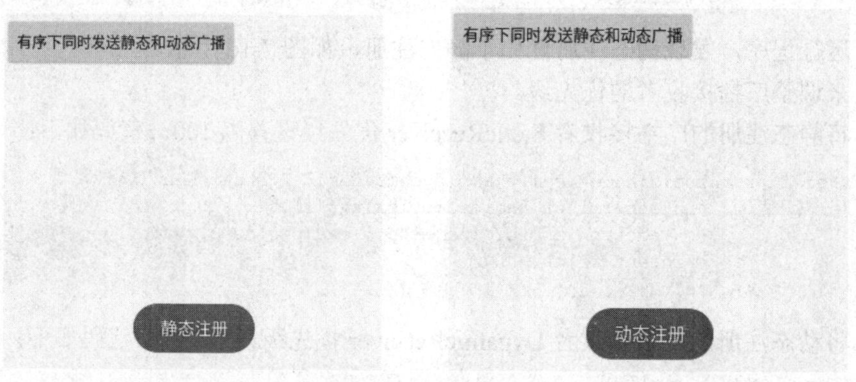

图 7-18 广播案例效果图 3　　　　图 7-19 广播案例效果图 4

【任务实施】

1. 任务分析

本任务将实现微动开关被触发关闭设备上红灯和橙灯的同时，发送广播消息，将界面上的红灯和橙灯也关闭。本任务在任务 1 完成后的代码 Project7_1 基础上实现。

(1) 在 Fragment 的基类 BaseFragment.java 中添加一个广播接收者，以动态方式注册。接收到广播后调用获取数据的抽象方法(在 BaseFragment 的实现类中实现)。

(2) 在 MyService 中创建 Intent，设置 Action 与广播过滤器的 Action 一致。微动开关被触发，在 Intent 中添加不同参数，分别表明要关闭界面上的红灯或橙灯，然后发送广播，在接收到广播的地方关闭对应的灯。

(3) 在 EntranceFragment.java 代码中接收广播消息，如果接收到要关闭橙灯的消息，并且当前橙灯是亮着的，关闭界面上橙灯。

(4) 在 MonitoringFragment.java 代码中实时更新红外对射的数据，如果有人并且红灯没亮，亮红灯；如果没人并且红灯亮，灭红灯。接收广播消息，如果接收到要关闭红灯的消息，并且当前红灯是亮着的，关闭界面上红灯。

2. 任务实现

(1) 在 Fragment 的基类 BaseFragment.java 中添加一个广播接收者，以动态方式注册。写一个获取数据的抽象方法，接收到广播后调用该方法。在 onDestroy()方法中注销广播。BaseFragment.java 类的代码如下：

```
1.  public abstract class BaseFragment extends Fragment {
2.      private MyDataReceiver receiver;
3.      public static SharePreferenceUtil sharePreferenceUtil;
4.      public Adam4150 adam4150;
5.      @Override
6.      public void onActivityCreated(@Nullable Bundle savedInstanceState) {
7.          super.onActivityCreated(savedInstanceState);
8.          if (adam4150 == null) {
9.              adam4150 = new Adam4150();
10.             adam4150.openPort(Constants.COMS[0], 9600);
11.         }
12.         if (sharePreferenceUtil == null) {
13.             sharePreferenceUtil = SharePreferenceUtil.getInstant(getActivity());
14.         }
15.         //注册接收数据的广播
16.         IntentFilter filter = new IntentFilter();
17.         filter.addAction("senddata");
18.         receiver = new MyDataReceiver();
19.         getActivity().registerReceiver(receiver, filter);
20.     }
21.
22.     /**
23.      * 获取广播发送过来的数据
24.      *
25.      * @param intent
26.      */
27.     public abstract void getData(Intent intent);
28.
29.     @Override
30.     public void onDestroy() {
31.         super.onDestroy();
32.         //注销广播
33.         getActivity().unregisterReceiver(receiver);
34.     }
35.
36.     //广播接收者
37.     private class MyDataReceiver extends BroadcastReceiver {
38.         @Override
39.         public void onReceive(Context context, Intent intent) {
40.             getData(intent);
41.         }
42.     }
43. }
```

由于第 27 行 getData()方法是一个抽象方法，所以继承 BaseFragment 的 3 个非抽象类 EnvironmentFragment.java、MonitoringFragment.java 和 CloudFragment.java 中要实现该方法，分别添加如下代码：

```
@Override
public void getData(Intent intent) {

}
```

(2) 在 MyService 的 onCreate()方法中创建 Intent，设置 Action 与广播过滤器的 Action 一致。如果微动开关被触发，在 Intent 中添加不同参数，分别表明要关闭界面上的红灯或橙灯，然后发送广播。

① 在 MyService 的 onCreate 方法 TODO 1 的位置创建 Intent，设置动作为 senddata，代码如下：

```
// TODO 1: 创建 Intent 设置动作为 senddata
intent = new Intent();
intent.setAction("senddata");
```

② 在本项目任务 1 任务实施的第二步中已经实现了如果微动开关被触发，这时如果红灯亮灭红灯，如果橙灯亮灭橙灯，并改变 SharePreferences 中对应灯的标记状态。任务 1 中灭红灯和橙灯的时候只是把设备上的灯关闭掉，并没有改变界面上灯的状态。所以这里要加上灭红灯和橙灯的代码，分别在 MyService 类 onCreate()方法 TODO 2 和 TODO 3 的位置添加如下代码：

```
// TODO 2: 设置 Intent 传递的红灯参数对应的值为 ture，表明要关闭界面上的红灯
intent.putExtra("closeRed", true);
// TODO 3: 设置 Intent 传递的橙灯参数对应的值为 ture，表明要关闭界面上的橙灯
intent.putExtra("closeOrange", true);
```

③ 在 MyService 的 onCreate 方法 TODO 4 的位置发送广播，代码如下：

```
// TODO 4: 发送广播
 sendBroadcast(intent);
```

(3) 在 EntranceFragment.java 代码的 getData()方法中接收广播消息，如果接收到的 closeOrange 为 true，关闭界面上的橙灯，在 getData()方法中添加如下代码：

```
@Override
public void getData(Intent intent) {
    //接收广播发送过来的数据，如果为 true，关闭界面上橙灯
    boolean isClose= intent.getBooleanExtra("closeOrange", false);
    if (isClose) {
        //关闭界面上橙灯
        cbSmoke.setChecked(false);
    }
}
```

(4) 在 MonitoringFragment.java 代码中实时更新红外对射的数据，如果有人并且红灯没亮，亮红灯；如果没人并且红灯亮，灭红灯。在 getData()方法中接收广播消息，如果接收到的 closeRed 为 true，则关闭界面上红灯。在 MonitoringFragment.java 中添加如下代码：

```
private TextView tvPerson;
//红灯对应的 CheckBox，true 对应亮灯图片，false 对应灭灯图片
private CheckBox cbRed;

//在 Activity 创建的时候绑定控件，并执行实时更新数据的 Runnable
@Override
public void onActivityCreated(@Nullable Bundle savedInstanceState) {
    super.onActivityCreated(savedInstanceState);
```

```
    tvPerson = getActivity().findViewById(R.id.tv_person);
    //执行 Runnable，实时更新红外对射的数据，如果有人并且红灯没亮，亮红灯；如果没人并且红灯亮，灭红灯
    handler.post(runnable);
}

Handler handler = new Handler();
//实时更新红外对射的数据，如果有人并且红灯没亮，亮红灯；如果没人并且红灯亮，灭红灯
Runnable runnable = new Runnable() {
    @Override
    public void run() {
        String person = adam4150.getInfrared(Constants.DIPORTS[1]);
        if (person.equals("有人") && !sharePreferenceUtil.getIsRedOn()) {
            //开红灯
            adam4150.openRelay(Constants.RELAYPORTS[3]);
            sharePreferenceUtil.setIsRedOn(true);
            cbRed.setChecked(true);
        } else if (person.equals("无人") && sharePreferenceUtil.getIsRedOn()) {
            //关红灯
            adam4150.closeRelay(Constants.RELAYPORTS[3]);
            sharePreferenceUtil.setIsRedOn(false);
            cbRed.setChecked(false);
        }
        tvPerson.setText(person);
        handler.postDelayed(runnable, 1000);
    }
};

@Override
public void getData(Intent intent) {
    //接收广播发送过来的数据，如果为 true，关闭界面上红灯
    boolean isClose = intent.getBooleanExtra("closeRed", false);
    if (isClose) {
        //关闭界面上红灯
        cbRed.setChecked(false);
    }
}
```

至此，便完成了使用广播触发三色灯状态的时时改变功能。完整代码可查阅随书资料中的 Project7_2 工程源代码。

3. 运行结果

程序运行结果如图 7-20 和图 7-21 所示。发现有烟雾或有人的时候，触发微动开关，设备上的红灯和橙灯灭了，界面上灯的状态也随着改变，说明发送广播成功。

图 7-20　微动开关触发时室内环境采集系统

图 7-21　微动开关触发时园区监控系统

项 目 总 结

本项目在服务中监测微动开关的状态，如果微动开关被触发，灭设备上的红灯和橙灯，亮绿灯，同时发送广播，关闭界面上的红灯和橙灯。读者要掌握服务的启动以及广播的发送方法。

思考与练习

7-1　简述标准广播与有序广播的区别。
7-2　简述静态注册和动态注册的区别。
7-3　简述启动式服务的生命周期。
7-4　简述绑定式服务的生命周期。

项目 8 媒体动画的实现

【项目描述】

手机，正在影响着人们的生活，听音乐、看新闻、看电影电视、跟友人聊天、随手拍拍照……这些都离不开手机强大的多媒体功能的支持，Android 也提供了一系列的 API 方便调用手机的多媒体资源。在物联网的应用中，手机作为智能终端，也需要利用多媒体呈现功能的状态结果，如预警警示、刷卡的开门动作、降温用的风扇转动等。

本项目以 4 个任务的形式讲解如何利用手机媒体动画编写出丰富多样的物联网应用程序。具体任务列表如图 8-1 所示。

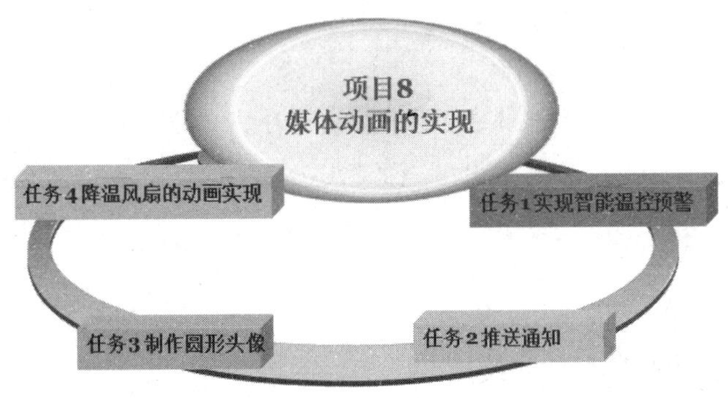

图 8-1　项目 8 任务列表

【学习目标】

知识目标：

- 会使用 MediaPlayer 播放音频。
- 会使用 VideoView 播放视频。
- 会推送消息。
- 会调用系统照相机拍照和从图库取图。
- 会处理动画。

技能目标：

- 能编程实现音视频播放。
- 能进行消息的推送。
- 能处理用户圆形头像。
- 能实现风扇的动画效果。

任务 1　实现智能温控预警

【任务描述】

扫码观看视频讲解

本任务模拟智能温控系统中的环境检测模块对家里的温度进行监测，可以在智能终端上实时查看温度的数值，当温度超过设定的预定值时发出“温度太高”的播报预警。任务清单如表 8-1 所示。

表 8-1　任务清单

任务课时	4 课时	任务组员数量	建议 3 人
任务组采用设备	有线温湿度传感器一个 ZigBee、四输入采集器一个 ZigBee 协调器一个 工控平板 PC 一台 电源线、串口线若干		

【拓扑图】

本任务拓扑图如图 8-2 所示。

注意：本任务温度传感器信号线接 IN2 口。

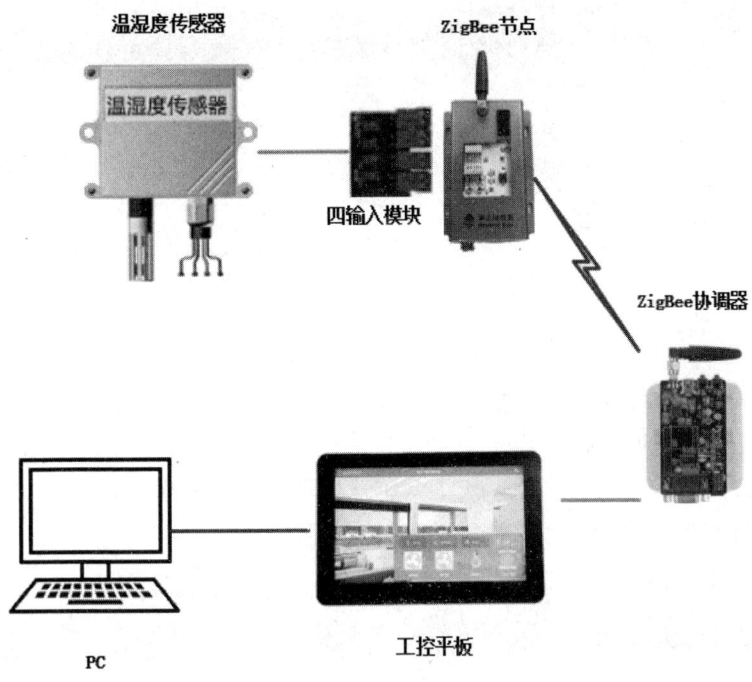

图 8-2 拓扑图

【知识解析】

1. 播放音频

在 Android 中用 MediaPlayer 类来播放音频，它可以播放多种格式的音频文件，也提供了控制播放的开始、暂停和停止方法。

用 MediaPayer 播放音频文件，有两种方法构建 MediaPlayer 对象。

播放 SD 上的音乐文件使用 new 的方式：

```
MediaPlayer mediaPlayer=new MediaPlayer();
```

播放资源中的音乐文件使用 create 方式：

```
MediaPlayer mediaPlayer=MediaPlayer.create(this,R.raw.音乐文件名);
```

要用 MediaPlayer 播放音乐，先创建出一个 MediaPlayer 对象，接着调用 setDataSource()方法设置音乐文件的路径，再调用 prepare()方法使 MediaPlayer 进入准备状态，然后调用 start()方法开始播放，调用 pause()方法暂停，调用 stop()方法停止播放。另外，eset()方法重置到刚刚创建的状态，seekTo()方法从指定的位置开始播放，release()方法可以释放资源，isPalying()方法判断当前是否正在播放资源，getDuration()方法获取音频文件的时长，etLooping(true)方法循环重复播放。

使用 create 方法创建的 MediaPlayer 对象不需要调用 setDataSource()和 prepare()。

1) 播放 raw 目录下的音频文件

例 8-1 播放 raw 目录下的音频文件。

第一步：创建工程 Demo8_1，在 res 下创建 raw 目录，把音频文件 light.mp3 放到目录下。创建 raw 目录的过程如图 8-3 所示。

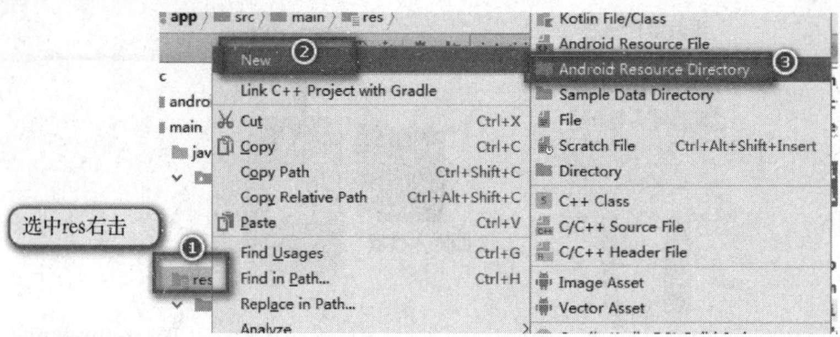

图 8-3　创建 raw 目录

在 Resource type 中选择 raw，然后把音频文件复制到 raw 下，如图 8-4 所示。

第二步：创建 BackMusicActivity，对应的布局 activity_music.xml 如下，3 个按钮用于播放、暂停、停止音乐，一个文本框用于显示要播放的音频文件名称，如图 8-5 所示。

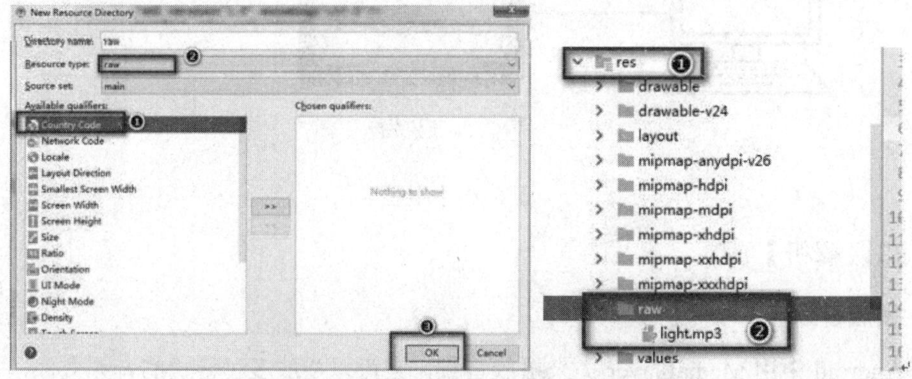

图 8-4　创建 raw 目录并把音频文件放进去

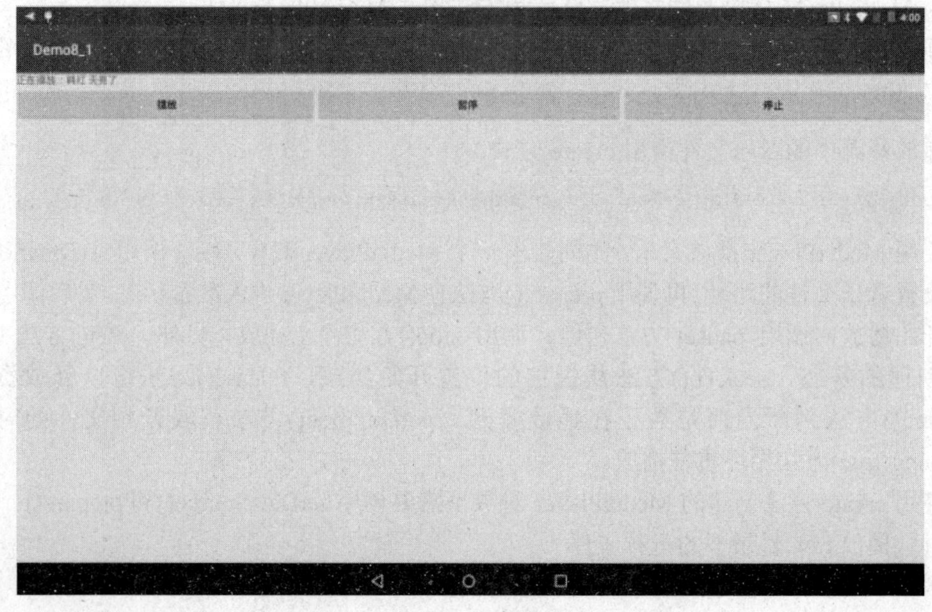

图 8-5　布局界面

```
<?xml version="1.0" encoding="utf-8"?>
<LinearLayout xmlns:android="http://schemas.android.com/apk/res/android"
    xmlns:app="http://schemas.android.com/apk/res-auto"
    xmlns:tools="http://schemas.android.com/tools"
    android:layout_width="match_parent"
    android:layout_height="match_parent"
    tools:context=".BackMusicActivity"
    android:orientation="vertical"
    >
    <TextView
        android:layout_width="match_parent"
        android:layout_height="wrap_content"
        android:id="@+id/path"
        android:hint="正在播放:韩红 天亮了.mp3"
        />
    <LinearLayout
        android:layout_width="match_parent"
        android:layout_height="wrap_content"
        android:orientation="horizontal"
        >
    <Button
        android:id="@+id/play"
        android:onClick="click"
        android:layout_width="0dp"
        android:layout_weight="1"
        android:layout_height="wrap_content"
        android:text="播放"/>
    <Button
        android:id="@+id/pause"
        android:onClick="click"
        android:layout_width="0dp"
        android:layout_weight="1"
        android:layout_height="wrap_content"
        android:text="暂停"/>
    <Button
        android:id="@+id/stop"
        android:onClick="click"
        android:layout_width="0dp"
        android:layout_weight="1"
        android:layout_height="wrap_content"
        android:text="停止"/>
    </LinearLayout>
</LinearLayout>
```

第三步：编写 BackMusicActivity 中的代码，如下所示：

```
public class BackMusicActivity extends AppCompatActivity {
    private MediaPlayer mediaPlayer;//音频播放器
    private int currPosition;//当前播放位置

    @Override
    protected void onCreate(Bundle savedInstanceState) {
        super.onCreate(savedInstanceState);
        setContentView(R.layout.activity_music);
    }

    public void click(View v) {
        switch (v.getId()) {
            case R.id.play://播放
                //1.实例化播放对象
                mediaPlayer = MediaPlayer.create(BackMusicActivity.this, R.raw.light);
```

```
16.                 //不停地重复播放
17.                   mediaPlayer.setLooping(true);
18.                   //2.开始播放
19.                  mediaPlayer.start();
20.                  //3.如果之前有播放过，就从上次停下来的位置继续播放
21.                 if (currPosition != 0) {
22.                     mediaPlayer.seekTo(currPosition);
23.                 }
24.                 break;
25.             case R.id.pause://暂停
26.                 if (mediaPlayer.isPlaying()) {
27.                     //获取当前的播放位置
28.                  currPosition = mediaPlayer.getCurrentPosition();
29.                     mediaPlayer.pause(); //暂停
30.                 }
31.                 break;
32.             case R.id.stop://停止
33.                 if (mediaPlayer.isPlaying()) {  //如果播放器正在播放中
34.                     currPosition = 0;  //把位置设置为0
35.                     mediaPlayer.stop();  //停止播放
36.                 }
37.                 break;
38.         }
39.     }
40.
41.     //释放播放器对象
42.     @Override
43.     protected void onDestroy() {
44.         super.onDestroy();
45.         if (mediaPlayer != null) {
46.             mediaPlayer.stop();   //停止播放
47.             mediaPlayer.release();   //释放播放器对象
48.         }
49.     }
50. }
```

上述代码在 11～39 行的点击事件中进行播放的控制。43～49 行是利用 Activity 的生命周期在退出 Activity 时进行播放器对象的释放。

运行程序，单击“播放”按钮音乐开始播放，单击“暂停”按钮音乐暂停了，再次单击“播放”按钮会从暂停的位置继续播放，单击“停止”按钮音乐停了。再次单击“播放”按钮音乐又开始播放了。

2) 播放 SD 卡目录下的音频文件

例 8-2 播放 SD 卡目录下的音频文件。

第一步：在例 8-1 的基础上新建 SDActivity，加载的也是布局文件 activity_music.xml。在清单文件中把 SDActivity 设置为主 Activity：

```
<activity android:name=".SDActivity">
    <intent-filter>
        <action android:name="android.intent.action.MAIN" />
        <category android:name="android.intent.category.LAUNCHER" />
    </intent-filter>
</activity>
```

第二步：因为要访问 SD 卡，需要在清单文件中添加访问 SD 卡的权限：

```
<uses-permission android:name="android.permission.READ_EXTERNAL_STORAGE"/>
<uses-permission android:name="android.permission.WRITE_EXTERNAL_STORAGE"/>
```

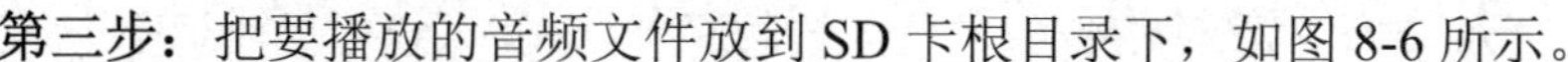

第三步：把要播放的音频文件放到 SD 卡根目录下，如图 8-6 所示。

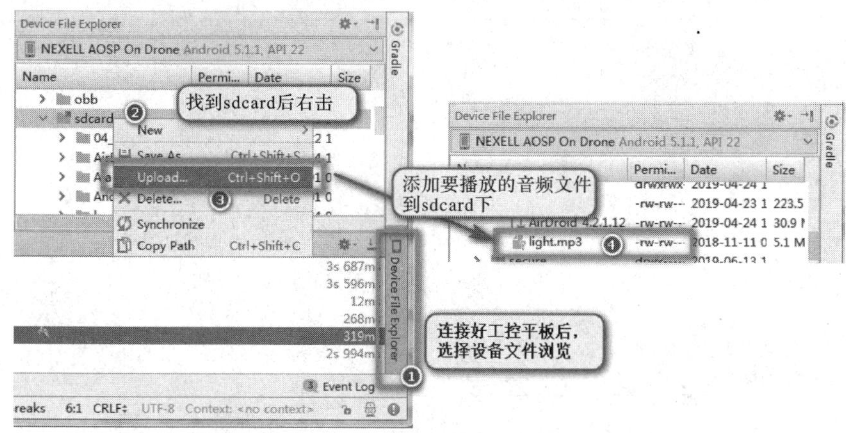

图 8-6 将音频文件放到 SD 卡

第四步：编写 SDActivity 中的代码实现播放 SD 卡上的音频文件，要注意处理 SD 卡的运行时访问权限。

```
public class SDActivity extends AppCompatActivity {
    private MediaPlayer mediaPlayer;//播放器对象
    private int currPosition;//当前播放位置
    private Button play, pause, stop;//播放 暂停 停止按钮

    @Override
    protected void onCreate(Bundle savedInstanceState) {
        super.onCreate(savedInstanceState);
        setContentView(R.layout.activity_music);
        play = findViewById(R.id.play);
        pause = findViewById(R.id.pause);
        stop = findViewById(R.id.stop);
        //检查 SD 卡的授权
        checkGrant();
    }

    @SuppressLint("NewApi")
    private void checkGrant() {
        //1.如果用户没有授权
        if (checkSelfPermission(Manifest.permission.WRITE_EXTERNAL_STORAGE) != PackageManager.PERMISSION_GRANTED) {
            //2.请求授权：参数一要将授权的权限放入 String 数组中，参数二是请求码
            requestPermissions(new String[]{Manifest.permission.WRITE_EXTERNAL_STORAGE}, 1);
        } else {
            //如果已有授权
            // 3.初始化媒体播放器
            initMediaPlayer();
            Log.i("TAG", "SD 卡可读可写");
        }
    }

    //授权监听的返回结果 参数： 请求码 授权结果的数组 申请授权的数组
    @Override
    public void onRequestPermissionsResult(int requestCode, String[] permissions, int[] grantResults) {
```

```
        super.onRequestPermissionsResult(requestCode, permissions, grantResults);
        if (requestCode == 1 && grantResults[0] == PackageManager.PERMISSION_GRANTED) {
            //如果授权成功，初始化媒体播放器
            initMediaPlayer();
            Log.i("TAG", "请求授权成功");
        } else {
            Toast.makeText(SDActivity.this, "没有访问 SD 卡权限", Toast.LENGTH_LONG).show();
            //如果用户不同意授权访问 SD 卡，则结束
            finish();
        }
    }

    //初始化音乐播放器
    private void initMediaPlayer() {
        //设置要播放的文件的路径是 SD 卡上的 light.mp3
        File file = new File(Environment.getExternalStorageDirectory(), "light.mp3");
        Log.i("path", file.getPath());
        if (file.exists() && file.length() > 0) {
            try {
                //1.初始化播放器
                mediaPlayer = new MediaPlayer();
                //2.设置要播放的文件在哪里
                mediaPlayer.setDataSource(file.getPath());
                //3.准备播放器
                mediaPlayer.prepare();
            } catch (IOException e) {
                e.printStackTrace();
            }
        } else {
            Toast.makeText(SDActivity.this, "播放文件出错", Toast.LENGTH_LONG).show();
        }
    }
    public void click(View v) {
        switch (v.getId()) {
            case R.id.play://播放
                play();
                break;
            case R.id.pause://暂停
                pause();
                break;
            case R.id.stop://停止
                stop();
                break;
        }
    }
    //释放播放器对象
    @Override
    protected void onDestroy() {
        super.onDestroy();
        if (mediaPlayer != null) {
            mediaPlayer.stop();//停止
            mediaPlayer.release();//释放
            mediaPlayer = null;
        }
    }

    private void pause() {
        if (mediaPlayer.isPlaying()) {
```

```
                //记下当前播放位置
                currPosition = mediaPlayer.getCurrentPosition();
                mediaPlayer.pause();
                pause.setText("继续");
                return;
            }
            if (mediaPlayer != null) {
                if (pause.getText().toString().equals("继续")) {
                    mediaPlayer.seekTo(currPosition);
                    mediaPlayer.start();
                    pause.setText("暂停");
                    return;
                }
            }
        }
    private void stop() {
        if (mediaPlayer.isPlaying()) {
            mediaPlayer.stop();
            play.setEnabled(true);
            pause.setEnabled(false);
            stop.setEnabled(false);
            initMediaPlayer();
            return;
        }
    }

    private void play() {
        if (!mediaPlayer.isPlaying()) {         //开始播放
            mediaPlayer.start();
            pause.setEnabled(true);
            stop.setEnabled(true);
            play.setEnabled(false);
        }
    }
}
```

运行程序，按以下步骤操作控制音频的播放、暂停、停止，如图 8-7 所示。

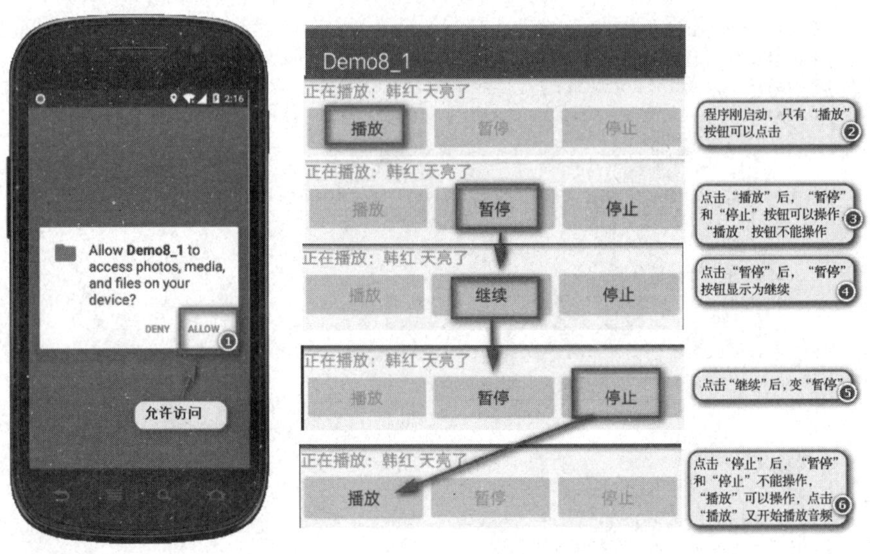

图 8-7 运行结果

2. 播放视频

播放视频主要使用 VideoView 类来完成，先创建一个 VideoView 对象，接着调用 setVideoPath()方法设置播放视频文件的路径，调用 start()方法开始播放，调用 pause()方法暂停，调用 suspend()方法挂起，调用 stop()方法停止播放，调用 resume()方法将视频从头开始播放，seekTo()方法从指定的位置开始播放，调用 isPalying()判断当前是否正在播放资源，调用 getDuration()方法获取视频文件的时长。

例 8-3 播放 SD 卡目录下的视频文件。

参考例 8-2 把视频文件 a.mp4 上传到 SD 卡目录下，复制例 8-2 中的布局文件，命名为 activity_video.xml，在布局中添加一个 VideoView 控件，代码如下：

```
<VideoView
        android:id="@+id/video_view"
        android:layout_width="match_parent"
        android:layout_height="wrap_content" />
```

同时，提示播放的内容修改信息：

```
<TextView
        android:layout_width="match_parent"
        android:layout_height="wrap_content"
        android:hint="正在播放：小熊" />
```

复制 SDActivity，命名为 VideoViewActivity，在清单文件中设置成为主 Activity：

```
<activity android:name=".VideoViewActivity">
           <intent-filter>
               <action android:name="android.intent.action.MAIN" />
               <category android:name="android.intent.category.LAUNCHER" />
           </intent-filter>
</activity>
```

编写程序实现视频的播放，代码如下：

```
public class VideoViewActivity extends AppCompatActivity {
    private VideoView videoView;//视频播放器对象
    private int currPosition;//当前播放位置
    private Button play, pause, stop;//播放、暂停、停止按钮

    @Override
    protected void onCreate(Bundle savedInstanceState) {
        super.onCreate(savedInstanceState);
        setContentView(R.layout.activity_video);
        //查找视频播放控件
        videoView = findViewById(R.id.video_view);
        play = findViewById(R.id.play);
        pause = findViewById(R.id.pause);
        stop = findViewById(R.id.stop);
        //检查 SD 卡的授权
        checkGrant();
    }

    @SuppressLint("NewApi")
    private void checkGrant() {
        //1.如果用户没有授权
        if (checkSelfPermission(Manifest.permission.WRITE_EXTERNAL_STORAGE) !=
PackageManager.PERMISSION_GRANTED) {
```

```
                //2.请求授权：参数一要将授权的权限放入 String 数组中，参数二是请求码
                requestPermissions(new String[]{Manifest.permission.WRITE_EXTERNAL_STORAGE}, 1);
            } else {
                //如果已有授权
                // 3.设置播放路径
                initVideoPath();
                Log.i("TAG", "SD 卡可读可写");
            }
        }

        //授权监听的返回结果  参数:  请求码  授权结果的数组  申请授权的数组
        @Override
        public void onRequestPermissionsResult(int requestCode, String[] permissions, int[] grantResults) {
            super.onRequestPermissionsResult(requestCode, permissions, grantResults);
            if (requestCode == 1 && grantResults[0] == PackageManager.PERMISSION_GRANTED) {
                //如果授权成功，设置播放路径
                initVideoPath();
                Log.i("TAG", "请求授权成功");
            } else {
                Toast.makeText(VideoViewActivity.this, "没有访问 SD 卡权限", oast.LENGTH_LONG).show();
                //如果用户不同意授权访问 SD 卡，则结束
                finish();
            }
        }

        //设置播放路径
        private void initVideoPath() {
          //设置要播放文件的路径是 SD 卡上的.mp4
            File file = new File(Environment.getExternalStorageDirectory(), "a.mp4");
            Log.i("path", file.getPath());
            if (file.exists() && file.length() > 0) {

              videoView.setVideoPath(file.getPath());
            } else {
                Toast.makeText(VideoViewActivity.this, "播放文件出错", Toast.LENGTH_LONG).show();
            }
        }
        public void click(View v) {
            switch (v.getId()) {
                case R.id.play://播放
                    if(!videoView.isPlaying()) {
                        videoView.start();//开始播放
                    }
                    break;
                case R.id.pause://暂停
                    if(videoView.isPlaying()) {
                        videoView.pause();//开始播放
                    }
                    break;
                case R.id.stop://停止
                    if(videoView.isPlaying()) {
                        videoView.resume();//开始播放
                    }
                    break;
            }
        }
```

```
    //释放播放器对象
    @Override
    protected void onDestroy() {
        super.onDestroy();
        if (videoView != null) {
           videoView.suspend();//挂起
            videoView = null;
        }
    }
}
```

运行程序，同意授权 SD 卡的访问权限后，控制播放结果如图 8-8 所示。

图 8-8　视频播放结果

至此，读者学会了播放音频和视频，VideoView 内部仍然是使用 MediaPlayer 对视频进行控制。同时需要注意的是，VideoView 在视频格式的支持以及播放效率方面都存在较大的不足，只适用于播放一些片头动画或视频宣传片。

【任务实施】

1. 任务分析

本任务要实现当采集到的温度超过一定值时进行音乐播放提示。

(1) 设备准备。

(2) 添加传感器数据采集库到工程中。

(3) 四输入 ZigBee 串口直连采集数据关键方法说明。

(4) 获取采集到的温度数据。

(5) UI 设计。

(6) 实现温度超过预设值进行音乐播报警示。

2. 任务实现

1) 设备准备

按接线图连接好设备，ZigBee 四输入模块烧写 Sensor Route2.3.hex，ZigBee 协调器烧写 collector2.3.hex，并用 Zigbee组网参数设置V2.0.exe 工具进行组网配置，确保 ZigBee 协调器通过串口正确连接到物联网移动终端的 COM2 口(物联网移动终端的 COM0 口一般用于调试)。

2) 添加传感器数据采集库到工程中

把随书资料中的库文件添加到工程中，其中 armeabi-v7a 文件是自定义的操作串口的.so 文件，在此不做讲解，hardware.jar 和 hardware-sources.jar 是操作物联网工程实训系统的硬件相关类。本任务主要用提供的方法获取 ZigBee 四输入中温度传感器的数据，需要熟悉库中的 Zigbee 类提供的相关方法。

新建工程 Project8_1，把 armeabi-v7a、hardware.jar 和 hardware-sources.jar 复制并粘贴到 libs 目录下，在工程 app 下的 build.gradle 文件的 android 闭包中添加以下代码：

```
compileOptions{
    sourceCompatibility 1.8
    targetCompatibility 1.8
 }
sourceSets {
    main {
        jniLibs.srcDirs = ['libs']
    }
}
```

接着单击“同步”按钮，这样程序就可以使用库中的方法了，如图 8-9 所示。

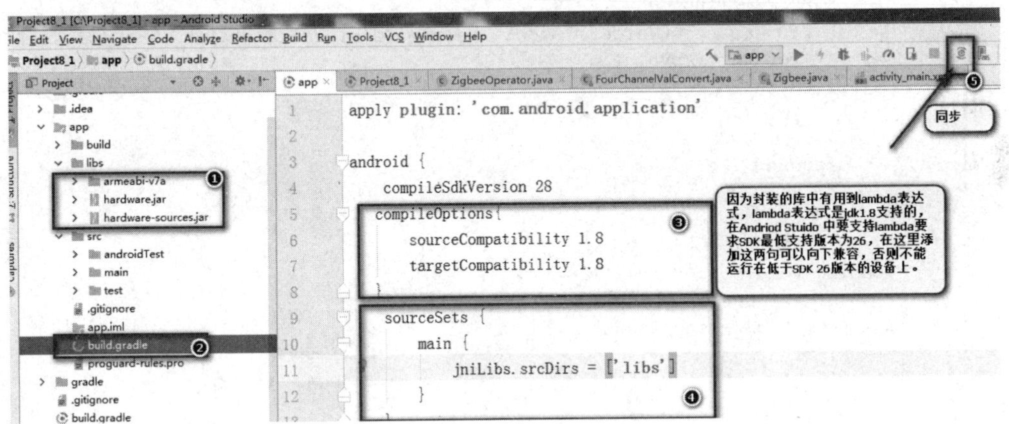

图 8-9　导入库并同步

3)　四输入 ZigBee 串口直连采集数据关键方法说明

```
//初始化
Zigbee zigBee = new Zigbee(DataBusFactory.newSerialDataBus(2, 38400));
//2：串口号， 38400：比特率
//获取四输入数据：
double[] vals = zigBee.getFourEnter();
/*vals 数组分别表示四输入模块从 1 到 4 的端口的电流值，需要根据实际接入情况转换为实际传感值，
使用 FourChannelValConvert 类的静态方法来转换，如温度接在 IN2 口时的转换：
(IN1-IN4 分别对应 vals[0]-vals[3])转换：
*/
public static double getTemperature(vals[1])
//湿度转换(接 IN3 口)：
public static double getHumidity(vals[2])
//噪声转换(接 IN4 口)：
public static double getNoice(vals[3])
//二氧化碳转换(接 IN1 口)：
public static double getCO2(vals[0])
```

4)　获取采集到的温度数据

新建 ZigBeeOperator.java 类，用来获取温度数据，代码如下：

```
public class ZigBeeOperator {
    private ZigBee zigBee;//ZigBee 对象，用于打开串口
    private double[] zigBeeVals;//存放四输入模块采集到的传感器数据
```

```
    private DecimalFormat df;//用于格式化数字

    //打开串口
    public void openPort(int com, int baudRate) {
        zigBee = new ZigBee(DataBusFactory.newSerialDataBus(com, baudRate));
        df = new DecimalFormat("#.##");
    }
    //获取温度
    public Double getTemp() {
        try {
            //获取从串口采集到的四输入模块传感器数据
            zigBeeVals = ZigBee.getFourEnter();
            if(zigBeeVals==null)
                return -1.0;
            //温度接 in2 口
            Double temp = FourChannelValConvert.getTemperature(zigBeeVals[1]);
            //保留两位小数
            Double tempValue = Double.valueOf(df.format(temp));
           return tempValue;
        } catch (Exception e) {
            e.printStackTrace();
        }
        return 0.0;
    }

    //关闭串口，释放资源
    public void closePort() {
        if (zigBee != null) {
            zigBee.stopConnect();
        }
    }
}
```

5) UI 设计

设计布局，在 activity_main.xml 中设置一个文本控件用于显示温度，当温度超过一定值时要播报预警声，所以简单设置一下，不做设置的界面跳转，如图 8-10 所示。

图 8-10　布局设计

布局不再展开，其中，设置温度的点击事件为 android:onClick="setTemp"，设置温度的控件 id 是 android:id="@+id/ed_temp"，显示采集到的温度传感器数据的控件 id 是 android:id="@+id/tv_temp"。

6) 实现温度超过预设值进行音乐播报警示

把声音文件 temp_is_h.wav 文件放到 res/raw 目录下，编程获取温度的最高值，当监测到的温度值超过设置的温度值时，启用媒体播放警报，可以设置一直循环播报。当监测到的温度值低于设置的温度值时，预警播报停止。

```
public class MainActivity extends Activity {
    private TextView tv_temp;
    private EditText ed_temp;//温度设置
    private ZigbeeOperator ZigBeeOperator;
    private double setTempVal = 10;//接收设置的温度值
    private static double data = 35.0;
    private MediaPlayer player;

    @Override
    protected void onCreate(Bundle savedInstanceState) {
        super.onCreate(savedInstanceState);
        setContentView(R.layout.activity_main);
        getId();

        //1.初始化对象
        zigbeeOperator = new ZigbeeOperator();
        //2.打开串口，将四输入模块接入 COM2 口，比特率是 38400
        zigbeeOperator.openPort(2, 38400);

        new Thread(new Runnable() {
            @Override
            public void run() {
                while (true) {
                    showTemp();
                    try {
                        Thread.sleep(1000);
                    } catch (InterruptedException e) {
                        e.printStackTrace();
                    }
                }
            }
        }).start();

    }

    public void showTemp() {
        Double data = zigBeeOperator.getTemp();
        Message msg = new Message();
        msg.what = 0;
        msg.obj = data;
        Log.i("handle data=", data + "");
        handler.sendMessage(msg);
    }

    //3.创建消息对象
    Handler handler = new Handler() {
        public void handleMessage(android.os.Message msg) {
            //4.使用 handleMessage 处理接收到的 msg
            data = (double) msg.obj;
            switch (msg.what) {
                case 0: //温度
```

```java
                    //5.对温度值进行操作
                    tv_temp.setText("温度感应:" + data);
                    // 判断如果大于预设温度值，则打开音乐
                    if (data >= setTempVal) {
                        if (player != null) {
                            //如果正在播放就不再播放
                            if (player.isPlaying())
                                return;
                        } else {
                            //1.实例化播放对象
                            player = MediaPlayer.create(MainActivity.this, R.raw.temp_is_h);
                            //不停地重复播放
                            player.setLooping(true);
                            //2.开始播放
                            player.start();
                        }
                    } else {
                        if (player != null) {
                            player.stop();
                            player = null;
                        }
                    }
                    break;
            }
        };
    };

    private void getId() {
        tv_temp = findViewById(R.id.tv_temp);
        ed_temp = findViewById(R.id.ed_temp);
    }

    //获取设置的温度值
    public void setTemp(View v) {
        String str = ed_temp.getText().toString().trim();
        if (!str.equals("") || str.length() != 0)
            setTempVal = Double.parseDouble(str);
        else
            setTempVal = 0.0;
    }

    @Override
    protected void onDestroy() {
        super.onDestroy();
        zigBeeOperator.closePort();
        if (player != null) {
            player.stop();
            player = null;
        }
    }
}
```

3. 运行结果

运行程序，可以看到当前获取到的温度是 48.12℃，设置温度值为 35℃，预警报警响起，当设置为 50℃时，报警声关闭了，如图 8-11 所示。

图 8-11　运行结果

至此，便完成了智能温控预警系统。完整代码可查阅随书资料中的 Project8_1 工程源代码。

任务 2　推 送 通 知

【任务描述】

扫码观看视频讲解

Notification 俗称通知，在 Android 中使用通知提示消息给用户，它展示在屏幕的顶端，表现为一个图标的形式，当用户向下滑动的时候，展示出通知具体的内容。物联网应用程序可以利用通知向用户推送一些预警信息。任务清单如表 8-2 所示。

本任务要求编程实现通知的发收。

表 8-2　任务清单

任务课时	6 课时	任务组员数量	建议 3 人
任务组采用设备	PC 一台 真机或模拟器或工控平板		

【知识解析】

App 中要能进行通知的收发，需要使用通知管理器类 NotificationManager。通过 NotificationManager.notify()发送通知，通过 NotificationManager.cancel()移除通知。下面就来学习如何进行通知的收发。

1. 通知管理器

通知一般通过通知管理器 NotificationManager 服务发送一个 Notification 对象来完成。NotificationManager 类是一个通知管理器类，这个对象是由系统维护的服务，应用程序可以通过它向系统发送全局的通知。在 Activity 中，可以使用 Activity.getSystemService(String) 方法获取某种服务管理器对象，在这里需要获取的是 NotificationManager 对象，所以直接传递 Context.NOTIFICATION_SERVICE 即可：

```
NotificationManager manager = (NotificationManager)
getSystemService(Context.NOTIFICATION_SERVICE);
```

2. Notification 对象

使用通知的时候，需要创建一个 Notification 对象来承载通知的内容，一般通过内部类 NotificationCompat.Builder 来实例化一个 Builder 对象，并设置通知的各项属性。最后通过 builder()方法得到一个 Notification 对象：

```
  Notification notification = new NotificationCompat.Builder(this, "通知渠道的id")
        .setContentTitle("小猪佩奇")//设置通知的标题
        .setContentText("哪里有美味?")//设置通知的内容
        .setWhen(System.currentTimeMillis())//发送通知的时间
        .setSmallIcon(R.mipmap.ic_launcher)//通知的小图标
        .setLargeIcon(BitmapFactory.decodeResource(getResources(), R.mipmap.ic_launcher_round))//通知的大图标
        .setAutoCancel(true)//是否可以取消通知，true 代表可能取消
        .build();
```

3. 发送通知

当获得 Notification 对象之后，就可以使用 NotificationManager.notify()方法发送通知。发送通知的时候，需要传递一个全局唯一的标识符，也就是通知的 id：

```
manager.notify(通知 id, notification);
```

4. 更新与移除通知

当通知展示在通知栏之后，如何取消呢？Android 提供两种方式移除通知。

(1) Notification 自己维护，使用 setAutoCancel(boolean flag)方法设置是否维护，传递一个 boolean 类型的数据。

(2) 使用 NotificationManager 通知管理器对象来维护，它通过 notify()发送通知的时候，有一个全局唯一的通知标识 id，可以使用 cancel(int id)来移除一个指定的通知，也可以使用 cancelAll()移除所有的通知。

```
manager.cancel(2);//移除 id 为 2 的通知
manager.cancelAll();//移除所有通知
```

5. 通知的单击效果

通常收到通知后会下意识地单击来看内容。但是上面的方式发送通知后，单击是没有任何效果的，如果要单击展开消息内容，还需要设置 PendingIntent。

我们知道 Intent 意图用于启动活动、启动服务、启动广播，它是立即执行某个动作，而 PendingIntent 是一个延迟意图，在某个指定的时间去执行某个动作，所以它是延迟执行的 Intent。

PendingIntent 通过静态方法获取实例对象，可以按照需求延迟启动活动、广播和服务。不管启动什么，方法接收的参数都是相同的，如图 8-12 所示。

参数三是一个 Intent，这里单击通知想看内容，要用一个新的 Activity 来呈现。所以，这个 Intent 是要跳转到新的 Activity 中。

设置好 PendingIntent 后，在构建 Notification 对象时通过 setContentIntent()设置，它的参数就是一个 PendingIntent，这样当用户单击通知后就会执行相应逻辑，如图 8-13 所示。

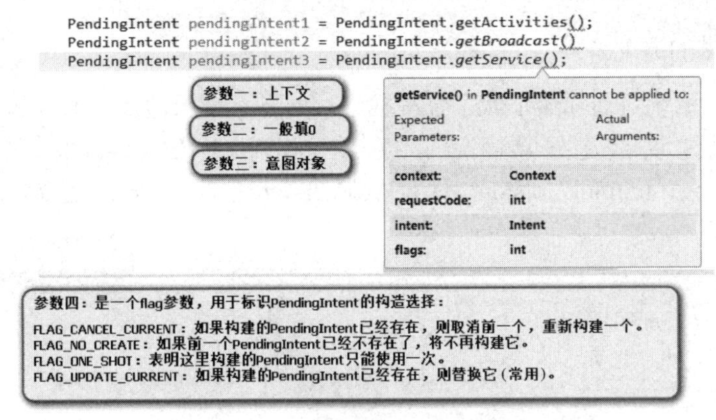

图 8-12　PendingIntent 的参数说明

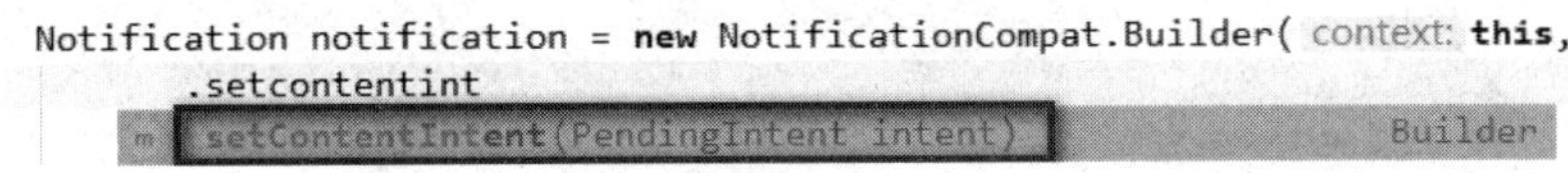

图 8-13　在通知中添加 PendingIntent 参数

6. 通知渠道

上面在构建通知对象时用了 new NotificationCompat.Builder(this, "通知渠道的 id")，这里的参数二是一个通知渠道，什么是通知渠道呢？

每个 App 都可以通过通知发信息给用户，但是通知太多，用户会天天被垃圾推送打扰，忍无可忍就会卸载掉 App。如果用户可以选择接收哪些通知，不理会哪些通知，这样用户对通知有了控制权，比如他可以设置只接收快递消息，不想收到美食通知信息，这时使用通知渠道就可以做这些事情了。

通知渠道就是每条通知都要属于一个对应的渠道。每个 App 可以随意创建通知渠道，但是这些通知渠道由用户掌控。如图 8-14 所示，依次打开“设置”→“应用管理”→“爱奇艺”→“通知管理”，用户可以自己选择这些通知渠道的重要程序，比如可以关闭或打开通知、是否在屏幕上显示、通知的铃声和振动设置、是否闪灯等。

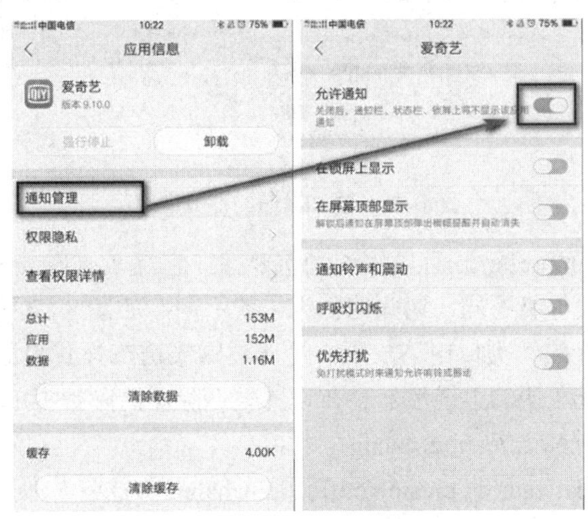

图 8-14　应用中的通知管理

7. 通知渠道的适配

Android 8.0 系统对通知渠道做了强制推广，如果不使用通知渠道，那么 App 通知将不能弹出，所以建议创建好项目后，打开 app 下的 build/gradle 文件检查，确保 compileSdkVersion、minSdkVersion、targetSdkVersion 已经是 26 或更高的版本，如图 8-15 所示。

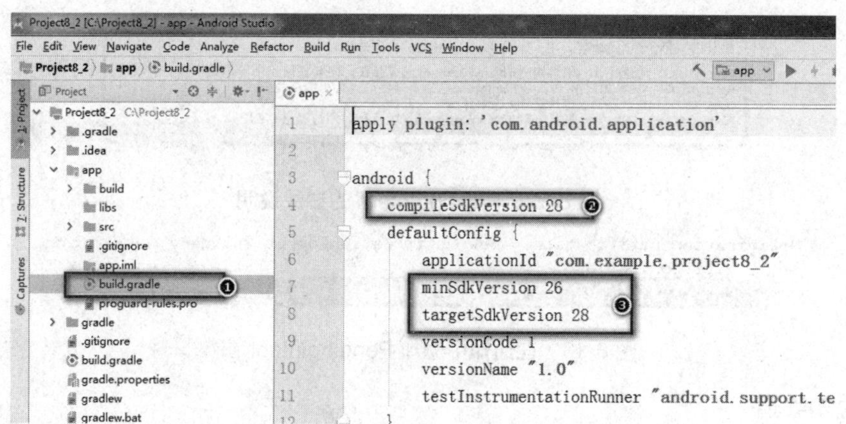

图 8-15　确认应用的版本能使用通知渠道

8. 通知渠道的使用

(1) 初始化通知渠道。

通知渠道用 NotificationChannel 创建，至少需要渠道 id、渠道名称和重要等级这 3 个参数，如图 8-16 所示。

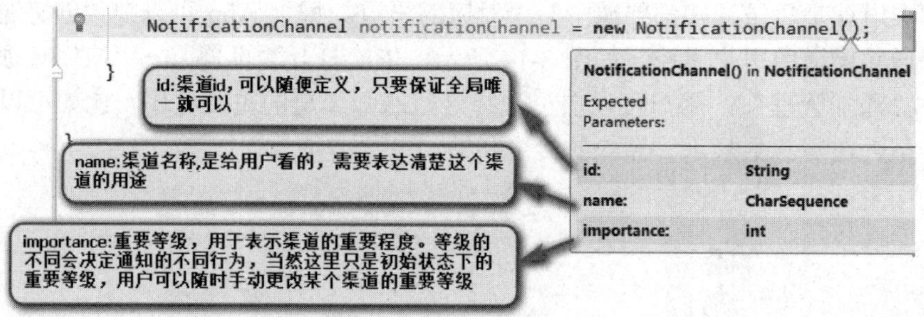

图 8-16　通知渠道的参数说明

参数三(重要等级 importance)有以下 6 种重要性，它是 NotificationManager 类中的静态常量，依次递增。重要等级不同，通知的行为也就不同，如图 8-17 所示。

渠道是向系统注册的，类似申请权限，注册过的渠道在设置里面都会看见，即使代码改变，再次运行，上次注册的渠道依然会在设置里，除非清除数据或者卸载重装应用。

(2) 将渠道添加到 NotificationManager 中。

使用 NotificationManager 的 createNotificationChannel 方法：

```
mNotificationManager.createNotificationChannel(notificationChannel);
```

(3) 构建通知时用有通知渠道的方法，参数二是通知渠道的 id:

```
NotificationCompat.Builder builder= new NotificationCompat.Builder(this,"1");
```

(4) 通知渠道的属性设置，在初始化时一起设置后添加到通知管理器中。

```
// 配置通知渠道的描述信息
notificationChannel .setDescription("渠道的描述");
// 设置通知出现时的闪灯
notificationChannel .enableLights(true);
notificationChannel .setLightColor(Color.GREEN);
// 设置通知出现时的振动(如静止 1 秒，振动 3 秒，静止 1 秒，振动 2 秒)
notificationChannel .enableVibration(true);
notificationChannel .setVibrationPattern(new long[]{1000, 3000, 1000,2000});
```

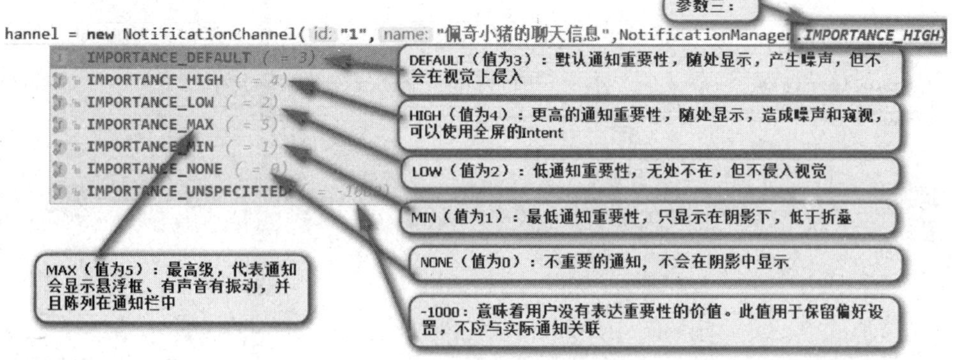

图 8-17 通知渠道的参数三说明

【任务实施】

1. 任务分析

(1) 新建工程 Project8_2，进行 UI 设计。

(2) 发送通知。

(3) 添加通知渠道。

(4) 取消通知。

(5) 实现通知的单击效果。

2. 任务实现

1) UI 设计

新建工程 Project8_2，在 MainActivity 对应的布局中添加两个按钮，用于发送和取消通知。

```
<?xml version="1.0" encoding="utf-8"?>
<LinearLayout xmlns:android="http://schemas.android.com/apk/res/android"
    xmlns:app="http://schemas.android.com/apk/res-auto"
    xmlns:tools="http://schemas.android.com/tools"
    android:layout_width="match_parent"
    android:layout_height="match_parent"
    tools:context=".MainActivity">

```

```
    <Button
        android:onClick="send"
        android:layout_width="0dp"
        android:layout_weight="1"
        android:layout_height="wrap_content"
        android:text="发通知"
        />
    <Button
        android:onClick="cancel"
        android:layout_width="0dp"
        android:layout_weight="1"
        android:layout_height="wrap_content"
        android:text="取消通知"
        />
</LinearLayout>
```

2) 发送通知

```
public class MainActivity extends AppCompatActivity {
private NotificationManager manager;//通知管理器
    @Override
    protected void onCreate(Bundle savedInstanceState) {
        super.onCreate(savedInstanceState);
        setContentView(R.layout.activity_main);
    }
    public void send(View v)
    {
    //1.获取通知管理器的实例对象
     manager = (NotificationManager) getSystemService(NOTIFICATION_SERVICE);
    //2.构建消息对象
        Notification notification=new NotificationCompat.Builder(this)
                .setContentTitle("订阅公众号")
                .setContentText("欢迎大家光临 newland 教育家园")
                .setWhen(System.currentTimeMillis())
                .setSmallIcon(R.drawable.notification)
      .setLargeIcon(BitmapFactory.decodeResource(getResources(),R.mipmap.ic_launcher))
                .build();
      //3.发送通知
         manager.notify(1,notification);
    }
}
```

3) 添加通知渠道

运行程序，发现单击之后没反应，通知并没有发出来，如图 8-18 所示。

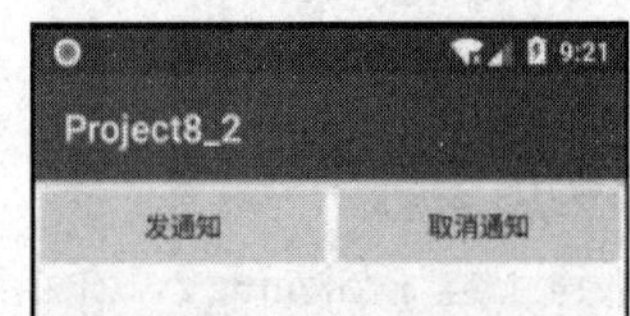

图 8-18 没有通知渠道的通知发送不出去

因为 Android 8.0 后发通知要添加通知渠道，在 MainActivity 中添加通知渠道的 id 和 name，在点击事件中创建通知渠道，并在 builder 的构建中使用通知渠道，示例代码如下：

```
public class MainActivity extends AppCompatActivity {
    private static final String CHINNEL_ID="订阅号";
    private static final String CHINNEL_NAME="物联网教育家园";
private NotificationManager manager;//通知管理器
    @Override
    protected void onCreate(Bundle savedInstanceState) {
```

```
        super.onCreate(savedInstanceState);
        setContentView(R.layout.activity_main);
    }
    @TargetApi(Build.VERSION_CODES.O)
    public void send(View v)
    {
        //1.获取通知管理器的实例对象
        manager = (NotificationManager) getSystemService(NOTIFICATION_SERVICE);

        //2.创建通知渠道
        NotificationChannel channel=new NotificationChannel(CHINNEL_ID,CHINNEL_NAME,NotificationManager.IMPORTANCE_HIGH);
        manager.createNotificationChannel(channel);
          //3.构建时使用参数二添加通知渠道
        Notification notification=new NotificationCompat.Builder(this,CHINNEL_ID)
                .setContentTitle("订阅公众号")
                .setContentText("欢迎大家光临 newland 教育家园")
                .setWhen(System.currentTimeMillis())
                .setSmallIcon(R.drawable.notification)
                .setLargeIcon(BitmapFactory.decodeResource(getResources(),R.mipmap.ic_launcher))
                .build();
       //4.发送通知,通知的 id 为 1
        manager.notify(1,notification);
    }
}
```

运行程序，可以看到单击“发送”按钮后通知可以发出来了。从下拉系统状态栏可以看到通知的详细信息，如图 8-19 所示。

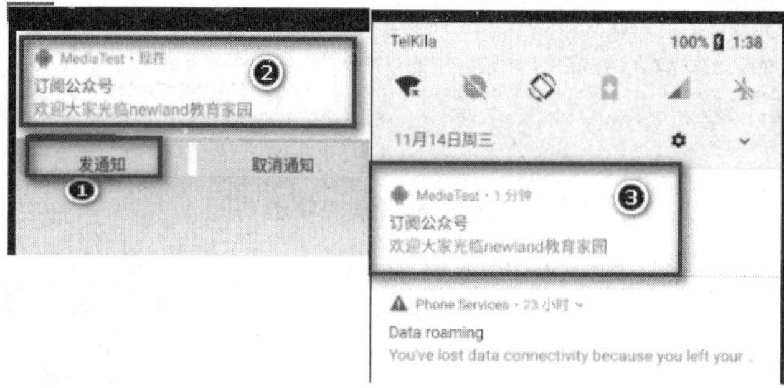

图 8-19　通知的详细信息

4)　取消通知

在布局中添加一个取消通知的按钮，在取消按钮的点击事件中取消通知，因为上面发通知时的通知 id 是 1，所以取消时填写的通知 id 也是 1。

```
布局中添加的按钮:
<Button
    android:onClick="cancel"
    android:layout_width="wrap_content"
    android:layout_height="wrap_content"
    android:text="取消通知"
    />
在 MainActivity 中按钮的点击事件:
public void cancel(View v)
```

```
{
    manager.cancel(1);
}
```

运行程序，测试通知的发送和取消，如图 8-20 所示。

图 8-20　可以单击的通知

5)　实现通知的单击效果

再次发送消息，单击通知后，发现没有任何反应。正常是应该可以单击的，之所以不能单击是因为没有实现通知的单击效果，需要加上 PendingIntent。按图 8-21 的方式添加延期意图。在发送通知的点击事件中加上以下代码，如图 8-21 所示。

```
//添加延期意图
Intent mIntent = new Intent( packageContext: this,MainActivity.class);
//这里需要设置Intent.FLAG_ACTIVITY_NEW_TASK属性，即设置点开通知时用一个新的task打开
mIntent.setFlags(Intent.FLAG_ACTIVITY_NEW_TASK);
PendingIntent contentIntent =PendingIntent.getActivity( context: this, requestCode: 0, mIntent,
        PendingIntent.FLAG_UPDATE_CURRENT);

//3.构建时使用参数二添加通知渠道
Notification notification=new NotificationCompat.Builder( context: this,CHINNEL_ID)
        .setContentTitle("订阅公众号")
        .setContentText("欢迎大家光临newland教育家园")
        .setContentIntent(contentIntent)
        .setWhen(System.currentTimeMillis())
        .setSmallIcon(R.drawable.notification)
        .setLargeIcon(BitmapFactory.decodeResource(getResources(),R.mipmap.ic_launcher))
        .build();
```

构建消息对象时添加延期意图

图 8-21　给通知添加单击功能

运行程序，通知可以单击了，如图 8-22 所示。

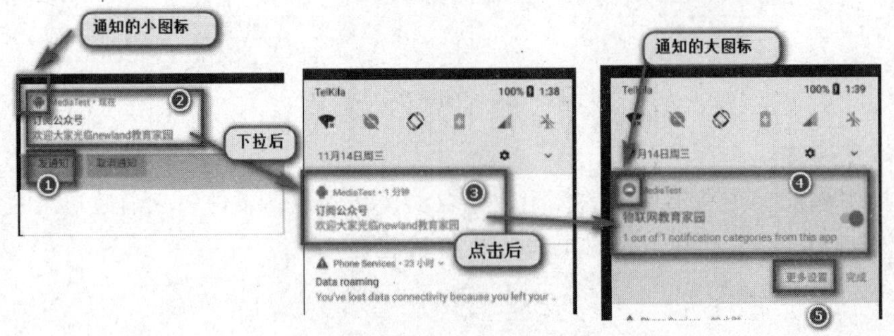

图 8-22　有单击功能的通知

单击图 8-22 中的“更多设置”按钮可以设置通知，如图 8-23 所示。

至此，通知的发送和取消就实现了，完整代码可查阅随书资料中的 Project8_2 工程源代码。

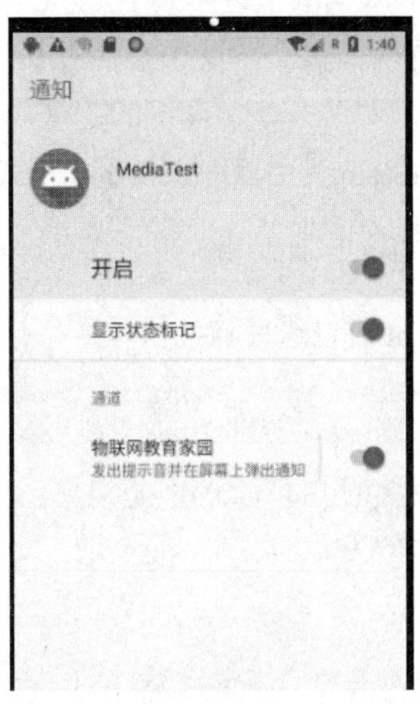

图 8-23 通知的更多设置界面

任务 3 制作圆形头像

【任务描述】

现在很多 App 的个人信息填写时都需要上传头像，头像可以拍照，也可以从现有图库(相册)中获取。本任务要求用户可以调用系统相机进行拍照或者从图库中选择图片，对图片进行裁剪处理后，利用第三方库 CircleImageView 制作成圆形头像。任务清单如表 8-3 所示。

表 8-3 任务清单

任务课时	4 课时	任务组员数量	建议 1 人
任务组采用设备	PC 一台 模拟器或手机或工控平板		

【知识解析】

要制作圆形头像，得先有照片，照片通常通过调用系统照相机进行拍照获得或者从系统图库中取图。在 App 中调用系统现有的应用，需要使用 Intent 指定要开启的应用的 Action，然后通过 startActivity(Intent)或者 startActivityForResult(Intent,int)开启指定的 Activity。

在开启系统相机拍照或从图库取图时，均需要处理图片，所以要使用 startActivityForResult()方法开启。

1. 调用系统照相机的 Intent

Android 系统从现有的相机应用中请求一张图片对应的 Action 如下：

```
android.media.action.IMAGE_CAPTURE
```

从现有的相机应用中请求一段视频对应的 Action 如下：

```
android.media.action.VIDEO_CAPTURE
```

上面两个参数，均在 MediaStore 类中以静态常量的形式定义好了，分别如下：

```
MediaStore.ACTION_IMAGE_CAPTURE(相机)
MediaStore.ACTION_VIDEO_CAPTURE(摄像机)
```

所以，调用系统相机的 Intent 可以这样写：

```
Intent intent = new Intent(MediaStore.ACTION_IMAGE_CAPTURE);
startActivityForResult(intent, CAMERA_REQUEST_CODE);
```

其中，请求码 CAMERA_REQUEST_CODE 是用来在回传结果中区别是调用相机的返回结果用的，它是一个普通整数即可。

2. 获取拍照回传的图片

使用 startActivityForResult() 方法开启系统相机后，需要重写 onActivityResult (int,int,Intent)并在其中处理拍照返回的图片。在方法中先判断请求码 requestCode 是否和调用系统相机时的请求码一致，再判断结果码 resultCode 是不是 RESULT_OK，如果是，再判断返回的数据 data 是否为空，不为空的时候再通过 data. getExtras()取回拍照回来的 Bundle 对象，凭关键字 data 取出基中的 Bitmap 图像，显示在控件上，代码如下：

```
@Override
protected void onActivityResult(int requestCode, int resultCode, Intent data) {
    if(requestCode == CAMERA_REQUEST_CODE)//拍照返回
    {
        if(resultCode==RESULT_OK) {
            if (data == null) {
                return;
            } else {
                Bundle extras = data.getExtras();
                if (extras != null) {
                    Bitmap bm = extras.getParcelable("data");
                    imageView.setImageBitmap(bitmap);// 显示在控件上
                }
            }
        }
    }
}
```

拍照的照片存放在 SD 卡上，需要处理 SD 卡的运行时权限。

3. 调用图库的 Intent

Android 系统中从现有的图库应用中请求一张图片对应的 Action 在 Intent 类中以常量的方式定义好了：public static final String ACTION_GET_CONTENT = "android.intent.action.GET_CONTENT"，所以调用图库的 Intent 如下：

```
Intent intent = new Intent(Intent.ACTION_GET_CONTENT);
intent.setType("image/*");
startActivityForResult(intent, IMAGE_REQUEST_CODE);
```

其中，请求码 IMAGE_REQUEST_CODE 用于返回结果时判断是否是图库取图返回的标志。

4. 获取图库回传的图片

从图库返回时，判断请求码 requestCode 是否和调用系统图库时的请求码一致，再判断结果码 resultCode 是不是 RESULT_OK，如果是，再判断返回的数据 data 是否为空，不为空的时候再通过 data.getData()取回图库回来的 Uri 对象，从 Uri 中获取对应的字节流 InputStream，用 BitmapFactory.decodeStream()把字节流的数据封装成 Bitmap 对象，最后显示在控件上。代码如下：

```
@Override
protected void onActivityResult(int requestCode, int resultCode, Intent data) {
     if(requestCode == IMAGE_REQUEST_CODE)//图库返回
    {
        if(data == null)
        {
            return;
        }
        Uri uri;
        //不裁剪和压缩时
         try {
             uri = data.getData();//得到图片所对应的Content类型的Uri
            //在这里对取回来的数据进行处理
          InputStream is = getContentResolver().openInputStream(uri);
          Bitmap bitmap = BitmapFactory.decodeStream(is);
           imageView.setImageBitmap(bitmap);// 显示在控件上
         }catch(Exception e)
         {

         }
    }
}
```

【任务实施】

1. 任务分析

本任务要完成制作圆形头像的功能。

(1) 创建工程，添加圆形头像的依赖库。

(2) 在布局中使用圆形头像控件。

(3) 自定义对话框的实现。

(4) 调用系统照相机拍照和从图库取图并对图片进行处理。

2. 任务实现

1) 创建工程并添加圆形头像的依赖库

新建工程 Projcet8_3，在工程中添加圆形头像的依赖库，添加完成后可以到工程 app 的 build.gradle 中查看是否添加成功，如图 8-24 和图 8-25 所示。

2) 在布局中使用圆形头像控件

在 res/drawable 目录下添加 add_pic.png 和 user_bd.jpg，如图 8-26 所示。

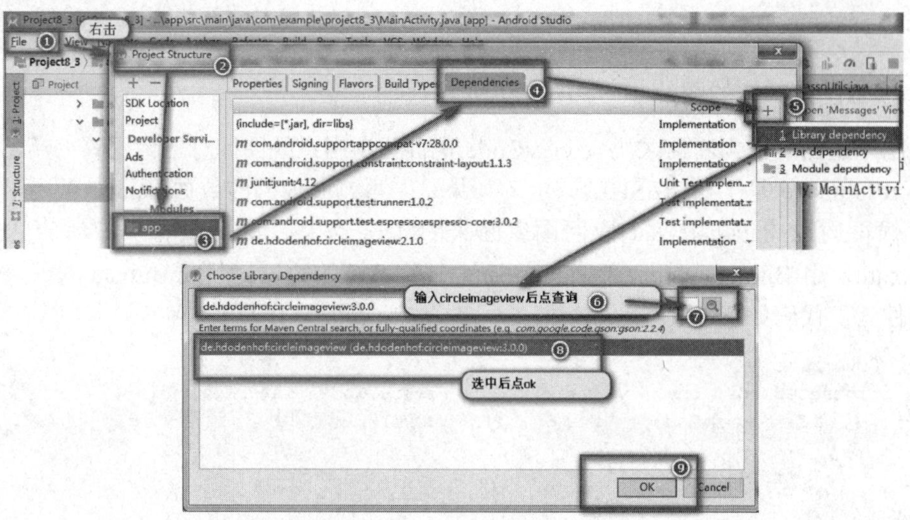

图 8-24　添加圆形头像依赖库

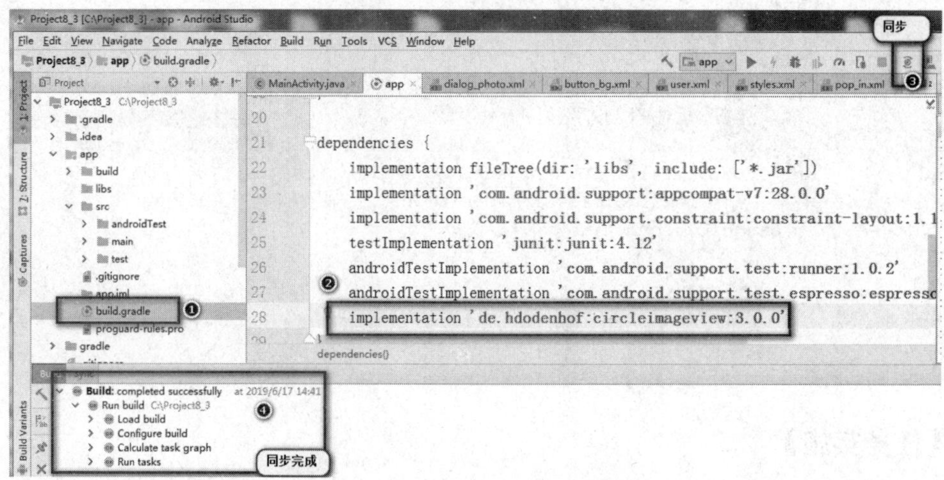

图 8-25　核查依赖库添加情况

在 MainActivity 文件对应的布局 user.xml 中使用圆形头像控件，使用时要写出完整包名 de.hdodenhof.circleimageview.CircleImageView，圆形头像控件中间是一个“+”的图形，后面通过单击它弹出自定义对话框，效果如图 8-27 所示，代码如下：

```
<?xml version="1.0" encoding="utf-8"?>
<LinearLayout xmlns:android="http://schemas.android.com/apk/res/android"
    xmlns:app="http://schemas.android.com/apk/res-auto"
    android:layout_width="match_parent"
    android:layout_height="match_parent"
    android:orientation="vertical">
    <LinearLayout
        android:layout_width="match_parent"
        android:layout_height="200dp"
        android:background="@drawable/user_bg"
        android:gravity="center"
        android:orientation="vertical">

```

```
        <de.hdodenhof.circleimageview.CircleImageView
            android:id="@+id/profile_image"
            android:layout_width="96dp"
            android:layout_height="96dp"
            android:src="@drawable/add_pic"
            app:civ_border_color="@color/colorPrimary"
            app:civ_border_width="2dp" />
    </LinearLayout>
</LinearLayout>
```

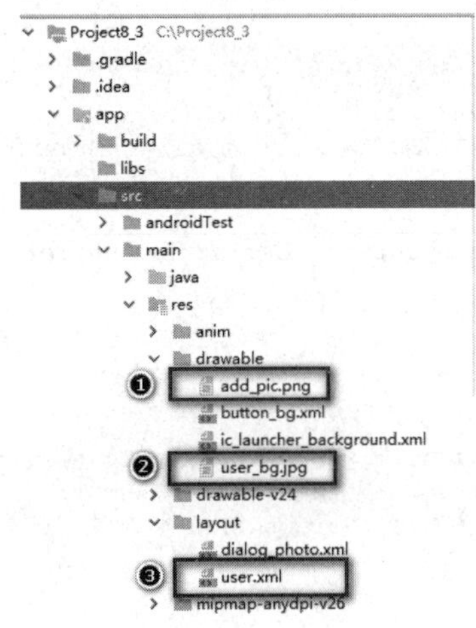

图 8-26　工程文件结构

图 8-27　圆形头像效果

3)　自定义对话框的实现

本任务当单击圆形头像时，会弹出自定义对话框，用于选择是拍照还是从图库中取图，自定义控件在前面的项目中已有所讲解，这里不再展开讲解，读者可以从随书资料中拿到相关资源，整合到本任务中，动画部分可先使用，在任务 4 中会展开讲解，涉及对话框的文件如图 8-28 所示。

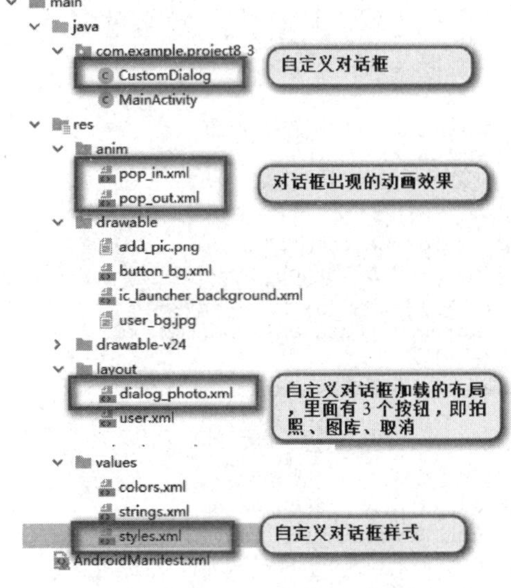

图 8-28　工程文件结构

第一步，在 values/styles.xml 中添加样式，代码如下：

```
<!--自定义 Dialog-->
   <style name="pop_anim_style">
       <item name="android:windowEnterAnimation">@anim/pop_in</item>
       <item name="android:windowExitAnimation">@anim/pop_out</item>
   </style>
```

第二步，编写自定义对话框类，代码如下：

```
public class CustomDialog extends Dialog{
    //定义模板
    public CustomDialog(Context context,int layout,int style) {
        this(context, WindowManager.LayoutParams.MATCH_PARENT,
                WindowManager.LayoutParams.WRAP_CONTENT,layout,style, Gravity.CENTER);
    }
    //定义属性
public CustomDialog(Context context,int width,int height,int layout,int style, int gravity,int anim){
        super(context,style);
        //设置属性
        setContentView(layout);
        Window window = getWindow();
        WindowManager.LayoutParams layoutParams = window.getAttributes();
        layoutParams.width = WindowManager.LayoutParams.MATCH_PARENT;
        layoutParams.height = WindowManager.LayoutParams.WRAP_CONTENT;
        layoutParams.gravity = gravity;
        window.setAttributes(layoutParams);
        window.setWindowAnimations(anim);
    }
    public CustomDialog(Context context,int width,int height,int layout,int style, int gravity){
        this(context,width,height,layout,style,gravity,R.style.pop_anim_style);
    }
}
```

第三步，编写自定义对话框加载的布局，代码如下：

```
<?xml version="1.0" encoding="utf-8"?>
<LinearLayout
    xmlns:android="http://schemas.android.com/apk/res/android"
    android:layout_width="match_parent"
    android:layout_height="match_parent"
    android:gravity="bottom"
    android:orientation="vertical"
    android:padding="10dp">

    <Button
        android:id="@+id/btn_camera"
        android:layout_width="match_parent"
        android:layout_height="wrap_content"
        android:background="@drawable/button_bg"
        android:text="拍照"
        android:textColor="@android:color/white"/>

    <Button
        android:id="@+id/btn_picture"
        android:layout_width="match_parent"
        android:layout_height="wrap_content"
        android:layout_marginBottom="10dp"
        android:layout_marginTop="10dp"
        android:background="@drawable/button_bg"
```

```
        android:text="图库"
        android:textColor="@android:color/white"/>


    <Button
        android:id="@+id/btn_cancel"
        android:layout_width="match_parent"
        android:layout_height="wrap_content"
        android:background="@drawable/button_bg"
        android:text="取消"
        android:textColor="@android:color/white"/>

</LinearLayout>
```

4) 调用系统照相机拍照和从图库取图并对图片进行处理

第一步，让 MainActivity 实现点击事件的接口 View.OnClickListener，在 MainActivity 类中定义控件、请求码和结果码、拍照保存的文件，在 onCreate()方法中查找控件、初始化控件、添加事件处理和处理运行时异常，同时注意要在清单文件中添加访问 SD 卡的权限。代码如下：

```
public class MainActivity extends AppCompatActivity implements View.OnClickListener{

    private CircleImageView profile_image;//圆形头像
    private CustomDialog dialog;//自定义对话框

    private Button btn_camera;//调用系统照相机按钮
    private Button btn_picture;//调用系统图库按钮
    private Button btn_cancel;//取消按钮


    //保存照片的文件和路径
    private File tempFile = null;
    public static final String PHOTO_IMAGE_FILE_NAME = "fileImg.jpg";
    //调用系统照相机的请求码
    public static final int CAMERA_REQUEST_CODE = 1;
    //调用系统图库的请求码
    public static final int IMAGE_REQUEST_CODE = 2;
    //结果码
    public static final int RESULT_REQUEST_CODE = 3;


    @Override
    protected void onCreate(Bundle savedInstanceState) {
        super.onCreate(savedInstanceState);
        setContentView(R.layout.user);
        //1.初始化圆形头像控件并添加事件
        profile_image = (CircleImageView) findViewById(R.id.profile_image);
         profile_image.setOnClickListener(this);
        //2.初始化 dialog
        dialog = new CustomDialog(MainActivity.this, 0, 0,
                R.layout.dialog_photo, R.style.pop_anim_style, Gravity.BOTTOM, 0);

        //3.设置对话框以外单击无效
        dialog.setCancelable(false);
        //4.查找对话框加载布局中的按钮并添加事件
        btn_camera = (Button) dialog.findViewById(R.id.btn_camera);
        btn_camera.setOnClickListener(this);
        btn_picture = (Button) dialog.findViewById(R.id.btn_picture);
        btn_picture.setOnClickListener(this);
```

```
40.         btn_cancel = (Button) dialog.findViewById(R.id.btn_cancel);
41.         btn_cancel.setOnClickListener(this);
42.
43.         //5.请求授权可以访问 SD 卡
44.         checkGrant();
45.     }
46.
47.     @SuppressLint("NewApi")
48.     private void checkGrant() {
49.         //如果用户没有授权则请求授权，且授权码为 1
50.         if (ContextCompat.checkSelfPermission(MainActivity.this, Manifest.permission.WRITE_EXTERNAL_STORAGE) != PackageManager.PERMISSION_GRANTED) {
51.             ActivityCompat.requestPermissions(MainActivity.this, new String[]{Manifest.permission.WRITE_EXTERNAL_STORAGE}, 1);
52.         } else {
53.             Log.i("TAG", "SD 卡可读可写");
54.         }
55.     }
56.     //授权监听的返回结果
57.     @Override
58.     public void onRequestPermissionsResult(int requestCode, String[] permissions, int[] grantResults) {
59.         super.onRequestPermissionsResult(requestCode, permissions, grantResults);
60.         if (requestCode == 1) {
61.             if (grantResults.length > 0 && (grantResults[0] == PackageManager.PERMISSION_GRANTED)) {
62.                 Log.i("TAG", "请求授权成功");
63.             } else {
64.                 Toast.makeText(MainActivity.this, "没有访问 SD 卡权限", Toast.LENGTH_LONG).show();
65.             }
66.         }
67.     }
68. //在此处进行处理
69. }
```

第二步，处理跳转到相机。

在上面的第 68 行代码处添加跳转到相机的处理，代码如下：

```
//跳转到相机
private void toCamera() {
    Intent intent = new Intent(MediaStore.ACTION_IMAGE_CAPTURE);
    //判断内存卡是否可用，可用的话就进行存储
    intent.putExtra(MediaStore.EXTRA_OUTPUT,
            Uri.fromFile(new File(Environment.getExternalStorageDirectory(), PHOTO_IMAGE_FILE_NAME)));
    startActivityForResult(intent, CAMERA_REQUEST_CODE);
    dialog.dismiss();
}
```

第三步，处理跳转到图库。

接着第二步添加跳转到图库的处理，代码如下：

```
//跳转到图库
private void toPicture() {
    Intent intent = new Intent(Intent.ACTION_PICK);
    intent.setType("image/*");
    startActivityForResult(intent, IMAGE_REQUEST_CODE);
    dialog.dismiss();
}
```

第四步，裁剪图片。

添加裁剪功能，调用系统自带的裁剪功能对拍照或从图库取回来的图片进行处理，代码如下：

```
1.  //裁剪图片
2.  private void startPhotoZoom(Uri uri) {
3.      if (uri == null) {
4.         Toast.makeText(MainActivity.this,"取图失败",Toast.LENGTH_LONG).show();
5.          return;
6.      }
7.      Intent intent = new Intent("com.android.camera.action.CROP");
8.      intent.setDataAndType(uri, "image/*");
9.      //设置裁剪
10.     intent.putExtra("crop", "true");
11.     //裁剪宽高比例
12.     intent.putExtra("aspectX", 1);
13.     intent.putExtra("aspectY", 1);
14.     //裁剪图片的质量
15.     intent.putExtra("outputX", 320);
16.     intent.putExtra("outputY", 320);
17.     //发送数据
18.     intent.putExtra("return-data", true);
19.     startActivityForResult(intent, RESULT_REQUEST_CODE);
20. }
```

第 18 行是把裁剪后的数据回传，用第 19 行的请求码 RESULT_REQUEST_CODE 表示是裁剪后的数据返回。

第五步，显示图片。

裁剪回来的数据显示在控件上。代码如下：

```
//设置图片
private void setImageToView(Intent data) {
    Bundle bundle = data.getExtras();
    if (bundle != null) {
        Bitmap bitmap = bundle.getParcelable("data");
        profile_image.setImageBitmap(bitmap);
    }
}
```

第六步，处理拍照、图库、裁剪回来的数据。

重写 onActivityResult 方法，处理拍照、图库、裁剪回来的数据。代码如下：

```
@Override
public void onActivityResult(int requestCode, int resultCode, Intent data) {
    if (resultCode != RESULT_CANCELED) {
        switch (requestCode) {

            case IMAGE_REQUEST_CODE: //相册数据
                startPhotoZoom(data.getData());
                break;

            case CAMERA_REQUEST_CODE:  //相机数据
                tempFile = new File(Environment.getExternalStorageDirectory(), PHOTO_IMAGE_FILE_NAME);
                startPhotoZoom(Uri.fromFile(tempFile));
                break;
            case RESULT_REQUEST_CODE:
                //有可能单击舍弃
```

```
                    if (data != null) {
                        //拿到图片，显示到控件上
                        setImageToView(data);
                        //既然已经设置了图片，就删除保存的文件
                        if (tempFile != null) {
                            tempFile.delete();
                        }
                    }
                    break;
            }
        }
    }
```

第七步，处理点击事件。

在 onClick 中处理跳转事件。代码如下：

```
@Override
public void onClick(View v) {
    switch (v.getId()) {
        case R.id.profile_image:
            dialog.show();//显示对话框
            break;
        case R.id.btn_cancel:
            dialog.dismiss();//取消对话框的显示
            break;
        case R.id.btn_camera:
            toCamera();//跳转到照相机
            break;
        case R.id.btn_picture:
            toPicture();//跳转到图库
            break;
    }
}
```

3. 运行结果

运行程序，拍照取图结果如图 8-29 所示。

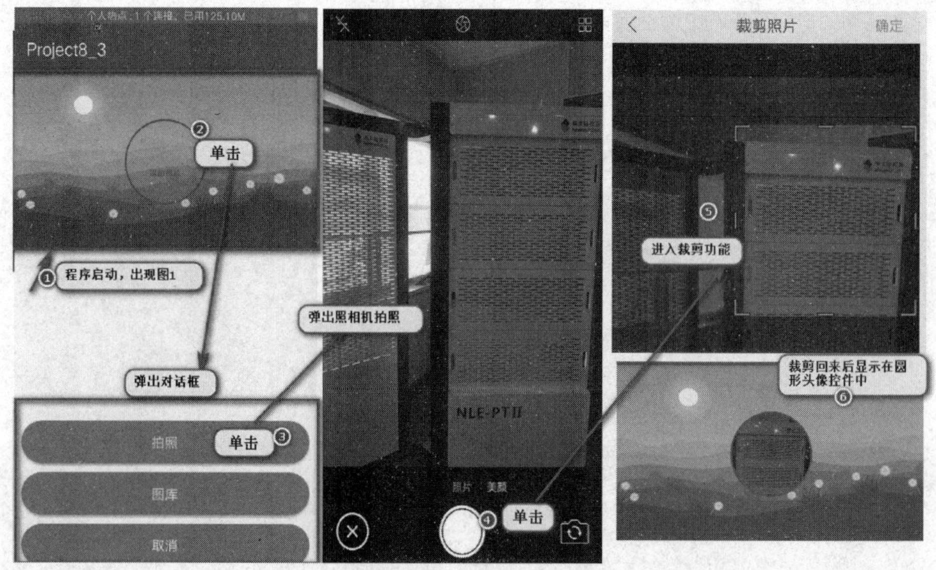

图 8-29　拍照裁剪和圆形头像

单击“图库”按钮后，进入图库选图，选好后会进行裁剪，裁剪完成也显示到了圆形头像控件上，如图 8-30 所示。

至此，圆形头像就实现了，完整代码可查阅随书资源中的 Project8_3 工程源代码。需要注意的是，本任务并没有对头像进行保存和读取，有兴趣的读者可以自行完成该功能。

图 8-30　从图库取图

任务 4　降温风扇的动画实现

【任务描述】

本任务要求在任务 1 的基础上，当监测到温度过高时开启风扇降温，风扇的转动用动画实现。任务清单如表 8-4 所示。

扫码观看视频讲解

表 8-4　任务清单

任务课时	2 课时	任务组员数量	建议 3 人
任务组采用设备	有线温湿度传感器一个 ZigBee、四输入采集器一个 ZigBee 协调器一个 工控平板 PC 一台		

【拓扑图】

本任务拓扑结构图如图 8-31 所示。

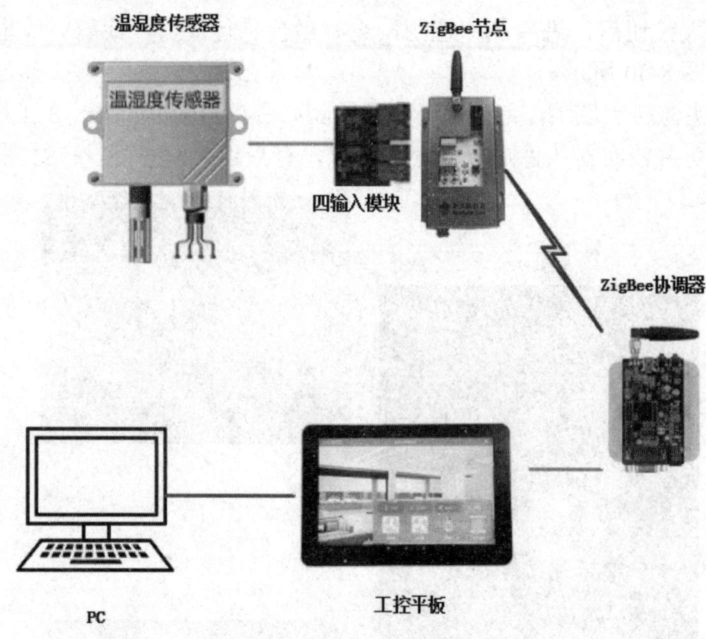

图 8-31 拓扑图

注意：本任务温湿度传感器信号线接 IN2 口。

【知识解析】

Android 的动画有补间动画、逐帧动画和属性动画，属性动画使用的设置比较多，这里只介绍补间动画和逐帧动画。

1. 逐帧动画

(1) 逐帧动画(frame-by-frame animation)的工作原理很简单，就是将一个完整的动画拆分成一张张单独的图片，然后再将它们连贯起来进行播放，类似于动画片的工作原理。

(2) 逐帧动画要使用动画列表文件，该文件只能放在 drawable 目录下。使用时要在 drawable 下新建动画文件 xml，创建动画文件 xml 按图 8-32 所示步骤操作。

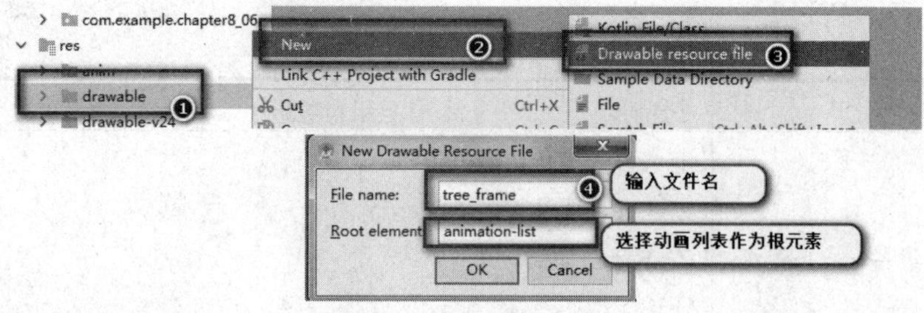

图 8-32 创建动画文件

(3) 把图片放到 minmap-hdpi 目录下，编辑 tree_frame.xml 文件，在列表项中设置图片来源以及每张图片播放的时长，如图 8-33 所示。android:oneshot="true"代表动画只做一次，false 代表动画可以一直持续。

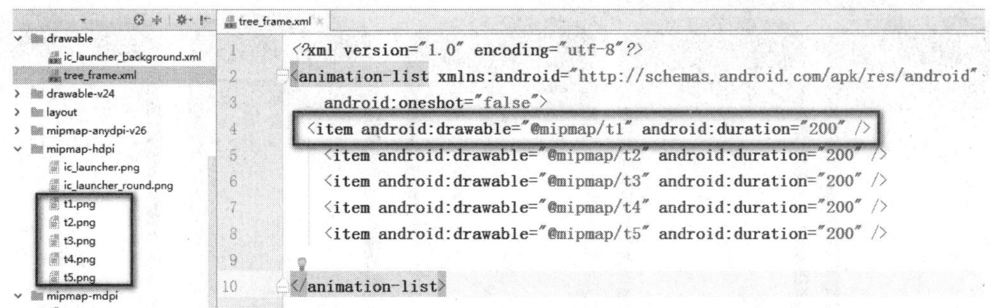

图 8-33　动画列表的编写

(4) 假设要把动画文件 tree_frame.xml 应用在控件 imageFrame 上，代码如下：

```
ImageView imageFrame = findViewById(R.id.image);
//将 Drawable 中的动画文件 xml 设置给 ImageView 控件
imageFrame.setImageResource(R.drawable.tree_frame);
//获取播放动画对象
AnimationDrawable
animation = (AnimationDrawable) imageFrame.getDrawable();
animation.start();//播放动画
animation.stop();//停止动画
```

例 8-4　逐帧动画的使用。

新建工程 Demo8_4，FrameActivity 对应的布局文件为 activity_frame.xml。参考上面的知识解析把 5 张树的成长图片放在 mipmap-hdpi 目录下，在 drawabe 下新建资源文件 tree_frame.xml，设计布局文件(此处不展开)，有两个按钮用于启动和停止动画(按钮的点击事件分别是 begin 和 end)，一个 ImageView 控件用于显示树的生长图片。

在 FrameActivity 中进行动画的启动和停止，代码如下：

```
public class FrameActivity extends AppCompatActivity {
   private ImageView image;
   private AnimationDrawable animation;
    @Override
    protected void onCreate(Bundle savedInstanceState) {
        super.onCreate(savedInstanceState);
        setContentView(R.layout.activity_frame);
        image = findViewById(R.id.image);
        image.setImageResource(R.drawable.tree_frame);
        //获取播放动画对象
        animation =(AnimationDrawable) image.getDrawable();
    }
    public void begin(View v )
    {
        animation.start();
    }

    public void end(View v )
    {
        animation.stop();
    }
}
```

运行程序，单击启动，动画效果如图 8-34 所示，单击停止，动画就停止了。

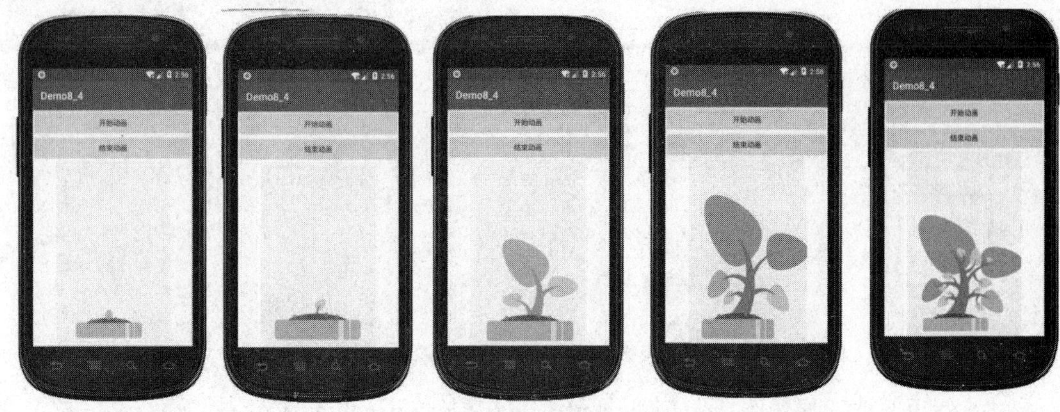

图 8-34 运行结果

2. 补间动画

(1) 补间动画(tweened animation)可以对 View 进行一系列的动画操作，包括淡入淡出、缩放、平移、旋转 4 种操作。相应地由 4 个子类来实现。

- 渐变透明度动画效果：AlphaAnimation。
- 画面转换位置移动动画效果：TranslateAnimation。
- 渐变尺寸伸缩动画效果：ScaleAnimation。
- 画面转移旋转动画效果：RotateAnimation。

比如位移的动画要这样使用：

```
TranslateAnimation animation = new TranslateAnimation(0,400,0,0);//创建动画对象
animation.setDuration(1000);//显示时长
animation.setFillAfter(true);//动画停留在移动后的位置
imageTween.startAnimation(animation); //设置哪个 View 使用动画
animation.start();//启动动画
```

其中，TranslateAnimation(float fromXDelta, float toXDelta, float fromYDelta, float toYDelta)方法的 fromXDelta 和 toXDelta 这两个参数的含义如下。

fromXDelta：动画开始的点离当前 View X 坐标上的差值。举个例子来理解一下，如果这里 fromXDelta=7，表示这个点的 X 轴坐标在自己本身位置的右边 7 个单位。fromXDelta=−7，表示的是这个点在自己本身位置的左边 7 个单位，也就是说，平移动画要从自身向左偏移 7 个单位开始动画。fromXDelta 也可以理解为从哪个位置开始动画，当然偏移量是相对于自身的坐标来说的。

toXDelta：动画结束的点离当前 View X 坐标上的差值。如果 toXDelta=7，表示这个点的位置距离动画物体位置 X 轴右边 7 个单位。这里 toXDelta 也可以理解为参照物在 X 轴方向向左平移 7 个单位。

所以，fromXDelta=−5，toXDelta=5 代表的是，假设有一个物体做 TranslateAnimation(−5, 5, 0, 0)的平移动画。这里表示物体需要自己从 X 轴左边方向 5 个单位移动到自己右边 X 轴方向 5 个单位。反正这个 fromXDelta 和 toXDelta 都是相对自己的距离。

但大部分情况下用 xml 动画文件设置动画属性。

(2) 动画文件放在 res/anim 目录下，anim 目录需要自己创建。

透明度动画文件如图 8-35 所示。

图 8-35　透明度动画文件

位移动画文件如图 8-36 所示。

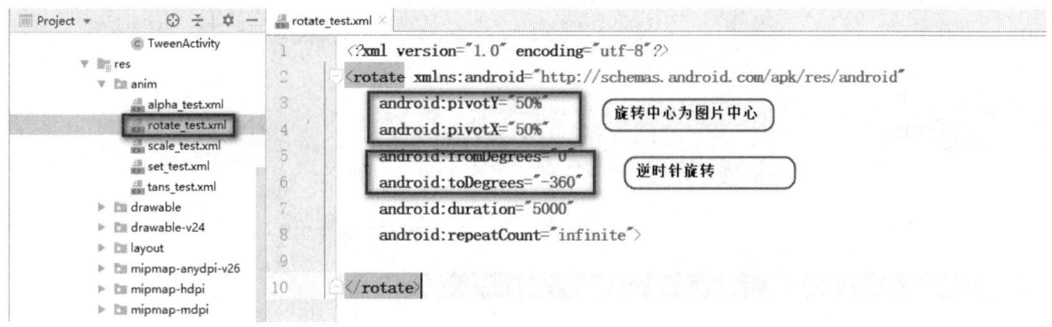

图 8-36　位移动画文件

缩放动画文件如图 8-37 所示。

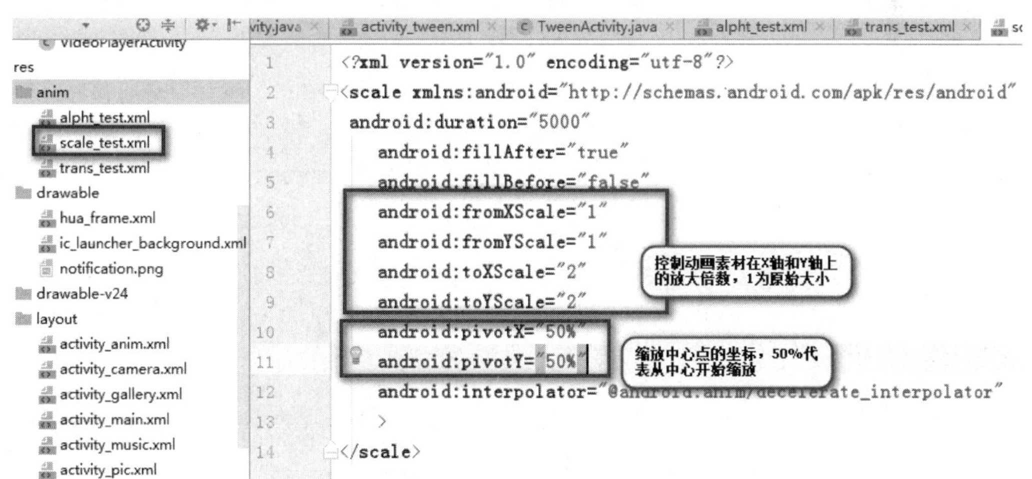

图 8-37　缩放动画文件

旋转动画文件如图 8-38 所示。

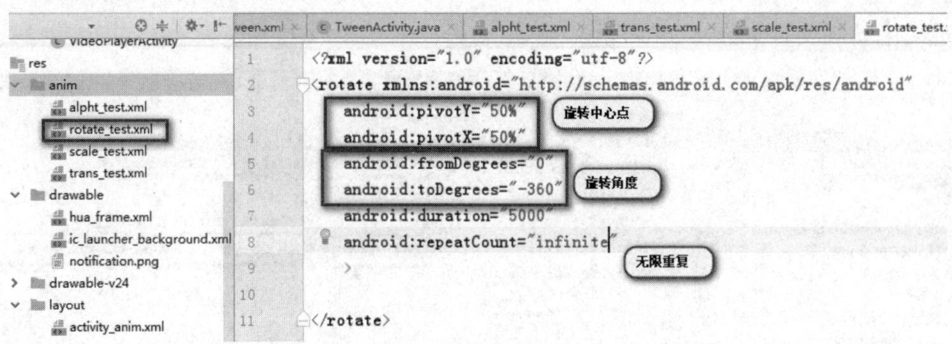

图 8-38　旋转动画文件

把上面的动画文件混合在一起要用 set 动画集，如图 8-39 所示。

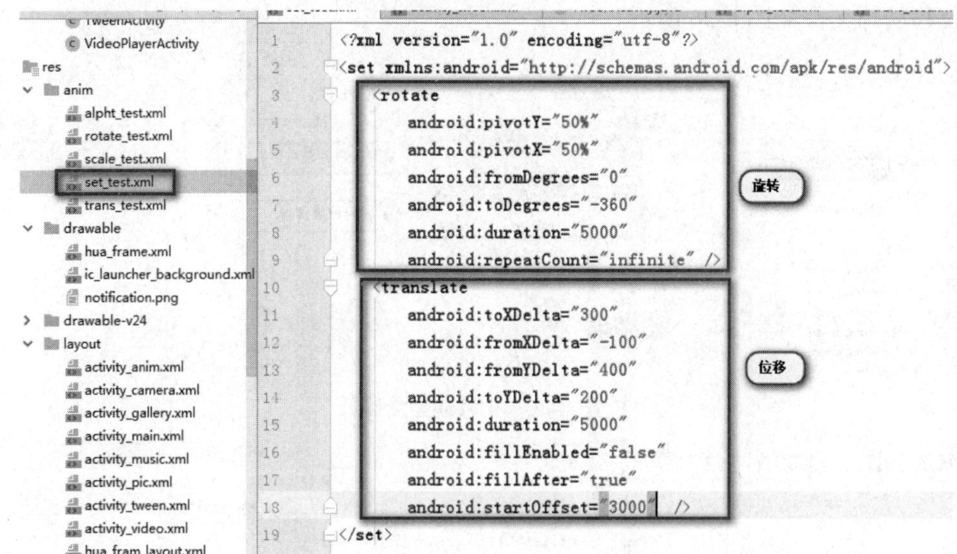

图 8-39　动画合集

从动画文件 xml 中生成 Animation 动画对象需要调用 Animation 类的 loadAnimation()方法，加载动画需要调用 startAnimation()方法，代码如下：

```
ImageView imageTween=findViewById(R.id.imageTween);
Animation anim=
 AnimationUtils.loadAnimation(getApplicationContext(),R.anim.set_test);
imageTween.startAnimation(anim);
```

例 8-5　实现补间动画。

在工程中新建 TweenActivity，对应的布局文件为 activity_tween，布局中有一个按钮用于启动动画，一个 ImageView 用于显示动画图片，默认图选 t5.png。

在 TweenActivity 中用两种方式实现动画，一种是平移的，另一种是上面 set_test.xml 中设置的 set 合集的。代码如下：

```
public class TweenActivity extends Activity implements View.OnClickListener {
    private ImageView imageTween;
    private Button startTween;
    @Override
```

```
    protected void onCreate(Bundle savedInstanceState) {
        super.onCreate(savedInstanceState);
        setContentView(R.layout.activity_tween);
        imageTween = (ImageView) findViewById(R.id.imageTween);
        startTween = (Button) findViewById(R.id.startTween);
        imageTween.setOnClickListener(this);
        startTween.setOnClickListener(this);
    }

    @Override
    public void onClick(View v) {
        switch (v.getId()){
            case R.id.imageTween:
                Toast.makeText(this,"图片移走了还能被点击，这是补间动画的缺陷，可以用属性动画弥补", Toast.LENGTH_SHORT).show();
                break;
            case R.id.startTween:
                 //第一种：平移
              TranslateAnimation animation = new TranslateAnimation(0,400,0,0); //创建动画对象
                //显示时长
                animation.setDuration(1000);
                //动画停留在移动后的位置
                animation.setFillAfter(true);
                imageTween.startAnimation(animation);
                //启动动画
                animation.start();
             //第二种：set 合集
            /*   imageTween=findViewById(R.id.imageTween);
                Animation anim=AnimationUtils.loadAnimation(getApplicationContext(),R.anim.set_test);
                imageTween.startAnimation(anim);
             */
                break;
            default:
                break;
        }
    }
}
```

在清单文件中把 TweenActivity 设置为主 Activity，运行程序，先执行第一种的，结果如图 8-40 左边所示，再执行第二种的，结果如图 8-40 右边所示。

图 8-40　运行结果

【任务实施】

1. 任务分析

本任务要在任务 1 的基础上，当采集到的温度超过预设的最大值时启动风扇转动的动画效果模拟降温；反之停止风扇的动画效果。

(1) 复制任务 1 的工程并重命名为 Project8_4。

(2) 添加动画图片和编写动画文件。

(3) 温度过高时开启动画，否则关闭动画。

2. 任务实现

(1) 复制任务 1 的工程并重命名为 Project8_4。

复制一份 Project8_1 的工程，重命名为 Project8_4，用 Android Studio 打开 Project8_4，使用工具栏 Build 中的 clear project 进行工程清理，接着再用 Build 中的 rebuild project 对工程进行重新构建，并把包名由 com.example.project8_1 改成 com.example.project8_4。在布局文件中添加一个 ImageView，用于显示风扇图片。

(2) 添加动画图片和编写动画文件。

先把风扇图片放在 drawable-hdpi 中，接着在 drawable 中添加动画文件 feng.xml，如图 8-41 所示。

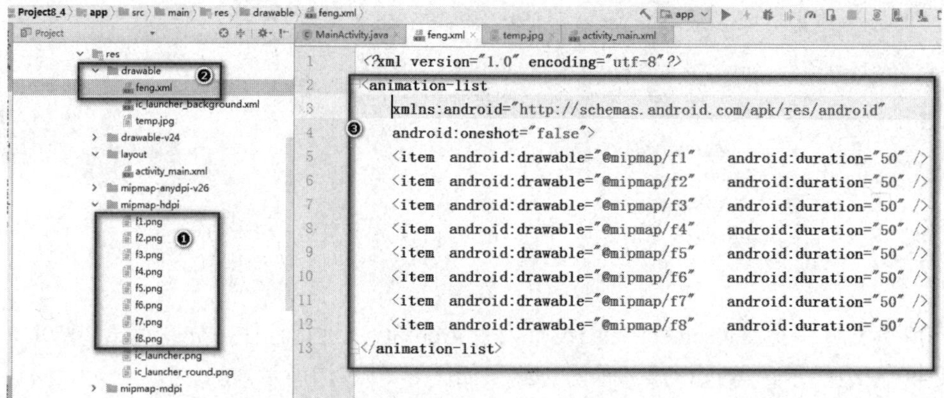

图 8-41 动画列表文件

(3) 温度过高时开启动画，否则关闭动画。

```java
public class MainActivity extends AppCompatActivity {
    private ImageView image;
    private AnimationDrawable animation;

     private TextView tv_temp;
     private EditText ed_temp;//温度设置
     private ZigBeeOperator zigBeeOperator;
     private double setTempVal = 10;//接收设置的温度值
     private static double data = 35.0;
     @Override
     protected void onCreate(Bundle savedInstanceState) {
         super.onCreate(savedInstanceState);
         setContentView(R.layout.activity_main);
```

```
        tv_temp = findViewById(R.id.tv_temp);
        ed_temp = findViewById(R.id.ed_temp);
        image = findViewById(R.id.image);
        //1.将 Drawable 中的动画 xml 文件设置给 ImageView 控件
        image.setImageResource(R.drawable.feng);
        //2.获取播放动画对象
        animation =(AnimationDrawable) image.getDrawable();

        //初始化对象
        zigBeeOperator = new ZigBeeOperator();
        //打开串口 将四输入接入 COM2    接的是 COM 口，比特率是 38400
        zigBeeOperator.openPort(2, 38400);

        //每隔 1 秒取一次数据
        new Thread(new Runnable() {
            @Override
            public void run() {
                while (true) {
                    showTemp();
                    try {
                        Thread.sleep(1000);
                    } catch (InterruptedException e) {
                        e.printStackTrace();
                    }
                }
            }
        }).start();
    }
   //取到数据后把数据封装到消息对象中发到 UI 线程
    public void showTemp() {
        Double data = zigBeeOperator.getTemp();
        Message msg = new Message();
        msg.what = 0;
        msg.obj = data;
        Log.i("handle data=", data + "");
        handler.sendMessage(msg);
    }
    Handler handler = new Handler() {
        public void handleMessage(android.os.Message msg) {
            //使用 handleMessage 处理接收到的 msg
            data = (double) msg.obj;
            switch (msg.what) {
                case 0: //温度
                    //对温度值进行操作
                    tv_temp.setText("温度感应:" + data);
                    // 判断如果大于预设温度值
                    if (data >= setTempVal) {
                        //3.开启动画
                        animation.start();
                    } else {
                        //4.停止动画
                        animation.stop();
                    }
                    break;
            }
        }        ;
    };

    //获取设置的温度值
    public void setTemp(View v) {
```

```
        String str = ed_temp.getText().toString().trim();
        if (!str.equals("") || str.length() != 0)
            setTempVal = Double.parseDouble(str);
        else
            setTempVal = 0.0;
    }
}
```

3. 运行结果

运行程序，结果如图 8-42 所示。当温度为 40℃，超过设定的最大值 25℃时，风扇启动，低于设定的温度值时，风扇停止。

图 8-42　运行结果

项 目 总 结

本项目主要讲解了 Android 中的媒体与动画实现，包括音频播放、视频播放、推送通知、调用系统照相机和从图库取图并利用第三方库制作圆形头像、逐帧动画和补间动画的实现，最后以智能温控预警系统和降温风扇为例讲解了多媒体在物联网应用中的使用。

思考与练习

8-1　简述播放音、视频的流程。

8-2　简述在工程中添加访问 ZigBee 四输入 jar 包的流程。

8-3　简述推送通知的流程。

8-4　简述调用系统相机拍照和从图库取图的流程。

8-5　简述逐帧动画和补间动画的使用流程。

项目9 室内环境采集系统和园区监控系统的实现

【项目描述】

本项目要实现室内环境采集系统和园区监控系统这两个项目。

室内环境采集系统要实时显示室内的温湿度、光照、烟雾数据，能根据温度高低自动判断是否需要开/关通风扇，根据湿度大小自动判断是否需要开/关排气扇，根据光照的强弱自动判断是否需要开/关照明灯，根据有无烟雾自动判断是否需要开/关三色灯-橙灯。有烟的时候发出警报提示，微动开关被触发时打开三色灯-绿灯，关闭三色灯-橙灯。

园区监控系统当红外对射传感器检测到人员由外进入园区时，三色灯-红灯亮。当开启监控的时候，打开摄像头，此时可以控制摄像头上、下、左、右转动，检测到有人则拍照并存储，可查看照片。微动开关被触发时打开三色灯-绿灯，关闭三色灯-红灯。

本项目通过对两个任务(图 9-1)的学习，使读者具备一定的综合项目开发能力。

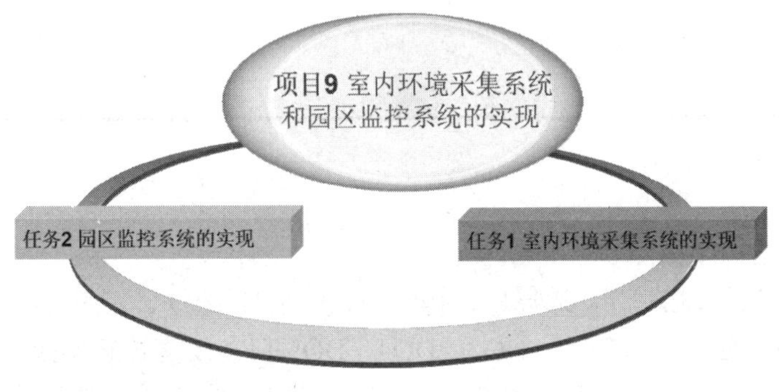

图 9-1　项目 9 任务图

【学习目标】

技能目标：

- 能综合应用数据存储技术。
- 能综合应用线程与消息机制。
- 能综合应用广播与服务。
- 能综合应用媒体动画。

任务 1 室内环境采集系统的实现

【任务描述】

实现温湿度、光照、烟雾数据的实时获取与显示。如果温度高于设定的阈值，开通风扇；如果湿度高于设定的阈值，开排气扇；如果光照高于设定的阈值，打开照明灯；如果有烟雾，三色灯-橙灯亮。任务清单如表 9-1 所示。

表 9-1 任务清单

任务课时	2 课时	任务组员数量	建议 3 人
任务组采用设备	ZigBee、四输入模块一个 ZigBee 协调器一个 温湿度传感器一个 光照传感器一个 烟雾传感器一个 ADAM-4150 模块一个 微动开关一个 三色灯一个 风扇两个 LED 灯一个 RS485 转 RS232 转接头一个 PC 一台 工控平板 公母串口线两条		

【拓扑图】

本任务拓扑结构图如图 9-2 所示。

注意：本任务的 LED 灯、通风扇、排气扇、三色灯-绿灯、三色灯-橙灯所接继电器分别接 ADAM-4150 的 DO0、DO1、DO2、DO4、DO5 口，烟雾传感器接 ADAM-4150 模块的 DI4 口，微动开关接 ADAM-4150 的 DI5 口，温湿度传感器、光照传感器分别接四输入模块的 IN2、IN3、IN1 口。

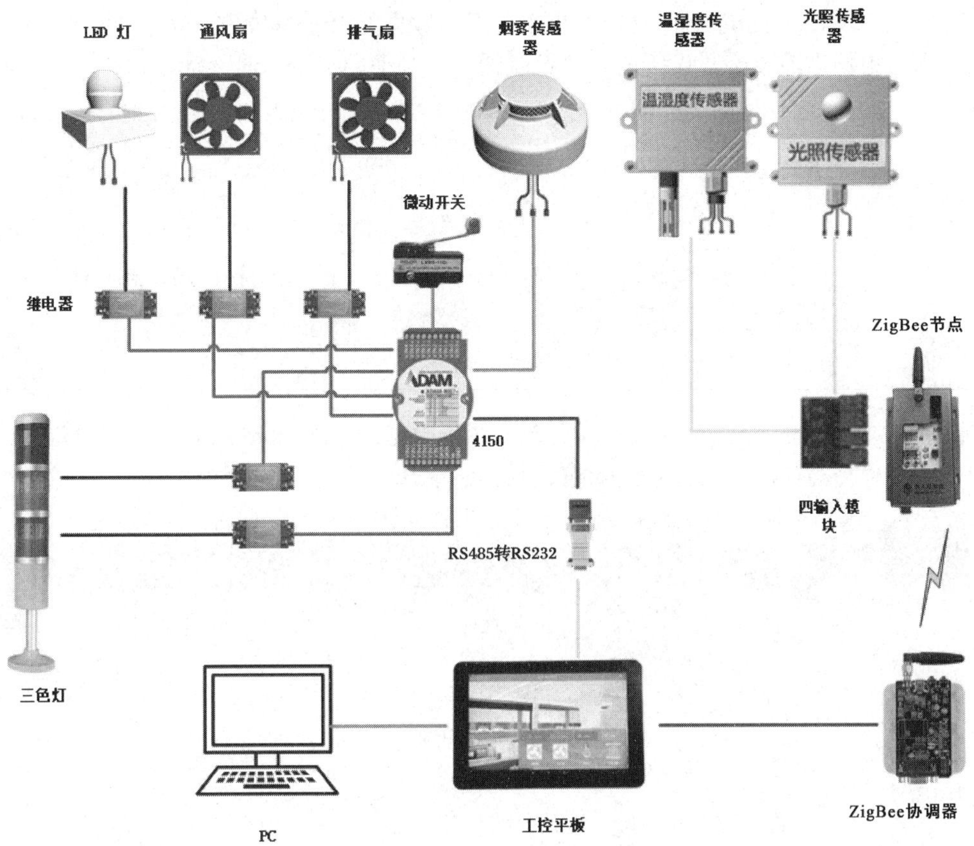

图 9-2　拓扑图

【任务实施】

1. 任务分析

本项目在 Project9_1Init 的基础上实现。初始项目 Project9_1Init 是项目 7 任务 2 完成后的代码。

(1)　在 EnvironmentFragment.java 中添加根据当前温度与设置的阈值控制通风扇开关、根据当前湿度与设置的阈值控制排气扇开关，以及根据当前光照与设置的阈值控制照明灯开关的代码。

(2)　新建 raw 文件夹，把报警提示音文件 alarm.wav 放到 raw 文件夹下。

(3)　创建通知工具类 NotificationUtils.java，设置通知的报警音乐，单击通知后通知自动消失，跳转到动作为 home、类别为 android.intent.category.DEFAULT 的页面中。

(4)　在 AndroidManifest.xml 文件的 HomeActivity 声明代码中，配置过滤器，动作为 home，类别为 android.intent.category.DEFAULT。

(5)　在服务中添加获取烟雾的业务代码，如果有烟雾则发送通知。

说明：Android 8.0 系统对通知渠道做了强制推广，如果不使用通知渠道，那么你的 App 通知将不能弹出。工控平板的 SDK 版本是 Android 5.1.1，所以项目中并没有用上通知渠道。

2. 任务实现

(1) 在 EnvironmentFragment.java 更新传感器数据的 Runnable 的 run()方法 TODO 1 的位置，添加根据当前温度与设置的阈值控制通风扇开关的代码，具体如下：

```
// TODO: 1 如果当前温度不小于温度阈值，并且通风扇没打开，打开通风扇并设置风扇状态为 true；
// 否则如果当前温度小于温度阈值，并且通风扇打开了，关闭通风扇并设置风扇状态为 false
            if(tempData>=sharePreferenceUtil.getTemp()&&!istempFanOpen){
                tempAnim.start();
                adam4150.openRelay(Constants.RELAYPORTS[0]);
                istempFanOpen=true;
            }else if(tempData<sharePreferenceUtil.getTemp()&&istempFanOpen){
                tempAnim.stop();
                adam4150.closeRelay(Constants.RELAYPORTS[0]);
                istempFanOpen=false;
            }
```

(2) 在 EnvironmentFragment.java 更新传感器数据的 Runnable 的 run()方法 TODO 2 的位置，添加根据当前湿度与设置的阈值控制排气扇开关的代码，具体如下：

```
// TODO: 2 如果当前湿度不小于湿度阈值，并且排气扇没打开，打开排气扇并设置风扇状态为 true；
// 否则如果当前湿度小于湿度阈值，并且排气扇打开了，关闭排气扇并设置风扇状态为 false
            if(humiData>=sharePreferenceUtil.getHumi()&&!isHumiFanOpen){
                humiAnim.start();
                adam4150.openRelay(Constants.RELAYPORTS[1]);
                isHumiFanOpen=true;
            }else if(humiData<sharePreferenceUtil.getHumi()&&isHumiFanOpen){
                humiAnim.stop();
                adam4150.closeRelay(Constants.RELAYPORTS[1]);
                isHumiFanOpen=false;

            }
```

(3) 在 EnvironmentFragment.java 更新传感器数据的 Runnable 的 run()方法 TODO 3 的位置，添加根据当前光照与设置的阈值控制照明灯开关的代码，具体如下：

```
// TODO: 3 如果当前光照不大于光照阈值，并且照明灯
// 没打开，打开照明灯并设置照明灯状态为 true；否则
// 如果当前光照大于光照阈值，并且照明灯打开了，关闭
// 照明灯并设置照明灯状态为 false
            if(lightData<=
sharePreferenceUtil.getLight()&&!isLampOpen){
                cbLamp.setChecked(true);
                adam4150.openRelay
(Constants.RELAYPORTS[2]);
                isLampOpen=true;

            }else if(lightData>
sharePreferenceUtil.getLight()&&isLampOpen){
                cbLamp.setChecked(false);
                adam4150.closeRelay
(Constants.RELAYPORTS[2]);
                isLampOpen=false;
            }
```

(4) 在 app→src→res 文件夹下新建 raw 文件夹，把报警提示音文件 alarm.wav 放到 raw 文件夹下，如图 9-3 所示。

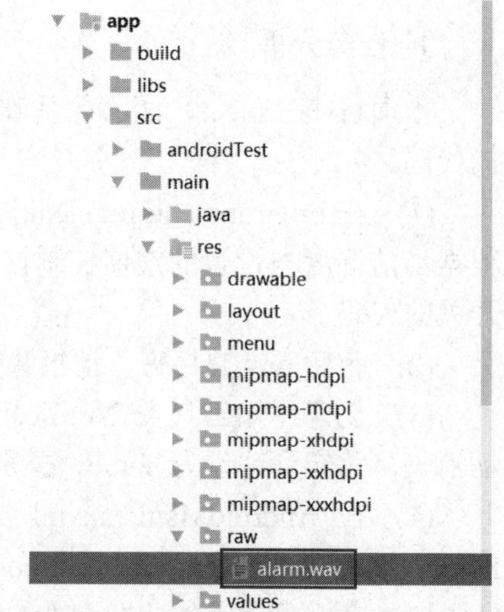

图 9-3 Project_1Init 代码目录

(5) 在 com.newland.smartpark.utils 包下创建通知工具类 NotificationUtils.java，设置通知的报警音乐，单击通知后通知自动消失，跳转到动作为 home、类别为 android.intent.category.DEFAULT 的页面中，代码如下：

```
public class NotificationUtils {

    private final Bitmap bitmap;
    public NotificationUtils(Context context, String content) {

        //点击事件真正想执行的动作，指定单击通知后跳转到activity
        Intent intent = new Intent();
        intent.setAction("home");
        intent.addCategory("android.intent.category.DEFAULT");
        bitmap = BitmapFactory.decodeResource(context.getResources(), R.mipmap.lamp_alarm_on);
        //当单击消息时就会向系统发送contentIntent意图
        PendingIntent contentIntent = PendingIntent.getActivity(context, 0, intent, PendingIntent.FLAG_UPDATE_CURRENT);
        //创建通知
        Notification notification = new NotificationCompat.Builder(context).setContentIntent(contentIntent).setContentTitle("室内烟雾报警").setLargeIcon(bitmap).setSmallIcon(R.mipmap.icon_smartpark).setContentText(content).build();

        //设置报警音乐
        notification.sound = Uri.parse("android.resource://" + context.getPackageName() + "/" + R.raw.alarm);

        //设置用户单击通知后，通知自动消失
        notification.flags = Notification.FLAG_AUTO_CANCEL;
        NotificationManager notificationManager = (NotificationManager) context.getSystemService(Context.NOTIFICATION_SERVICE);
        notificationManager.notify(0, notification);
        ;
    }
}
```

(6) 在 AndroidManifest.xml 文件的 HomeActivity 声明的代码中，配置过滤器，动作为 home，类别为 android.intent.category.DEFAULT，代码如下：

```
<!--配置过滤器，单击通知栏消息通知，可以跳入主界面-->
        <intent-filter>
            <action android:name="home" />
            <category android:name="android.intent.category.DEFAULT" />
        </intent-filter>
```

这样，单击通知栏的消息通知就可以跳转到 HomeActivity 了。

(7) 如果有烟雾发送通知的代码，在 MyService.xml 文件的 onCreate 方法的线程中 TODO 4 的位置添加获取烟雾数据，代码如下：

```
// TODO 4: 获取烟雾数据，如果有烟雾发送通知
        //获取烟雾数据
        String smoke = adam4150.getSmoke(Constants.DIPORTS[0]);
        //如果有烟，发送通知
        if (smoke == "有烟") {
            //发送通知
            NotificationUtils noti = new NotificationUtils(MyService.this, "烟雾警报");
        }
```

至此，便完成了室内环境采集系统。完整代码可查阅随书资料中的 Project9_1 工程源代码。

3. 运行结果

运行程序，设置温湿度光照阈值分别为 28℃、48%RH、3024lx，结果如图 9-4 所示。室内环境采集系统效果如图 9-5 所示，温湿度均高于设置的阈值，光照低于光照阈值，所以两个风扇和灯泡都开启了，检测到有烟雾，因此三色灯-橙灯亮，发送通知栏警报提示，如图 9-6 所示。

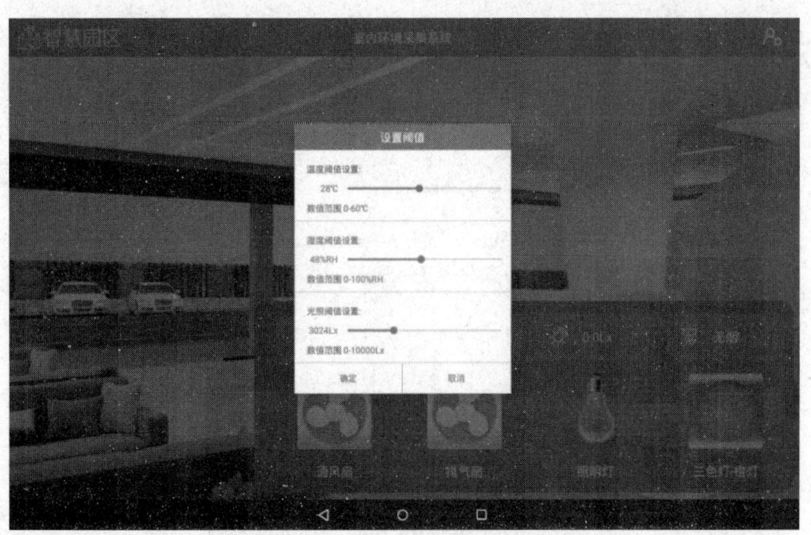

图 9-4　设置温湿度光照阈值

图 9-5　室内环境采集系统

图 9-6 通知栏室内烟雾警报消息通知

任务 2 园区监控系统的实现

【任务描述】

本任务在初始代码 Project9_2Init 的基础上实现，Project9_2Init 已经实现了“设置摄像头地址”和“查看截图”的页面及功能，以及有人时开启三色灯-红灯，没人时关闭红灯，微动开关被触发时关闭红灯。本任务要实现监控的开启和关闭以及控制摄像头上、下、左、右转动的功能，监控开启的时候如果有人则拍照，并实现单击“设置摄像头地址”和“查看截图”菜单项的时候弹出相应的对话框。效果如图 9-7 至图 9-9 所示。任务清单如表 9-2 所示。

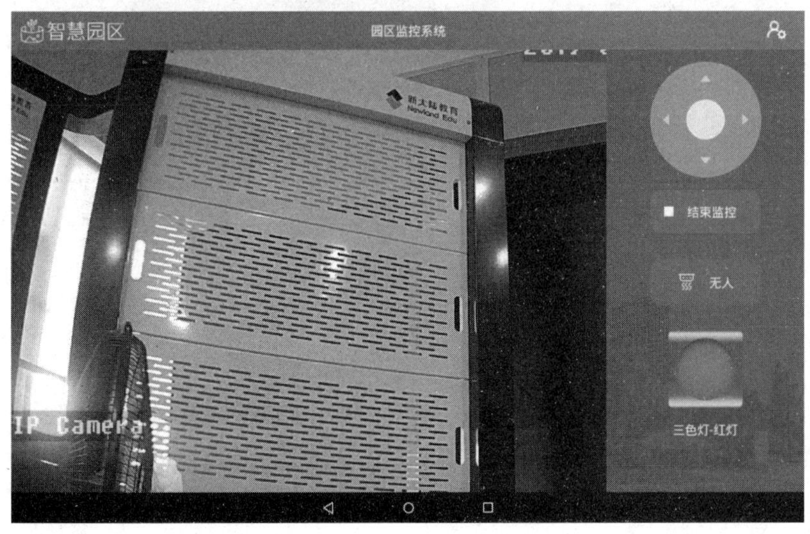

图 9-7 园区监控页面效果图

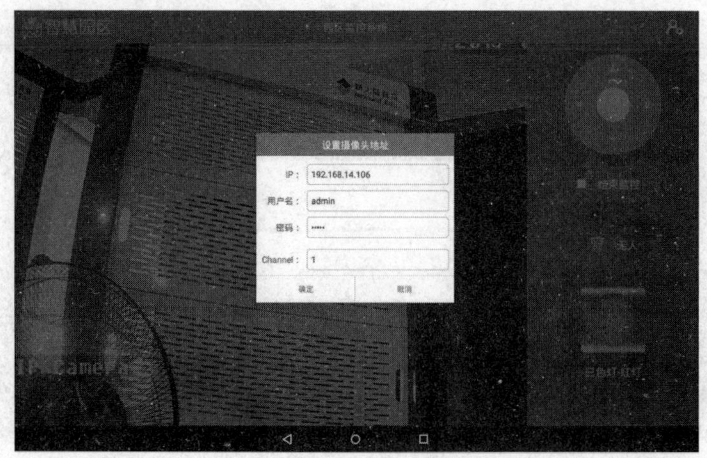

图 9-8　摄像头地址设置界面

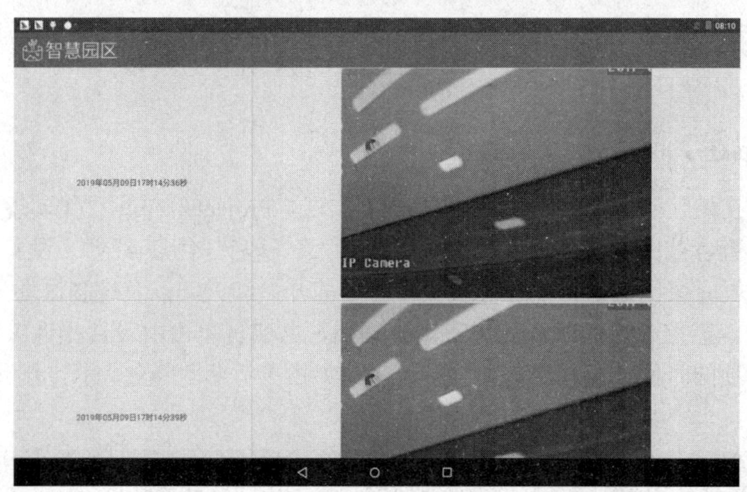

图 9-9　截图查看界面

表 9-2　任务清单

任务课时	2 课时	任务组员数量	建议 1 人
任务组采用设备	ADAM-4150 模块一个 微动开关一个 三色灯一个 风扇两个 红外对射传感器一对 RS485 转 RS232 转接头一个 PC 一台 工控平板 网络摄像头一个 路由器一个 公母串口线一条		

【拓扑图】

本任务拓扑结构图如图 9-10 所示。

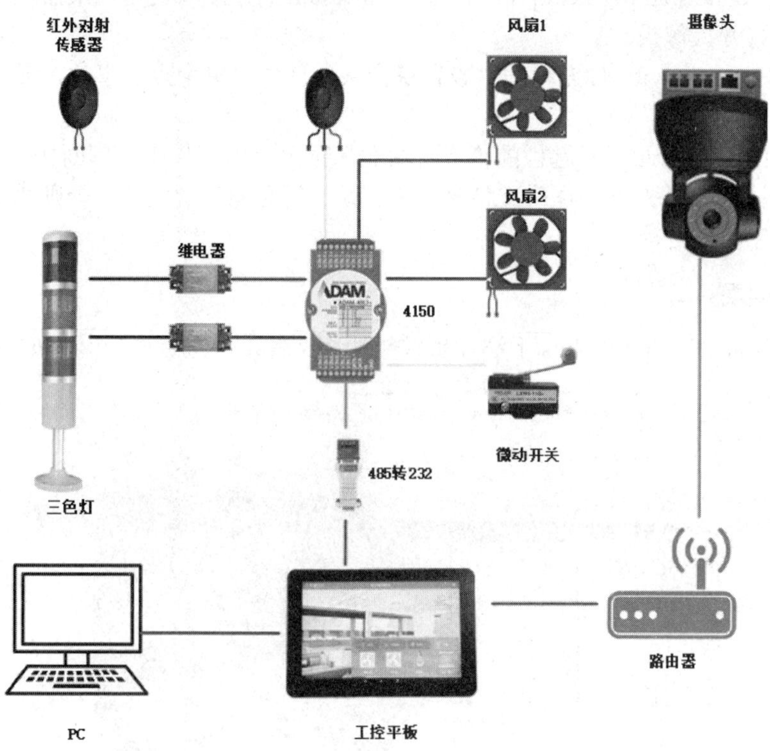

图 9-10　拓扑图

注意：本任务的三色灯-红灯、三色灯-绿灯所接继电器分别接 ADAM-4150 的 DO3、DO4，红外对射传感器、烟雾传感器、微动开关分别接 ADAM-4150 模块的 DI2、DI4、DI5 口，温湿度传感器、光照传感器分别接四输入模块的 IN2、IN3、IN1 口。

【任务实施】

1. 任务分析

本项目在 Project9_2Init 的基础上实现。

(1) 在 Project9_2Init 的 libs 下添加摄像头的库文件，并进行同步。

(2) 在清单文件中添加访问网络权限。

(3) 在 MonitoringFragment.java 中判断摄像头地址信息是否为空，为空则弹出提示并返回；否则打开摄像头。

(4) 在 MonitoringFragment.java 中添加“开启监控”ToggleButton 的状态监听，如果是“开启监控”状态，打开摄像头；如果摄像头不为空，释放摄像头。

(5) 在 MonitoringFragment.java 的控制摄像头上、下、左、右转动按钮的触摸事件中，判断如果监控未开启，弹出开启监控的提示，返回 false；如果监控已开启，移动按钮或手抬起时控制摄像头转动。

(6) 在 MonitoringFragment.java 中，如果有人，判断摄像头地址信息是否为空，为空弹出提示，不为空且监控是开启的，则拍照。

(7) 在 MonitoringFragment.java 中添加 onResume()方法，判断如果监控按钮是开启状态，0.3 秒后打开摄像头。

(8) 在 HomeActivity 标题栏的“设置摄像头地址”菜单中点击事件，显示设置摄像头地址菜单对话框。

(9) 在 HomeActivity 标题栏的“查看截图”菜单点击事件中，判断如果手机内置 SD 卡，根目录下 pic 文件夹不存在或文件夹为空，弹出暂无图片的提示；否则跳转到显示截图页面。

2. 任务实现

(1) 在 Project9_2Init 的 libs 下添加摄像头的库文件①，进行同步②，如图 9-11 所示。

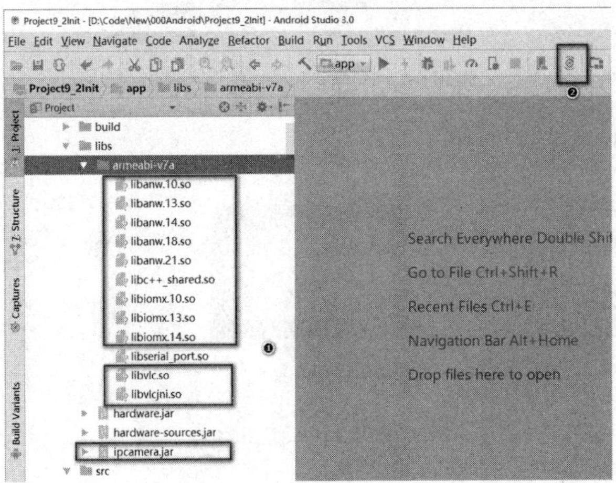

图 9-11　添加库文件并同步

(2) 要访问摄像头，所以要在清单文件中添加访问网络的权限：

```
<uses-permission android:name="android.permission.INTERNET" />
```

(3) 在 MonitoringFragment.java TODO 1 的位置添加名为 openCameraRunnable 的 Runnable，判断摄像头地址信息是否为空，为空弹出提示并返回；否则打开摄像头，代码如下：

```
// TODO: 1  添加名为 openCameraRunnable 的 Runnable，判断摄像头地址信息是否为空，为空弹出
// 提示并返回，否则打开摄像头
    Runnable openCameraRunnable = new Runnable() {
        @Override
        public void run() {
            //获取 Sharepreferences 中保存的摄像头地址信息
            getCameraData();
            if (ip.isEmpty() || userName.isEmpty() || pwd.isEmpty() ||
channel.isEmpty()) {
                Toast.makeText(getActivity(), "请单击右上角图标设置摄像头的地址信息",
Toast.LENGTH_SHORT).show();
                return;
            }
            cameraManager = CameraManager.getInstance();
```

```
            cameraManager.setupInfo(textureView, userName, pwd, ip, channel);
            cameraManager.openCamera();
        }
    };
```

(4) 在 MonitoringFragment.java 的 onActivityCreated()方法的 TODO 2 中，添加“开启监控”ToggleButton 的状态监听，如果是“开启监控”状态，则打开摄像头；如果摄像头不为空，释放摄像头，代码如下：

```
// TODO: 2 添加"开启监控"ToggleButton 的状态监听，如果是“开启监控”状态，打开摄像头；// 如果摄像头不为空，释放摄像头
    tbOpenOrClose.setOnCheckedChangeListener(new CompoundButton.OnCheckedChangeListener() {
        @Override
        public void onCheckedChanged(CompoundButton compoundButton, boolean open) {
            if (open) {
                //打开摄像头
                handler.post(openCameraRunnable);
            } else {
                if (cameraManager != null) {
                    //释放摄像头
                    cameraManager.releaseCamera();
                }
            }
        }
    });
```

(5) 在 MonitoringFragment.java 的控制摄像头上、下、左、右转动的按钮触摸事件中的 onTouch()方法的 TODO3 中添加代码，判断如果监控未开启，弹出开启监控的提示，返回 false；如果监控已开启，移动按钮或手抬起时控制摄像头转动。代码如下：

```
// TODO: 3 在控制摄像头上、下、左、右转动的按钮的触摸事件中判断：如果监控未开启，弹出开启监控的提示，返回 false；如果监控已开启，移动按钮或手抬起时控制摄像头转动
        try {
            int action = arg1.getAction();
            PTZ ptz = null;
            if (!tbOpenOrClose.isChecked()) {
                if (action == MotionEvent.ACTION_UP) {//如果未开始监听，手抬起时弹出提示
                    Toast.makeText(getActivity(), "请先开启监控", Toast.LENGTH_SHORT).show();
                    return false;
                }
            } else {
                //移动按钮或手抬起时控制摄像头转动
                if (action == MotionEvent.ACTION_CANCEL || action == MotionEvent.ACTION_UP) {
                    ptz = PTZ.Stop;
                } else if (action == MotionEvent.ACTION_DOWN) {
                    int viewId = arg0.getId();
                    switch (viewId) {
                        case R.id.ibTop:
                            ptz = PTZ.Up;
                            break;
                        case R.id.ibBottom:
                            ptz = PTZ.Down;
                            break;
                        case R.id.ibLeft:
                            ptz = PTZ.Left;
```

```
                                break;
                            case R.id.ibRight:
                                ptz = PTZ.Right;
                                break;
                        }
                    }
                }
                cameraManager.controlDir(ptz);
            } catch (Exception e) {
                e.printStackTrace();
            }
```

(6) 在 MonitoringFragment.java 名为 runnable 任务的 TODO 4 位置，添加如果有人，判断摄像头地址信息是否为空，为空时弹出提示，不为空时监控如果是开启的，则拍照，代码如下：

```
// TODO 4：如果有人，判断摄像头地址信息是否为空，为空时弹出提示，不为空时监控如果是开启的，拍照
            if (person.equals("有人")) {
                //获取 Sharepreferences 中保存的摄像头地址信息
                getCameraData();
                if (ip.isEmpty() || userName.isEmpty() || pwd.isEmpty() || channel.isEmpty()) {
                    Toast.makeText(getActivity(), "请单击右上角图标设置摄像头的地址信息", Toast.LENGTH_SHORT).show();
                } else {
                    if (tbOpenOrClose.isChecked()) {
                        //如果有人并且监控是开启的，拍照
                        capture();
                    }
                }
            }
```

(7) 在 MonitoringFragment.java 的 TODO 5 位置，添加从其他页面回到园区监控页面的时候，如果监控按钮是开启状态，0.3 秒后打开摄像头，代码如下：

```
// TODO 5：从其他页面回到园区监控页面的时候，如果监控按钮是开启状态，0.3 秒后打开摄像头
@Override
public void onResume() {
    super.onResume();
    if (tbOpenOrClose.isChecked()) {
        //如果开启监控，0.3 秒后打开摄像头
        handler.postDelayed(openCameraRunnable, 300);

    }
}
```

(8) 在 HomeActivity 标题栏菜单的点击事件 onOptionsItemSelected()方法 TODO 6 的位置，添加显示设置摄像头地址菜单对话框，代码如下：

```
// TODO: 6 显示设置摄像头地址菜单对话框
                //实例化对话框
                SettingCameraDialog dialog = new SettingCameraDialog(this);
                //设置区域外不可取消
                dialog.setCanceledOnTouchOutside(false);
                dialog.show();
```

(9) 在 HomeActivity 标题栏菜单的点击事件 onOptionsItemSelected()方法 TODO 7 的位置，添加判断如果手机内置 SD 卡根目录下 pic 文件夹不存在或文件夹为空，弹出暂无图片的提示；否则跳转到显示截图的 ShowActivity，代码如下：

```
// TODO: 7 如果手机内置SD卡根目录下pic文件夹不存在或文件夹为空，弹出暂无图片的提示；否则跳转
// 到显示截图的ShowActivity
            String path = Environment.getExternalStorageDirectory().getPath() + "/pic/";
            File f = new File(path);
            if (f.list() != null) {
                if (f.list().length == 0) {
                    ShowMessage("暂无图片");
                    break;
                } else {
                    //跳转到监控图的详情页
                    android.content.Intent intent = new android.content.Intent(HomeActivity.this, ShowActivity.class);
                    startActivity(intent);
                }
            } else {
                ShowMessage("暂无图片");
            }
```

至此，便完成了园区监控系统。完整代码可查阅随书资料中的 Project9_2 工程源代码。

3. 运行结果

将程序运行到工控平板，确保摄像头和工控平板连接的是同一个网络，单击园区监控页面右上角的“设置”图标，跳到设置摄像头地址页面，如图 9-12 所示，设置完后，单击“开启监控”按钮，可以看到监控界面，如图 9-13 所示。开启监控后发现有人就会进行拍照，拍照后会以列表的方式进行展示，如图 9-14 所示。

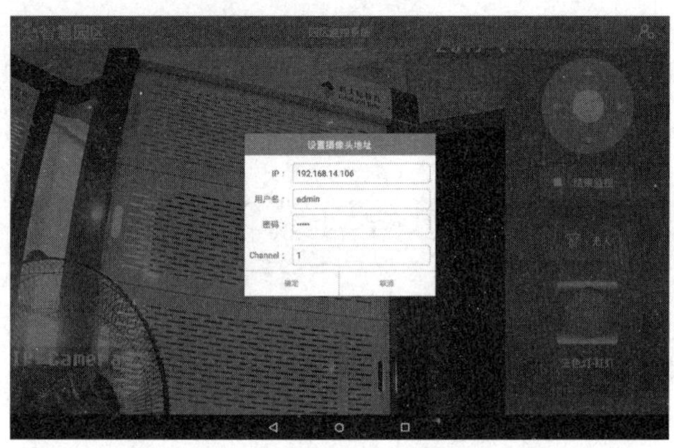

图 9-12　摄像头地址设置界面

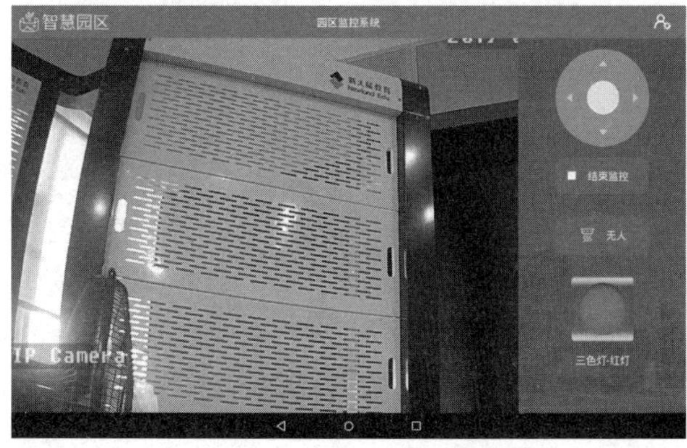

图 9-13　园区监控页面效果

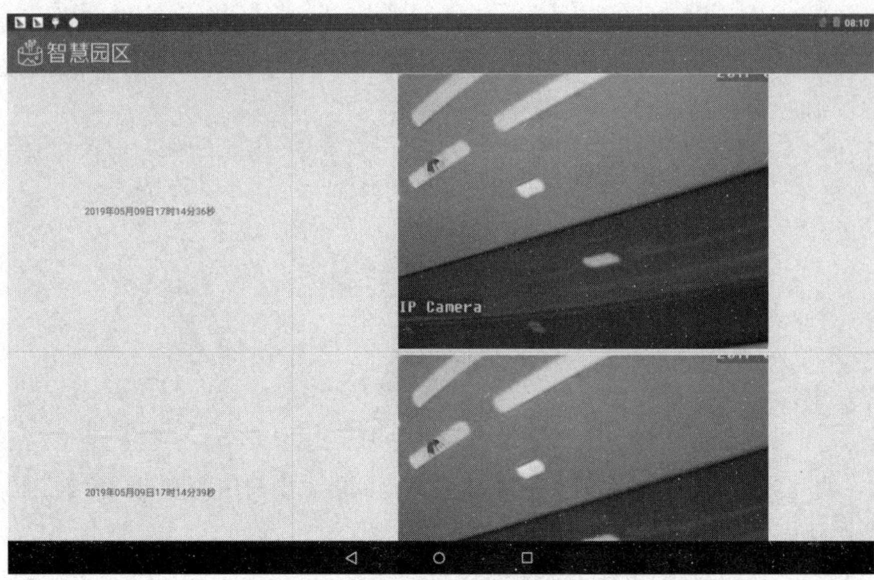

图 9-14　截图查看界面

项 目 总 结

本项目主要讲解了室内环境采集系统和园区监控系统的实现，利用采集到的数据进行判断和控制相应的设备，从而让读者明白传感数据采集在物联网中的应用。

思考与练习

9-1　简述报警通知的发送流程。

9-2　简述摄像数据的采集过程

9-3　简述拍照截图的实现过程。

项目10 网络与定位技术的使用

【项目描述】

物联网就是物物相连的互联网，其核心和基础仍然是互联网。不管是手机、电视、空调、汽车等，几乎都具备连接互联网的功能。在物联网云平台架构中，物联网平台可以实现底层终端设备的“管、控、营”一体化，为上层提供应用开发和统一接口，构建终端设备和业务的端到端通道。手机作为智能终端，也需要考虑如何使用网络技术连接物联网平台和从物联网平台获取数据。同时，万物互联时代中的数据，在很多时候，需要有对应的位置信息才更有价值，因此也需要考虑物联网应用中的定位技术是如何实现的。

本项目以 4 个任务的形式讲解终端如何使用 TCP 协议、HTTP 协议和物联网云平台进行网络交互，并对返回的数据进行解析，同时讲解了利用定位终端模块和地图 SDK 实现快速和精准地定位位置，如图 10-1 所示。

本项目所使用的物联网云平台是 NLECloud，访问地址是 api.nlecloud.com。终端接入和云平台服务是物联网运营的关键技术，终端是直接与用户接触的使用界面，而平台则是承载服务的核心系统，终端接入和平台服务保障了物联网应用端到端的服务质量。物联网云平台 NLECloud 是一个以设备为核心，针对物联网教育和科研而推出的云教学平台，如图 10-2 所示。

本教程所使用的云平台均指 NLECloud 物联网云平台，更多的使用可参考 api.nlecloud.com 上的相关说明。

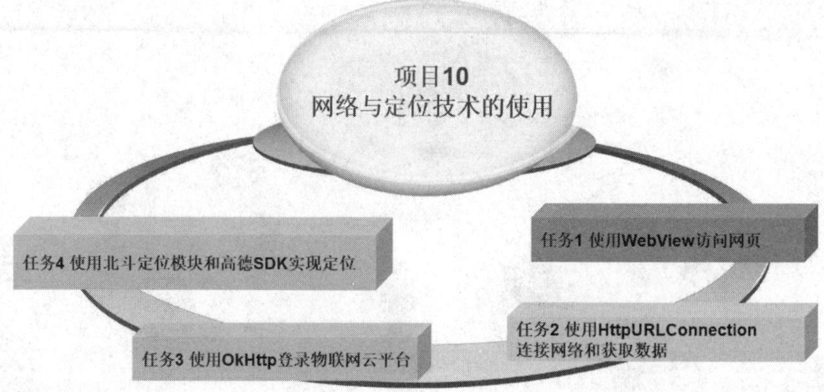

图 10-1 项目 10 任务列表

图 10-2 NLECloud 物联网云平台

【学习目标】

知识目标：

- 会使用手机显示网页。
- 会连接网络和获取数据。
- 会使用第三方库连接网络。
- 会使用定位技术。

技能目标：

- 能用 WebView 打开网页。
- 能用 HttpURLConnection 访问网络。
- 能使用 OkHttp 访问网络。
- 能使用定位模块和地图 SDK 实现定位。

任务 1 使用 WebView 访问网页

【任务描述】

扫码观看视频讲解

本任务要求使用 Android 控件 WebView 打开指定的网页。任务清单如表 10-1 所示。

表 10-1 任务清单

任务课时	2 课时	任务组员数量	建议 1 人
任务组采用设备	PC 一台 工控平板或模拟器或手机		

【知识解析】

以下我们来介绍 WebView 的用法。通常，加载和显示网页用浏览器完成，但是，如果在不用浏览器的情况下又想访问网页，就可以利用 WebView 控件来实现。WebView 控件可以在应用程序中嵌入一个浏览器，从而可以轻松地展示网页。

使用 WebView 时只需要在布局中添加 WebView 控件，代码如下。

```
<WebView
        android:id="@+id/webview"
        android:layout_width="match_parent"
        android:layout_height="match_parent" />
```

接着进行一些设置：

```
1.  WebView webView = findViewById(R.id.webview);
2.  webView.getSettings().setJavaScriptEnabled(true);
3.  webView.setWebViewClient(new WebViewClient());
4.  webView.loadUrl("要打开的网址");
```

代码解释如下。

第 1 行先通过控件 id 获取 WebView 的实例对象。

第 2 行调用 webView.getSettings()方法用来设置浏览器的一些属性，其中 setJavaScriptEnabled(true)设置了 WebView 支持 JavaScript 脚本。

第 3 行设置是指不打开系统浏览器，网页还运行在当前 WebView 中。也就是当需要从一个网页跳转到另一个网页时，仍然在当前 WebView 中显示。

第 4 行调用 loadUrl()，传入要打开的网址，就可以显示相应的内容了。

需要注意的是，WebView 需要连接网络，在 AndroidManifest.xml 清单文件中要添加使用网络的权限 <uses-permission android:name="android.permission.INTERNET"/> 和允许 HTTP 明文传输的安全策略 android:usesCleartextTraffic="true"。

【任务实施】

1. 任务分析

用 WebView 访问云平台首页的过程如下。

(1) 创建工程，在布局中添加 WebView 控件。

(2) 设置 WebView 相关属性。

(3) 在清单文件中添加网络权限和安全策略。

2. 任务实现

(1) 新建工程 Project10_1，在对应的布局文件中添加 WebView 控件，代码如下。

```
<?xml version="1.0" encoding="utf-8"?>
<LinearLayout xmlns:android="http://schemas.android.com/apk/res/android"
    xmlns:app="http://schemas.android.com/apk/res-auto"
    xmlns:tools="http://schemas.android.com/tools"
    android:layout_width="match_parent"
    android:layout_height="match_parent"
    tools:context=".MainActivity">
    <WebView
        android:id="@+id/webview"
        android:layout_width="match_parent"
        android:layout_height="match_parent"
        />
</LinearLayout>
```

(2) 在 MainActivity 中查找 WebView 控件并进行设置，代码如下。

```
public class MainActivity extends AppCompatActivity {
    @Override
    protected void onCreate(Bundle savedInstanceState) {
        super.onCreate(savedInstanceState);
        setContentView(R.layout.activity_main);
        WebView webView = findViewById(R.id.webview);
        webView.getSettings().setJavaScriptEnabled(true);
        webView.setWebViewClient(new WebViewClient());
        webView.loadUrl("http://api.nlecloud.com");
    }
}
```

(3) 在清单文件中添加网络权限和安全策略，代码如下。

```
<manifest xmlns:android="http://schemas.android.com/apk/res/android"
    package="com.example.project10_4">
<uses-permission android:name="android.permission.INTERNET"/>
    <application
        省略......
        android:usesCleartextTraffic="true"
        >
        <activity 省略......        </activity>
    </application>

</manifest>
```

3. 运行结果

在运行之前，确保手机或模拟器或物联网工控平板是联网的，运行后可以看到云平台的首页已经展示出来了，还可以进行网页的跳转，如图 10-3 所示。

WebView 的使用还有很多更高级的用法，这里不一一展开，有兴趣的读者可以自行研究。通过 WebView 的使用过程可以看到：加载网页时用 webView.loadUrl (http://api.nlecloud.com)，其中用到了 HTTP 协议，客户端(手机)向服务器(云平台 Web 站点)发出一条 HTTP 请求，服务器收到后分析出客户想要访问的是首页，于是把首页的数据返回给客户端，客户端的 WebView 控件内部的 WebKit (一个开源的浏览器引擎)调用浏览器对数据进行解析，最终将页面显示出来，也就是说，WebView 在后台处理了发送 HTTP 请求、接收服务器响应、解析返回数据以及页面展示的工作，这些工作过程就是 HTTP 的请求和响应过程。

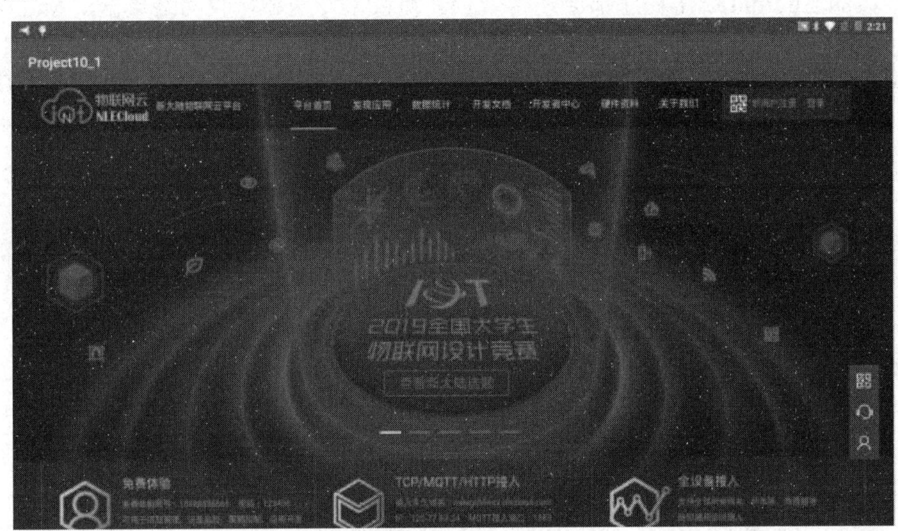

图 10-3　程序运行结果

任务 2　使用 HttpURLConnection 连接网络和获取数据

【任务描述】

本任务要使用 HttpURLConnection 进行网络连接，并通过获取云平台 Logo 的过程来演示使用 HttpURLConnection 如何连接网络和获取数据的过程。任务清单如表 10-2 所示。

扫码观看视频讲解

表 10-2　任务清单

任务课时	2 课时	任务组员数量	建议 1 人
任务组采用设备	PC 一台 工控平板或模拟器或手机		

【知识解析】

1. HTTP 协议

HTTP(HyperText Transfer Protocol，超文本传输协议)是 Internet 上目前使用最广泛的应用层协议，它基于传输层的 TCP 协议进行通信，工作原理很简单，就是客户端向服务器发起 HTTP 请求，服务器收到后返回数据给客户端，客户端进行数据解析和处理。

HTTP 协议支持客户/服务器模式，客户向服务器请求服务时，只需传送请求方法和路径。请求方法常用的有 GET、POST。GET 表示希望从服务器那里获取数据，而 POST 则表示希望提交数据给服务器。

大多数的 Android 应用程序都会使用 HTTP 协议来发送和接收网络数据，Android 中主要提供了两种方式来进行 HTTP 操作，即 HttpURLConnection 和 HttpClient。但是由于 HttpClient 存在对应的 API 过多等缺点，在 Android 6.0 后 HttpClient 的功能被正式弃用了，所以现在官方建议使用 HttpURLConnection 进行发送 HTTP 的请求。

使用 HttpURLConnection 建立与服务器的连接后，服务器会返回状态码，出现下列状态码，代表请求已成功被服务器接受、理解并接收。

200 OK：200(成功)服务器已成功处理了请求。通常，这表示服务器提供了请求的网页。

404 Not Found：404(未找到)服务器找不到请求的网页。

500：(服务器内部错误)服务器遇到错误，无法完成请求。

使用 HttpURLConnection 一般按以下步骤进行。

第一步，要先通过 new 一个 URL 对象传入网址，然后调用 openConnection()方法获得 HttpURLConnection 实例对象 conn，代码如下。

```
//新建一个 URL 对象
URL url = new URL("https://www.baidu.com/");
// 打开一个 HttpsURLConnection 连接
HttpsURLConnection conn = (HttpsURLConnection) url.openConnection();
```

第二步，有了实例对象 conn 后，可以设置 HTTP 请求所使用的方法及进行其他设置(根据实际要求设置)，然后通过返回的 code 值判断是否成功，代码如下。

```
//设置请求方式，可以不设置，默认就是 GET 请求
conn.setRequestMethod("GET");
// 设置连接主机超时时间，单位为毫秒
conn.setConnectTimeout(5000);
// 设置从主机读取数据超时时间
conn.setReadTimeout(5000);
// 判断请求是否成功，成功码为 200
 if (conn.getResponseCode()==200) {
}
```

第三步，接着就可以调用 getInputStream()获取服务器返回的输入流对象，通过输入流对象可以进行数据的读取，代码如下。

```
InputStream input =conn.getInputStream();
```

第四步，数据读取完毕，可以通过调用 disConnect()方法将 HTTP 连接关闭，代码如下。

```
conn.disConnect();
```

如果想提交数据给服务器，需要把请求方式改成 POST，并在获取输入流之前把要提交的数据写出去，每条数据以键值对的形式存在，数据与数据之间用&符号隔开。比如要向服务器提交用户名和密码，代码如下。

```
conn.setRequestMethod("POST");
OutputStream out=conn.getOutputStream();
out.write("username=admin&password=123456".getBytes());
```

2. 添加网络安全策略允许当前应用使用 htttp 明文请求

从 Andorid 9.0(API28)系统开始，限制了明文 HTTP 的网络请求，非加密的请求都会被系统禁止，因此，如果当前应用的请求是 HTTP 请求，而非 HTTPS，就会导致系统禁止当前应用的请求。因为 HttpURLConnection 使用的是 HTTP，需要进行安全策略配置以支持 HTTP 的连接，否则会报以下错误：java.net.UnknownServiceException: CLEARTEXT communication to wanandroid.com not permitted by network。

解决方法共以下四种。

(1) 如果一定要使用明文通信的话，则可以打开 AndroidManifest.xml 文件，在 application 标签中添加：android:usesCleartextTraffic="true"。

需要注意以下三点：①为了安全，不建议使用明文的通信方式，不过明文方法可以作为一种临时的通信策略。②如果声明不使用明文通信，则可以在 application 标签中添加：android:usesCleartextTraffic="false"，此声明指示该应用不使用明文网络通信，并使 Android Marshmallow 的平台网络堆栈禁止该应用中的明文通信。例如，如果应用意外尝试通过 HTTP 明文请求登录用户，该请求将被阻止，该用户的身份和密码信息不会泄露到网络上。③如果要临时使用明文通信策略，代码如下。

```
<manifest xmlns:android="http://schemas.android.com/apk/res/android"
    package="com.example.project10_4">
<uses-permission android:name="android.permission.INTERNET"/>
    <application
        省略......
        android:usesCleartextTraffic="true"
        >
        <activity 省略......        </activity>
    </application>

</manifest>
```

(2) 项目改用 https 请求。

(3) 项目的 targetSdkVersion 改为 28 以下。

(4) 在 res 的 xml 目录下，新建一个 xml 文件(名称自定义，如 network_security_config.xml)，内容如下。

```
<?xml version="1.0" encoding="utf-8"?>
<network-security-config>
    <base-config cleartextTrafficPermitted="true" />
</network-security-config>
```

在 manifest 清单文件配置 application 标签。

```
<?xml version="1.0" encoding="utf-8"?>
<manifest ... >
```

```
    <application
       android:networkSecurityConfig="@xml/network_security_config"
    ... >
    ...
    </application>
</manifest>
```

本任务采用的是第一种，在 application 标签中添加：android:usesCleartextTraffic="true"。

【任务实施】

1. 任务分析

本任务使用 HttpURLConnection 连接网络，并配合使用多线程技术，通过下载云平台 Logo 并更新到 UI 的过程讲解网络数据是如何获取的，过程如下。

(1) 创建工程并设计布局。

(2) 编写下载类。

(3) 使用线程下载 Logo 并更新 UI。

(4) 在清单文件中添加网络权限和安全策略。

2. 任务实现

(1) 创建工程并设计布局。创建新工程 Project10_2，对应的布局文件为 activity_main.xml，使用一个 ImageView 用来放下载好的图片，一个按钮用于单击下载，代码如下。

```
<?xml version="1.0" encoding="utf-8"?>
<LinearLayout xmlns:android="http://schemas.android.com/apk/res/android"
    xmlns:app="http://schemas.android.com/apk/res-auto"
    xmlns:tools="http://schemas.android.com/tools"
    android:layout_width="match_parent"
    android:layout_height="match_parent"
    android:orientation="vertical"
    tools:context=".MainActivity">

    <ImageView
        android:id="@+id/image"
        android:layout_width="match_parent"
        android:layout_weight="1"
        android:layout_height="0dp"
        android:src="@mipmap/ic_launcher" />
    <Button
        android:onClick="getPic"
        android:text="下载云平台 Logo"
        android:layout_width="match_parent"
        android:layout_height="wrap_content" />
</LinearLayout>
```

(2) 编写下载类。新建 MyDownload 类，在类里声明一个接口 DownListener。接口中有两个方法，一个是成功下载，一个是失败下载。在类中编写方法 downPic()，传入两个参数，一个是要下载数据的路径，另一个是接口的实现对象，代码如下。

```
public class MyDownload {
    public static void downPic(String path,DownListener downListener){
        ByteArrayOutputStream out=null;
        HttpURLConnection conn=null;
        try {
```

```
6.              URL url = new URL(path);
7.             conn=(HttpURLConnection) url.openConnection();
8.              conn.setReadTimeout(5000);
9.              conn.setConnectTimeout(5000);//设置连接超时
10.             conn.setRequestMethod("GET");//用 GET 的方式发送请求
11.             conn.setRequestProperty("accept", "*/*");
12.             conn.setRequestProperty("connection", "Keep-Alive");
13.            int code=conn.getResponseCode();
14.            if(code==200) {
15.                InputStream input = conn.getInputStream();
16.                out = new ByteArrayOutputStream();
17.                byte[] b = new byte[1024];
18.                int n = 0;
19.                while ((n = input.read(b)) > 0) {
20.                    out.write(b, 0, n);
21.                }
22.                downListener.success(out.toByteArray());  //如果成功下载，把数据传出去
23.            }
24.        } catch (Exception e) {
25.            downListener.error("图片下载失败");  //下载失败的提示
26.            e.printStackTrace();
27.        }finally{
28.            if(conn!=null)
29.                conn.disconnect();
30.        }
31.    }
32.    interface DownListener{
33.        //如果成功下载，则把数据存放到 b 中
34.        public void success(byte[] b);
35.        //如果下载失败，则提示出错信息
36.        public void error(String error);
37.    }
38. }
```

代码解析如下。

第 13 行获取响应码，如果是 200 则说明成功建立连接。

第 15～21 行从输入流中读取数据。

第 22 行把读取到的数据转换成字节数组，并调用接口中的 success()传递给调用者。

第 25 行如果下载失败，把出错信息用接口中的方法 error()传递出去。

第 29 行断开 http 的连接。

(3) 使用线程下载 Logo 并更新 UI。编写 MainActivity 类，因为访问网络是一个耗时动作，所以在按钮的点击事件 getPic()方法中开启线程进行网络的耗时操作，在线程的 run()方法中调用 downPic()方法进行下载，通过接口回调方法 success()获取下载好的数据，再通过 Android 提供的 runOnUiThread 进行 UI 的更新，代码如下。

```
public class MainActivity extends AppCompatActivity {
    private ImageView image;
    //云平台 Logo 的路径
   private String path="http://api.nlecloud.com/images/nlogo_blue_s.png";
    @Override
    protected void onCreate(Bundle savedInstanceState) {
        super.onCreate(savedInstanceState);
        setContentView(R.layout.activity_main);
        image=findViewById(R.id.image);
    }
    public void getPic(View v )
```

```
    {
        //1.开启线程，在线程中发送网络请求
        new Thread(new Runnable() {
            @Override
            public void run() {
                //2.调用下载方法，传入网址和回调接口实例对象，获取到的数据通过接口回调值byte[]b 返回
            MyDownload.downPic(path,new MyDownload.DownListener(){
                @Override
                public void success(final byte[] b) {
        //3.因为是在线程中连接网络，而更新UI 只能在主线程中进行，用Android的runOnUiThread来更新
                    runOnUiThread(new Runnable() {
                        @Override
                        public void run() {
                            //4.从返回的字节数组数据中构建Bitmap 对象
                            Bitmap bm = BitmapFactory.decodeByteArray(b,0,b.length);
                           //5.显示在控件上
                            image.setImageBitmap(bm);
                        }
                    });
                }
                @Override
                public void error(String error) {
                    Toast.makeText(MainActivity.this,"下载图片失败", Toast.LENGTH_LONG).show();
                }
            });
            }
        }).start();
    }
}
```

(4) 在清单文件中添加网络权限和安全策略。请参考任务 1 完成网络权限与安全策略的添加。

3. 运行结果

运行程序，结果如图 10-4 所示。

图 10-4　下载 Logo 的结果

任务3　使用 OkHttp 登录物联网云平台

【任务描述】

扫码观看视频讲解

本任务使用第三方通信库 OkHttp 进行网络连接，通过演示登录云平台的过程，讲解如何使用云平台提供的 RESTfulAPI 与云平台进行数据交互。任务清单如表 10-3 所示。

表 10-3　任务清单

任务课时	4 课时	任务组员数量	建议 3 人
任务组采用设备	PC 一台 工控平板或模拟器或手机		

【知识解析】

1. RESTfulAPI 简介

REST 是 Representational State Transfer 的缩写，如果一个架构符合 REST 原则，就称它为 RESTful 架构。RESTful 架构可以充分地利用 HTTP 协议的各种功能，是 HTTP 协议的最佳实践。

RESTful 用 URL 表示要操作的资源，用不同的 HTTP 请求(GET、POST、PUT、DELETE)描述对资源的操作，通过 HTTP 的状态码来判断此次对资源操作的结果，这就是 RESTful 风格。RESTful 风格是对请求的一次解耦，提高了 URL 的重用性。

客户端通过 HTTP 协议访问服务端资源，通过 HTTP 协议里面定义好的四个动词(GET、POST、PUT、DELETE)来对资源进行状态转化。GET 用来获取资源，POST 用来新增资源，PUT 用来更新资源，DELETE 用来删除资源。

RESTful 有什么意义呢？网络上所有的信息体都被看作是一种资源，对网络资源的某种操作，都是通过 METHOD 来确定的。对于现在的数据或者资源提供方，对外透露的接口一般是 RESTful 风格的，有利于不同系统之间的资源共享，而且只需要遵守规范，不需要做另外的配置就能达到资源共享。

API(Application Programming Interface，应用程序编程接口)是一些预先定义的函数，目的是提供应用程序与开发人员基于某软件或硬件的以访问一组例程的能力，而又无须访问源码，或理解内部工作机制的细节。符合 RESTful 风格的 API 就是 RESTfulAPI，它是一种软件架构风格、设计风格，可以让软件更清晰、更简洁、更有层次，可维护性更好。

打开 api.nelcloud.com，注册账号并成功登录云平台后，打开 api.nelcloud.com/doc/api，可以看到详细列表中有各种 RESTfulAPI，如图 10-5 所示。

2. 解析 JSON 格式的数据

当客户端通过 GET 方式或 POST 方式向服务器提交 HTTP 请求后，数据会传递到服务器，服务器也会根据请求返回数据给客户端，那么这些数据到底以什么格式在网络上传输呢？服务器收到数据后怎么知道客户端需要什么数据呢？

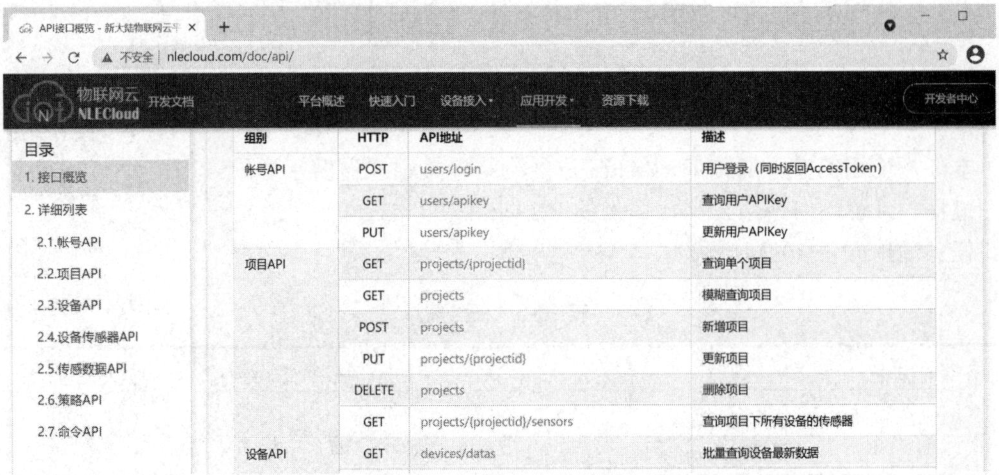
API接口概览 - 新大陆物联网云平
nlecloud.com/doc/api/
物联网云 NLECloud 开发文档
平台概述
快速入门
设备接入
应用开发
资源下载
开发者中心
目录
1. 接口概览
2. 详细列表
2.1.帐号API
2.2.项目API
2.3.设备API
2.4.设备传感器API
2.5.传感数据API
2.6.策略API
2.7.命令API
组别 | HTTP | API地址 | 描述
帐号API | POST | users/login | 用户登录（同时返回AccessToken）
GET | users/apikey | 查询用户APIKey
PUT | users/apikey | 更新用户APIKey
项目API | GET | projects/{projectid} | 查询单个项目
GET | projects | 模糊查询项目
POST | projects | 新增项目
PUT | projects/{projectid} | 更新项目
DELETE | projects | 删除项目
GET | projects/{projectid}/sensors | 查询项目下所有设备的传感器
设备API | GET | devices/datas | 批量查询设备最新数据

图 10-5　NLECloud 云平台提供的部分 RESTfulAPI

通常在网络上传输的数据都需要进行格式化，数据会有一定的结构和语义，当另一方收到数据后可以按照相同的结构和语义进行解析，从而取出想要的数据。

在网络上传输数据时常用的格式有两种，即 XML 和 JSON，因为云平台上用的是 JSON 格式的，所以在这里只针对 JSON 格式数据进行讲解。

JSON(JavaScript Object Notation)是一种轻量级的数据交换格式。因为易于阅读和编写，同时也易于机器解析和生成，故而被广泛应用于网络数据交换。解析 JSON 格式的数据可以使用官方的 JSONObject，也可以使用谷歌的开源库 GSON，一些第三方的开源库如 FastJson、Jackson 也很不错。

FastJson：阿里巴巴开发的 JSON 库，性能十分优异(本任务选用)。

Jackson：在社区十分活跃且更新速度很快。

常用的 JSON 格式数据有以下 5 种。

① JSON 数字：JSON 数字可以是整型或者浮点型：{ "age":30 }。

② JSON 对象：JSON 对象在大括号({})中书写，对象可以包含多个名称/值对：{ "name":"温度" , "value":30 }。

③ JSON 数组：JSON 数组在中括号中书写，数组可包含多个对象：

```
{
    "devices": [
        { "name":"温度" , "value":25.55 },
        { "name":"湿度" , "value":50.00},
        { "name":"光照" , "value":1234.56 }
    ]
}
```

在上面的例子中，对象"devices"是包含三个对象的数组。每个对象代表一条关于某个传感器(name、value)的数据。

④ JSON 布尔值：JSON 布尔值可以是 true 或者 false：{ "flag":true }。

⑤ JSON null：JSON 可以设置 null 值：{ "device":null }。

下面以 NLECloud 云平台的 JSON 数据做分析。

单击图 10-5“账号 API”中的 POST 用户登录 API，可以查看请求方式和地址、包体请求参数、请求示例，其中请求示例就是 JSON 格式的数据，如图 10-6 所示。

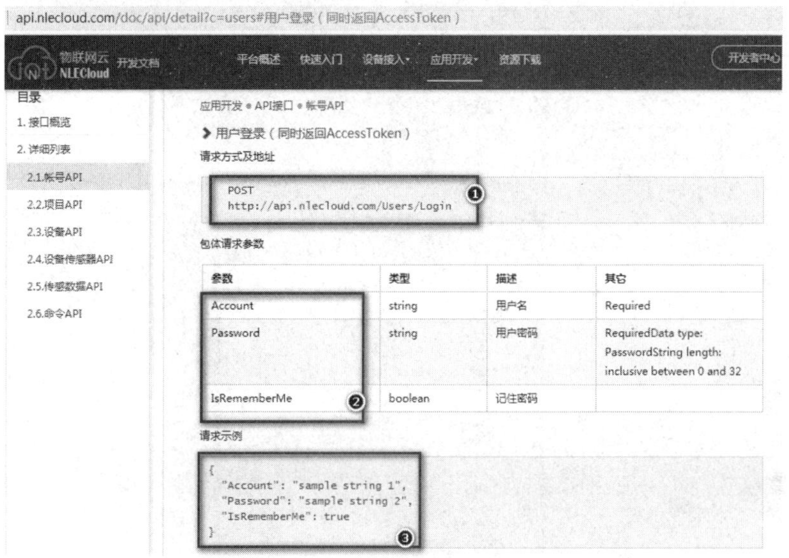

图 10-6　JSON 格式的请求参数

把请求示例中的 JSON 格式字符串复制出来，写一个 Java 程序来测试，如图 10-7 所示。


```
import com.alibaba.fastjson.JSON;

public class JSONTest {

    public static void main(String[] args) {
        String jsonstr="{\r\n" +
                "  \"Account\": \"sample string 1\",\r\n" +
                "  \"Password\": \"sample string 2\",\r\n" +
                "  \"IsRememberMe\": true\r\n" +
                "}";

        //解析JSON格式的字符串
         JSONObject jsonObj = JSON.parseObject(jsonstr);
         //凭关键字Account获取值
         String account = jsonObj.getString("Account");
         String password = jsonObj.getString("Password");
         String isRememberMe = jsonObj.getString("IsRememberMe");

        System.out.println("account = " +account+" password="+password
                +" "+" isRememberMe"+isRememberMe);
    }
```

```
Console  Problems  Debug Shell
<terminated> JSONTest [Java Application] C:\Program Files\Java\jdk-10.0.2\bin\javaw.exe (2019年6月21日 下午4:09:32)
account = sample string 1 isRememberMetrue password=sample string 2
```

图 10-7　解析 JSON 格式的数据结果

可以看到 JSON 的数据解析非常简单。用 JSON.parseObject(JSON 格式的字符串)可以把字符串解析为一个 JSONObject 对象 jsonObj，再由 jsonObj.getString(关键字)取出对应的值即可。

再继续看账号 API 的响应参数和响应示例，如图 10-8 所示。

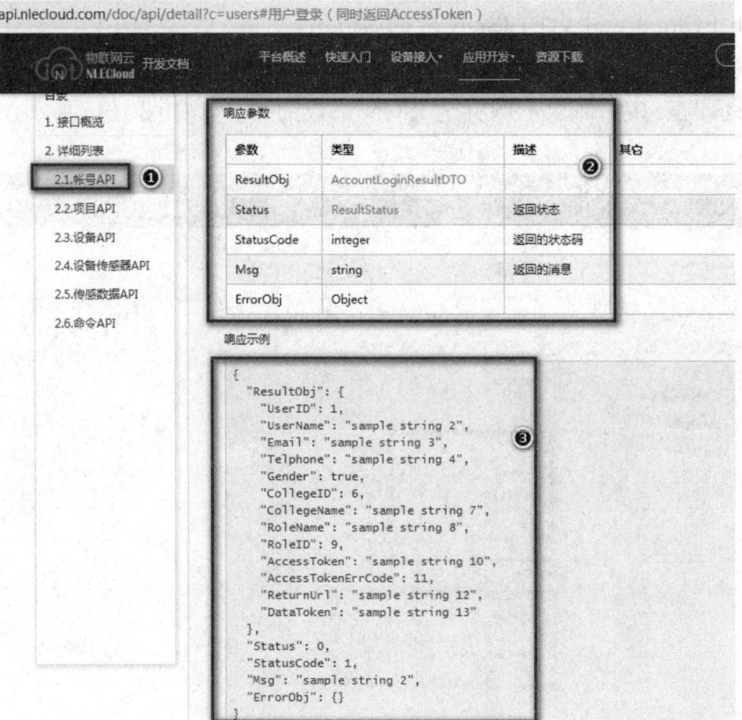

图 10-8 JSON 格式的响应

响应示例 ResultObj 中还有数据，这又要如何解释呢？接下来再写一个 Java 程序来测试，如图 10-9 所示。

JSONTest2.java

```
    public static void main(String[] args) {
        String jsonstr="{\r\n" +
                "  \"ResultObj\": {\r\n" +
                "    \"UserID\": 1,\r\n" +
                "    \"UserName\": \"sample string 2\",\r\n" +
                "    \"Email\": \"sample string 3\",\r\n" +
                "    \"Telphone\": \"sample string 4\",\r\n" +
                "    \"Gender\": true,\r\n" +
                "    \"CollegeID\": 6,\r\n" +
                "    \"CollegeName\": \"sample string 7\",\r\n" +
                "    \"RoleName\": \"sample string 8\",\r\n" +
                "    \"RoleID\": 9,\r\n" +
                "    \"AccessToken\": \"sample string 10\",\r\n" +
                "    \"AccessTokenErrCode\": 11,\r\n" +
                "    \"ReturnUrl\": \"sample string 12\",\r\n" +
                "    \"DataToken\": \"sample string 13\"\r\n" +
                "  },\r\n" +
                "  \"Status\": 0,\r\n" +
                "  \"StatusCode\": 1,\r\n" +
                "  \"Msg\": \"sample string 2\",\r\n" +
                "  \"ErrorObj\": {}\r\n" +
                "}\r\n" +
                "";
        JSONObject jsonObj = JSON.parseObject(jsonstr);
        JSONObject innerJson = jsonObj.getJSONObject("ResultObj");
        String accessToken = innerJson.getString("AccessToken");
        System.out.println("AccessToken = " +accessToken);
        String status = jsonObj.getString("Status");
        System.out.println("status = " +status);
    }
}
```

Console

```
<terminated> JSONTest2 [Java Application] C:\Program Files\Java\jdk-10.0.2\bin\javaw.exe (2019年6月21日 下午4:26:31)
AccessToken = sample string 10
status = 0
```

图 10-9 解析 JSON 格式的响应

可以看到，如果 JSON 格式数据的对象中还有对象，就用 getJSONObject()取出对象，接着再按关键字取对应的数据即可。

3. OkHttp 发送 GET 和 POST 请求

OkHttp 是一个开源的、很出色的网络通信库，用于替代原生的 HttpUrlConnection 和 Apache HttpClient。OkHttp 在接口封装上做得简单易用，现在已经成为 Android 开发首选的网络通信库。

(1) 添加依赖。在使用 OkHttp 之前，要先在项目中添加 OkHttp 库的依赖。在 app/build.gradle 文件的 dependencies 闭包中添加如下内容。

```
dependencies {
    implementation fileTree(include: ['*.jar'], dir: 'libs')
    implementation 'com.squareup.okhttp3:okhttp:4.0.0-RC1'
......
}
```

OkHttp 内部依赖另一个开源库 Okio，所以添加上述依赖时会自动下载两个库，一个是 OkHttp 库，一个是 Okio 库，Okio 是 OkHttp 的通信基础，OkHttp 的最新版本读者可以通过 OkHttp 的项目主页 https://github.com/square/okhttp 来查看。

(2) 使用 OkHttp 发送 GET 请求。使用 OkHttp 发送 GET 请求的步骤如下。

```
//1.创建 OkHttpClient 实例
   OkHttpClient client = new OkHttpClient();
//2.创建一个 Request 对象，用于发送 HTTP 请求,默认是 GET 方式
   Request request = new Request.Builder().build();
```

上述代码只是创建了一个空的 Request 对象，要让它连接网络发请求还需要在 build()方法前连缀其他方法，比如设置要连接的 url 等，代码如下。

```
Request request = new Request.Builder()
             .url("api.nlecloud.com")
             .build();
```

接着通过 client.newCall()方法来创建一个 call 对象，传入 request 对象，调用它的 execute()方法来发送请求并获得从服务器返回的数据 response，代码如下。

```
try {
    Response response = client.newCall(request).execute();
   String data =  response.body().string();
} catch (IOException e) {
    e.printStackTrace();
}
```

注意，Response 最低要求的 API level 为 19。

(3) 使用 OkHttp 发送 POST 请求。

使用 OkHttp 发送 POST 请求，代码如下。

```
 //1.声明媒体类型为 JSON，媒体类型在一些协议的消息头中叫作"Content-Type"，用 utf-8 编码格式
public static final MediaType JSON
    = MediaType.get("application/json; charset=utf-8");
//2.创建 OkHttpClient 实例
OkHttpClient client = new OkHttpClient();
//3.创建一个 RequestBody 对象，用来存放待提交的数据，其中 json 就是要提交的 JSON 格式数据
RequestBody requestBody = RequestBody.create(JSON,json);
```

```
//4.创建 Request 对象，调用 post 方法，并将 RequestBody 对象传入
Request request = new Request.Builder()
        .url("api.nlecloud.com")
        .post(requestBody)
        .build();
/*5.调用一个 OkHttpClient 的 newCall()方法创建 call 对象，传入 request 对象
调用它的 execute()方法来发送请求并获得从服务器返回的数据
*/
try {
    Response response = client.newCall(request).execute();
    String data = response.body().string();
} catch (IOException e) {
    e.printStackTrace();
}
```

上述 url 地址要根据实际地址进行更改。

【任务实施】

1. 任务分析

本任务使用 OkHttp 网络通信库访问云平台，通过封装 OkHttp 的 GET 请求和 POST 请求讲解 OkHttp 的使用过程，同时用登录功能进行测试，过程如下。

(1) 创建工程，添加 OkHttp 依赖库。

(2) 封装 OkHttp 的 GET 请求和 POST 请求方法。

(3) 登录云平台，记下 AccessToken。

(4) 在清单文件中添加网络权限和安全策略。

2. 任务实现

(1) 创建工程，添加 OkHttp 依赖库。创建新工程 Project10_3，在 app/build.gradle 文件中添加 OkHttp 的依赖库，同时修改最低支持的 API 为 19，添加完后要进行同步，如图 10-10 所示。

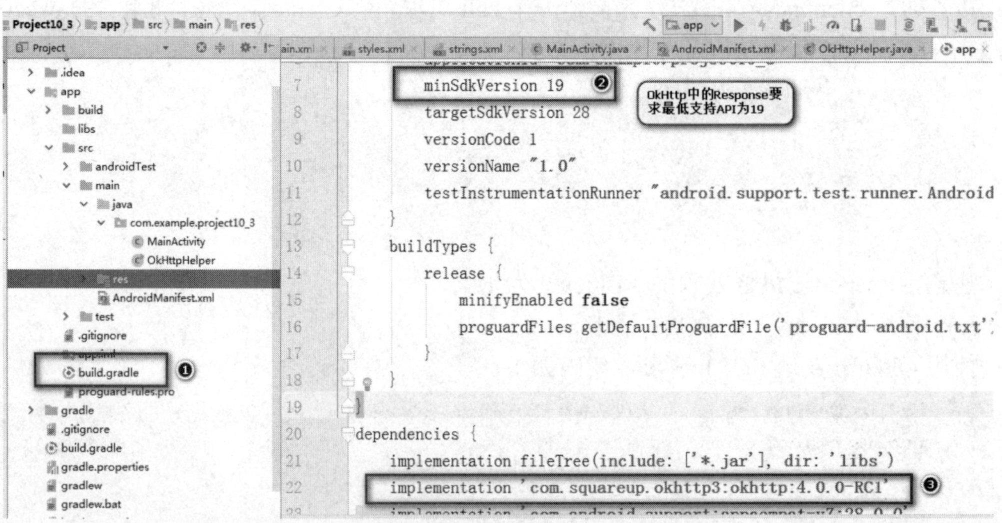

图 10-10　添加 OkHttp 依赖

(2) 封装 OkHttp 的 GET 请求和 POST 请求方法。新建类 OkHttpHelper，封装 GET 请求和 POST 请求方法，代码如下。

```
package com.example.project10_3;
import java.io.IOException;
import okhttp3.MediaType;
import okhttp3.OkHttpClient;
import okhttp3.Request;
import okhttp3.Request.Builder;
import okhttp3.RequestBody;
import okhttp3.Response;
public class OkHttpHelper {
    //声明媒体类型为 JSON
    private static final MediaType JSON = MediaType.get("application/json; charset=utf-8");
    private static final String TOKEN_KEY = "AccessToken";
    //使用 Okhttp
    public static OkHttpClient client = new OkHttpClient();
    //保存 AccseeToken 值(登录一次后，获取的 AccessToken 值可以重复多次使用)
    public static String accessToken;

    public static String get(String url) throws IOException {
        Request.Builder builder = new Request.Builder();
        if (accessToken != null)
            builder.addHeader(TOKEN_KEY, accessToken);

        Request request = builder.url(url).build();
        Response response = client.newCall(request).execute();
        return response.body().string();
    }

    public static String post(String url, String json) throws IOException {
        RequestBody requestBody = RequestBody.create(JSON, json);
        Builder builder = new Request.Builder();
        if (accessToken != null)
            builder.addHeader(TOKEN_KEY, accessToken);
        Request request = builder.url(url).post(requestBody).build();
        Response response = client.newCall(request).execute();
        return response.body().string();
    }
}
```

上述代码中的：

```
if (accessToken != null)
    builder.addHeader(TOKEN_KEY, accessToken);
```

用来在后续的 HTTP 请求中添加 AccessToken。因为云平台服务器接受登录请求后返回的 JSON 格式数据中有 AccessToken，以后的每次请求都需要带上 AccessToken，所以当登录成功得到 AccessToken 后，要在请求头里加上 AccessToken。

(3) 登录云平台，记下 AccessToken。在 MainActivity 中编写代码，对应的布局中有一个按钮，点击事件为 loginYun，代码如下。

```
package com.example.project10_3;
import android.os.Bundle;
import android.support.v7.app.AppCompatActivity;
import android.view.View;
import java.io.IOException;
```

```
6.  import com.alibaba.fastjson.JSON;
7.  import com.alibaba.fastjson.JSONObject;
8.  public class MainActivity extends AppCompatActivity {
9.      @Override
10.     protected void onCreate(Bundle savedInstanceState) {
11.         super.onCreate(savedInstanceState);
12.         setContentView(R.layout.activity_main);
13.     }
14.
15.     public void loginYun(View view) {
16.         new Thread() {
17.             @Override
18.             public void run() {
19.                 try {
20.                     //登录云平台，云平台账号和密码要按真实情况进行填写
21.                     String accessToken = login("云平台的账号", "云平台的密码");
22.                  System.out.println("AccessToken = " +accessToken);
23.                 } catch (Exception e) {
24.                     e.printStackTrace();
25.                 }
26.             }
27.         }.start();
28.     }
29.
30.     public String login(String loginName, String password) throws IOException {
31.         String json = "{'Account':'"+loginName+"','Password':'"+password+"',
'IsRememberMe':'true'}";
32.         String url = "http://api.nlecloud.com/Users/Login";
33.         //云平台上规则的登录用post
34.         String content = OkHttpHelper.post(url, json);
35.         System.out.println(content);
36.         JSONObject jsonObj = JSON.parseObject(content);
37.         JSONObject innerJson = jsonObj.getJSONObject("ResultObj");
38.         String accessToken = innerJson.getString("AccessToken");
39.
40.         String status = jsonObj.getString("Status");
41.         System.out.println("status = " +status);
42.         return accessToken;
43.     }
44. }
```

代码解析如下。

第 15～29 行因为访问网络是一个耗时动作，所以在点击事件中使用一个线程处理登录。

第 31 行是按云平台上的 JSON 数据请求格式，把用户名和密码组装成 JSON 格式字符串。

第 32 行是云平台上规定的账号 API 的 url 格式。

第 34 行是云平台上规定的登录采用 POST 方式。

第 36～42 行用 JSON 解析 AccessToken 和登录是否成功的状态值。Status 的返回值为 0 代表成功。

因为登录云平台提供的 API，所以需要注意确保“开发设置”中 API 密钥的期限，如果过期了要重新生成并确认提交，如图 10-11 所示。

(4) 在清单文件中添加网络权限和安全策略。请参考任务 1 完成网络权限与安全策略的添加。

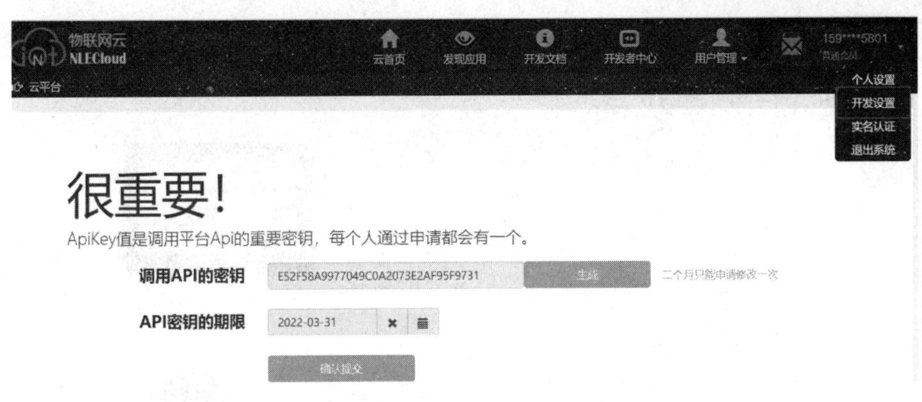

图 10-11　API 密钥的期限

3. 运行结果

运行程序， 结果如图 10-12 所示，成功登录了云平台并解析出 AccessToken 的值。

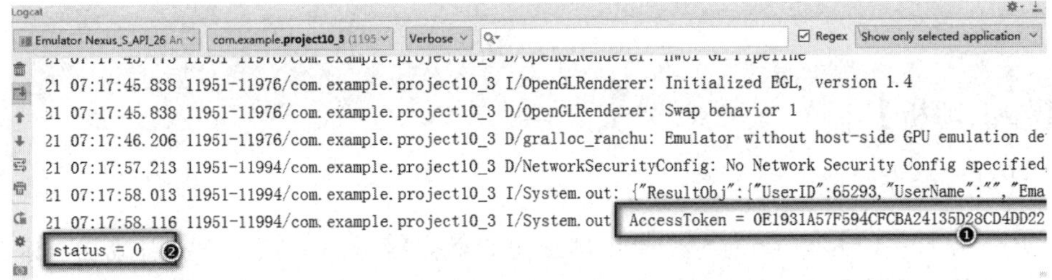

图 10-12　登录成功返回了 AccessToken

解析出 AccessToken 后，与云平台的 HTTP 连接请求都需要带 AccessToken 协议头，这些我们在封装的 GET 和 POST 方法中已经进行了处理，更多访问云平台的用法会在项目 11 中进行讲解。

任务 4　使用北斗定位模块和高德 SDK 实现定位

【任务描述】

本任务使用北斗定位模块进行经纬度数据的采集，通过 TCP 协议把经纬度数据上传到云平台，Android 程序使用 HTTP 协议从云平台获取经纬度数据并配合高德地图 SDK 实现定位。任务清单如表 10-4 所示。

表 10-4　任务清单

任务课时	2 课时	任务组员数量	建议 3 人
任务组采用设备	北斗定位模块一个 RS485 转 RS232 模块一个 PC 一台 工控平板或模拟器或手机		

【拓扑图】

本任务拓扑结构图如图 10-13 所示。

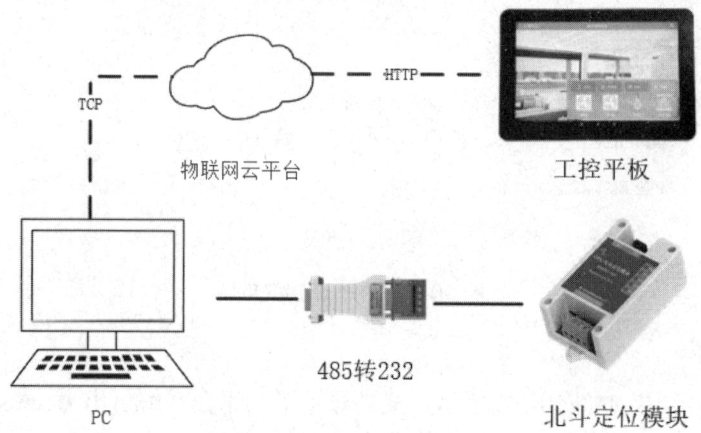

图 10-13　拓扑图

【知识解析】

1. 北斗导航系统简介

2020 年 6 月 13 日，我国北斗三号全球卫星导航系统最后一颗卫星发射成功，标志着北斗三号卫星导航系统共 35 颗卫星全球组网完毕，可对全球任何一个点进行定位、导航、测速、授时和短语通信，如图 10-14 所示。

图 10-14　北斗卫星导航系统

2012年，交通运输部做出规定，12吨以上固定参数和用途的车辆，如货运车辆和半挂牵引车辆等，必须安装北斗行车记录仪。其实不只大货车，按照国家规定，“两客一危”，即从事旅游的包车，三类以上班线的客车和运输危险化学品、烟花爆竹、民用爆炸品的专用车辆也都必须安装北斗行车记录仪。关于北斗行车记录仪，国家还出台了专门的国标，编号是GB-T19056-2012，尾号2012说明该国标是由2012年开始实施的。

北斗行车记录仪是北斗+GPS双模的，北斗系统有53颗卫星，GPS系统有35颗卫星，这是全世界规模最大的两个全球定位导航系统。北斗导航系统只具有导航定位功能，天上的北斗卫星只负责广播自己的精确位置和时间，地面接收机只要能看到4颗以上北斗卫星，就能确定自己的位置和时间。北斗导航系统就好像广播一样，卫星和地面接收装置一般没有双向联系，而车载终端是一个有通信功能的设备，在收到卫星导航信号后，还要和某些网络联系并标记位置，也就是说通信不是北斗导航系统的事情，北斗只管定位。

2. GPS/北斗定位模块简介

在随书提供的资源“./项目10 网络和定位技术的使用/”下有：GPS/北斗定位模块使用说明书.pdf。从说明书得知，HS6601 GPS/北斗定位模块(以下简称：HS6601定位模块)是一款具有GPS定位和北斗定位的双模定位终端，可以快速、精确定位位置，并且将定位信息以RS485接口和ModBus协议的方式提供给用户使用，可以通过PC设置软件或串口命令轻松控制，使用方便快捷。

HS6601定位模块读取定位数据的命令帧和响应帧如图10-15所示。

读取定位数据（RMC）

命令帧：01 03 00 05 00 23 14 12

地址	功能码	寄存器起始地址	寄存器个数	CRC 校验
0x01	0x03	0x00 0x05	0x00 0x23	0x14 0x12

响应帧：

地址	功能码	数据长度	数据	CRC 校验
0x01	0x03	0x46	70 字节数据	两字节校验

图10-15 HS6601定位模块读取定位数据的命令帧和响应帧

读取定位数据返回的70字节数据的ASCII显示如下：$GNRMC,072905.00,A,3640.46260,N,11707.54950,E,000.0,000.0,050119,OK*24，各数据含义解析如图10-16所示。

【任务实施】

1. 任务分析

Android程序利用经纬度数据实现定位的过程如下。

(1) 北斗定位模块采集经纬度数据并使用TCP协议发送到云平台。

(2) Android程序从云平台获取经纬度数据。

(3) 使用高德地图SDK配合经纬度数据实现定位。

字段	符号	含义	取值范围	举例	备注
1	$	语句起始符			
2	GNRMC	RMC协议头			RMC协议头, GNRMC表示联合定位
3	hhmmss.ss	UTC 时间	时时分分秒秒.秒秒	072905.00	北京东八区需要时+8
4	A	定位状态	A/V		A-有效，V-无效
5	ddmm.mmmmm	纬度	度度分分.分分分分分	3640.46260	计算要转为度： 36度 + 40.46260分。 40.46260/60=0.67438度,所以为36.67438度
6	a	纬度方向	N/S		N-北纬，S-南纬
7	dddmm.mmmmm	经度	度度度分分.分分分分分	11707.54950	计算要转为度： 117度 + 07.54950分。 07.54950/60=0.12583度,所以为117.12583度
8	a	经度方向	E/W		E-东经，W-西经
9	x.xxx –xxx.x	对地速度	节	123.2	地速率 节单位 地面速率 000.0~999.9节，Knot
10	x.xxx –xxx.x	对地航向	度	000.0~359.9	地面航向（000.0~359.9度，以正北为参考基准）
11	xxxxxx	日期	日月年	050119	2019 年那 1 月 5 日
12	aa	天线状态	OK/OP/OR		OK 代表天线正常 OK；OP 代表开路 OPEN；OR 代表天线短路 SHORT
13	*	语句结束符			
14	24	校验和	对“$”和“*”之间的数据（不包括这两个字符）按字节进行异或运算，用十六进制数值表示		

图 10-16 响应数据解析

2. 任务实现

(1) 北斗定位模块采集经纬度数据并使用 TCP 协议发送到云平台。

一般来说，要把采集的经纬度数据上传到云平台，需要在云平台上做设备接入。NLECloud 云平台的设备接入网址为 nlecloud.com/doc/devicedev，如图 10-17 所示。

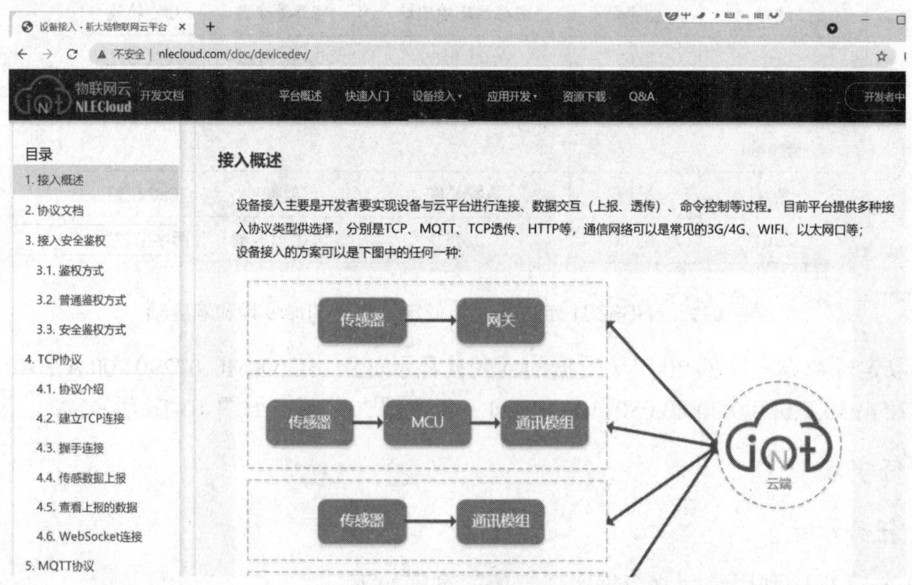

图 10-17 设备接入

从图 10-17 可以看出，我们可以把北斗定位模块采集的数据通过网关进行接入。网关的作用是把北斗定位模块的标准 ModBus 通信协议转成 TCP 协议，再按云平台的 TCP 接入协议进行接入。TCP 协议的设备接入过程可以用硬网关实现，也可以用软网关实现。接入过程可以参考：http://www.nlecloud.com/doc/devicedev/#TCP 协议。这里按照 TCP 接入协议和利用 JavaSE 相关技术实现了一个定位数据上报工具，相当于软网关的功能，针对从串口获取北斗定位模块采集的数据，解析出经纬度数据后，采用 TCP 协议上报到云平台的过程进行了详细的讲解，涉及网络编程、多线程编程、集合、TCP 协议数据上报及响应等的综合应用，有兴趣的读者可以在随书资料源码包：“./项目 10 使用网络和定位技术/”中找到相关源码工程 GPSTool，详读源码进行了解，在《基于 Java 的物联网基础应用开发》这本书项目 11 的任务实施部分也进行了详细的介绍，限于篇幅这里不再展开。

使用 eclipse 打开或者用 IntelliJ IDEA 导入 GPSTool 项目后，运行程序，如图 10-18 所示。(图中示例是用 IEDA 导入的)

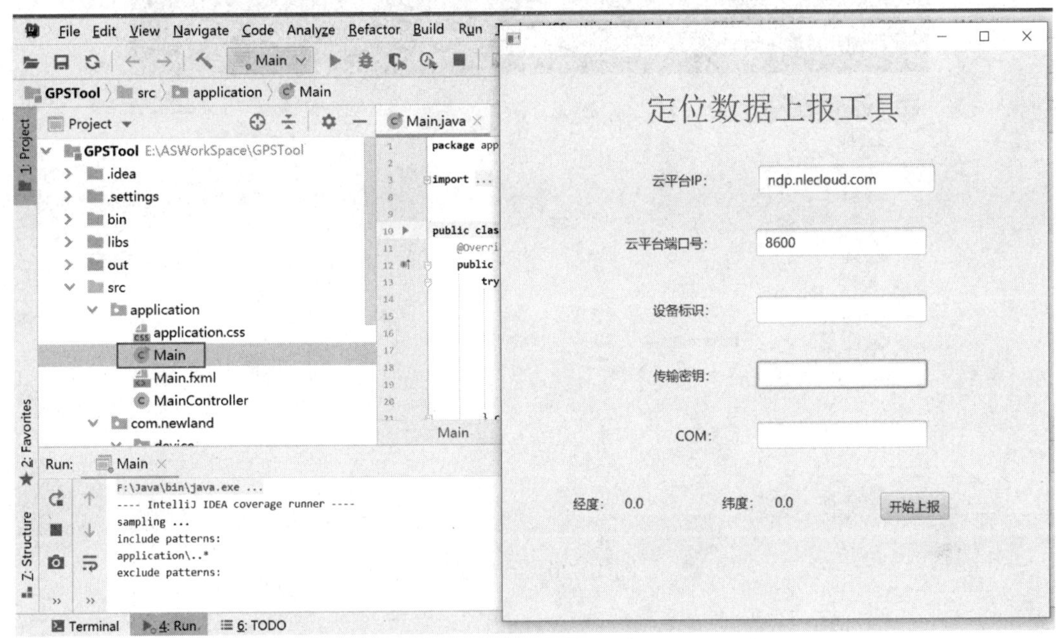

图 10-18　使用 IDEA 导入 GPSTool 项目并运行

从图 10-18 可以看到，需要填写设备标识、传输密钥和 COM 口号后单击“开始上报”按钮，才可以把采集的经纬度数据上报到云平台。

要获得设备标识、传输密钥这两个参数，需要在云平台上完成设备的创建，步骤如图 10-19 所示。

接下来，使用任务 3 注册好的账号登录云平台，切换到“开发者中心”界面，切换到“新增项目”选项卡，在弹出的“添加项目”界面中填写：项目名称为“经纬度数据上报”(可以自定义)；行业类别为“智慧城市”；联网方案为 WIFI，然后单击“下一步”按钮，如图 10-20 所示。

在弹出的“添加设备”界面中填写：设备名称为 BEIDOU(可以自定义)；通信协议为 TCP；设备标识为 PKR123456(可以自定义)，单击“确定添加设备”按钮，如图 10-21 所示。

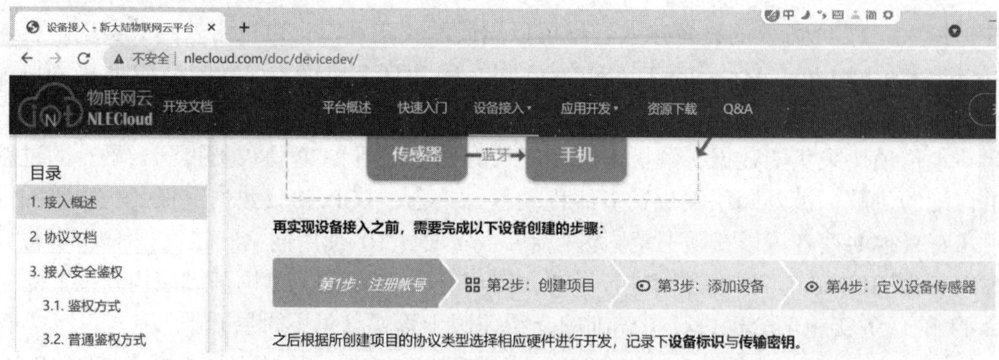

图 10-19　定位数据上报工具

图 10-20　添加项目

添加设备

*设备名称

BEIDOU　支持输入最多15个字符

*通讯协议

TCP　MQTT　CoAP　HTTP　LWM2M　ModbusTCP　TCP透传

*设备标识

PKR123456　英文、数字或其组合6到30个字符 解绑被占用的设备

数据保密性

公开(访客可在浏览中阅览设备的传感器数据)

数据上报状态

马上启用（禁用会使设备无法上报传感数据）

确定添加设备　关闭

图 10-21　添加设备

设备添加完成后，就创建好了“经纬度数据上报”项目，如图 10-22 所示。

单击图 10-22 中的“1 个设备”，可以看到项目信息，如图 10-23 所示。记下设备标识、传输密钥以及设备 ID，在后面的程序编写中需要这些信息。单击图 10-23 中的设备名称 BEIDOU，进入“添加传感器”界面，按图 10-24、图 10-25 中的信息添加经度和纬度传感器，传感器类型为“自定义”，标识名必须是 longitude 和 latitude。

图 10-22　经纬度数据上报项目

图 10-23　项目信息

图 10-24　添加经度传感器

图 10-25　添加纬度传感器

把北斗定位模块用 RS485 转 RS232 模块接到 PC 的 COM 口，按物联网工程实训相关知识和图 10-13 进行线路连接，记下识别出来的 COM 口号。打开串口助手 XCOM(也可以使用别的串口工具)，参数按图 10-26 进行设置，在发送窗口输入：01 03 00 05 00 23 14 12，单击“发送”按钮，可以看到接收区域显示了北斗定位模块获取的定位数据，如图 10-26 所示。

在定位数据上报工具中，填写识别出来的 COM 口号和设备标识、传输密钥后，再单击“开始上报”按钮，就可以看到云平台上设备上线了，开启实时数据后，经纬度数据实时上报到了云平台，同时定位数据上报工具中也显示了同步的经纬度数据，如图 10-27 所示。

(2) Android 程序从云平台获取经纬度数据。经纬度数据传送到云平台后， 接下来 Android 程序就可以从云平台获取经纬度数据。

① 创建 Project10_4 项目。本任务在任务 3 工程 Project10_3 的基础上实现，复制 Project10_3 并重命名为 Project10_4，进入 Project_4 文件夹，修改 Project10_3.iml 为 Project10_4.iml，修改 app/build.gradle 中的包名为 project10_4，然后使用 Rebuild Project 重构项目，如图 10-28 所示。

图 10-26　串口助手接收北斗定位数据

图 10-27　上报到云平台的经纬度数据

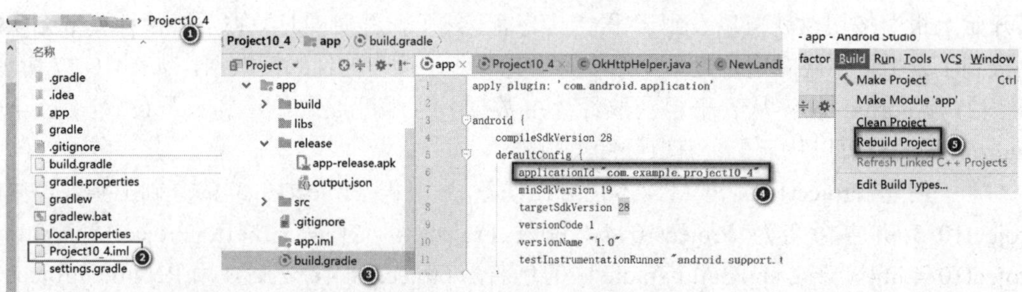

图 10-28　构建 Project10_4 项目

② 封装物联网云服务平台请求操作类。本任务要完成从云平台取经纬度数据，使用到两个API，一个是登录云平台(已在任务3中做了说明)，一个是批量查询设备最新数据，需要的参数是devids(设备ID)，返回的传感器数据在Msg中，如图10-29所示。

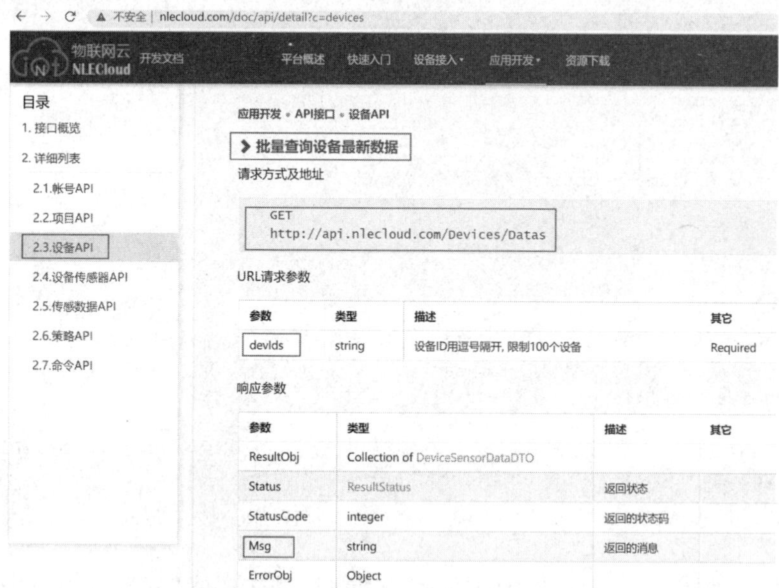

图10-29 批量查询设备最新数据的API使用说明

在包com.example.project10_4中新建NewLandECloud.java，用于实现云平台的相关操作，按登录和批量查询设备最新数据的API地址、请求参数、响应参数，封装的代码如下所示。

```
public class NewLandECloud {
 private static Logger logger = Logger.getLogger("NewLandECloud");
 private final String baseURL = "http://api.nlecloud.com";
 private static final String LOGIN = "/Users/Login";
 private static final String DEVICES = "/Devices";
private static final String DATAS = "/Datas";

 public  static final String Account = "XXXXXXXXXXX";  //填入你自己的云平台账号
 public  static final String Password = "XXXXXX";      //填入你自己的云平台密码
 public static String devIds = "229674";               //填入你自己的云平台项目的设备ID

 public NewLandECloud(){    }

 /**
  * 登录物联网云服务平台
  * @param loginname
  * @param password
  * @return
  * @throws RuntimeException
  * @throws IOException
  */
 public boolean login(String loginname, String password) throws RuntimeException,
IOException {
  String url = baseURL + LOGIN;
  String json = "{'Account':'"+loginname+"','Password':'"+password+"','IsRememberMe':
'true'}";
```

```
  String response = OkHttpHelper.post(url, json);
  logger.info(response);

  JSONObject jsonObj = JSONObject.parseObject(response);
  Integer status = jsonObj.getInteger("Status");
  if(status.equals(0)) {
      OkHttpHelper.accessToken = jsonObj.getJSONObject("ResultObj").getString("Access
Token");
      logger.info("accessToken = " + OkHttpHelper.accessToken);
      return true;
  }else {
   throw new RuntimeException("连接云平台失败！");
  }
 }

 /**
  * 批量查询设备最新数据
  * @param deviceId 多个设备编号，如果只查一个，则只传一个设备 ID 即可
  * @return 存放多个传感器键值对关系的 Map
  * @throws IOException
  */
 public Map<String, String> getDevicesData(String ... deviceId) throws IOException {
  if(deviceId == null || deviceId.length == 0) {
   return null;
  }

  StringBuilder url = new StringBuilder(baseURL).append(DEVICES).append(DATAS);
  url.append("?devIds=").append(deviceId[0]);
  for (int i = 1; i < deviceId.length; i++) {
   url.append(",").append(deviceId[i]);
  }

  String response = OkHttpHelper.get(url.toString());
  logger.info(response);
  System.out.println("传感器数据："+response);
  JSONObject jsonObj = JSONObject.parseObject(response);
  Integer status = jsonObj.getInteger("Status");
  if(!status.equals(0)) {
   throw new RuntimeException("您的请求操作失败，请检查您传入的参数");
  }
  //保存采集到的数据
  Map<String, String> data = new HashMap<>();
  //解析云平台返回的 JSON 数据
  JSONArray deviceAry = jsonObj.getJSONArray("ResultObj");
  int deviceSize = deviceAry.size();
  //遍历多个设备
  for (int i = 0; i < deviceSize; i++) {
   JSONObject deviceObj = deviceAry.getJSONObject(i);
   JSONArray sensorAry = deviceObj.getJSONArray("Datas");
   int sensorSize = sensorAry.size();
   //遍历获取每个设备中多个传感器的值，并保存到 Map 类型的对象中
   for (int j = 0; j < sensorSize; j++) {
    JSONObject sensorObj = sensorAry.getJSONObject(j);
    String apiTag = sensorObj.getString("ApiTag");
    String value = sensorObj.getString("Value");
    data.put(apiTag, value);
   }
  }
  return data;
 }
}
```

③ 登录云平台并获取经纬度数据。在布局文件 activity_main.xml 中编写代码，实现如图 10-30 所示的界面，其中经纬度的值可以从云平台获取并显示在控件上，也可以输入经纬度在地图上查看该位置，底部预留作为后续的地图显示使用。

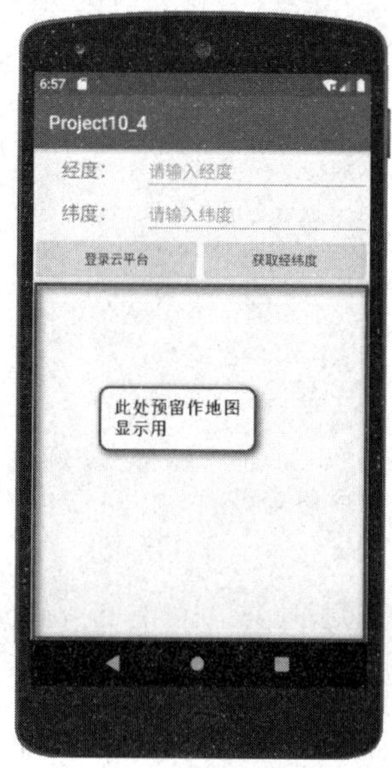

图 10-30 界面

布局文件 activity_main.xml 代码如下。

```
<?xml version="1.0" encoding="utf-8"?>
<LinearLayout xmlns:android="http://schemas.android.com/apk/res/android"
    xmlns:app="http://schemas.android.com/apk/res-auto"
    xmlns:tools="http://schemas.android.com/tools"
    android:layout_width="match_parent"
    android:layout_height="match_parent"
    android:orientation="vertical"
    tools:context=".MainActivity">
<LinearLayout
    android:layout_width="match_parent"
    android:layout_height="wrap_content"
    android:orientation="horizontal"
    >
    <TextView
        android:layout_width="0dp"
        android:layout_weight="1"
        android:layout_height="wrap_content"
        android:text="经度: "
        android:gravity="center"
        android:textSize="20sp"
        />
        <EditText
            android:id="@+id/edtJingdu"
            android:layout_width="0dp"
            android:layout_height="wrap_content"
            android:layout_weight="2"
            android:hint="请输入经度"

            />
```

```
</LinearLayout>
    <LinearLayout
        android:layout_width="match_parent"
        android:layout_height="wrap_content"
        android:orientation="horizontal"
        >
        <TextView
            android:layout_width="0dp"
            android:layout_weight="1"
            android:layout_height="wrap_content"
            android:text="纬度："
            android:gravity="center"
            android:textSize="20sp"
            />
        <EditText
            android:id="@+id/edtWeidu"
            android:layout_width="0dp"
            android:layout_weight="2"
            android:layout_height="wrap_content"
            android:hint="请输入纬度"

            />
    </LinearLayout>

    <LinearLayout
        android:layout_width="match_parent"
        android:layout_height="wrap_content">
        <Button
            android:layout_width="0dp"
            android:layout_weight="1"
            android:layout_height="wrap_content"
            android:id="@+id/loginYunButton"
            android:onClick="loginYun"
            android:text="登录云平台"
            />
        <Button
            android:layout_width="0dp"
            android:layout_weight="1"
            android:layout_height="wrap_content"
            android:id="@+id/btnFindButton"
            android:text="获取经纬度"
            android:onClick="getData"
            />
    </LinearLayout>
    <LinearLayout
        android:layout_width="match_parent"
        android:layout_height="0dp"
        android:layout_weight="1"
        >
        <-- 此处预留作地图显示用-->
    </LinearLayout>
</LinearLayout>
```

布局文件中登录按钮对应的点击事件处理方法为 loginYun，获取经纬度数据的按钮对应的点击事件处理方法为 getData，在 MainActivity.java 中编写代码，使用线程处理登录事件，登录成功或失败的结果通过 handler 消息对象传回主界面并进行相应提示。登录成功后，使用线程从云平中获取经纬度数据，需要传入的参数是设备 ID(在 NewLandECloud.devids 中已记录好)，获取数据后也是使用 handler 传回主界面，代码如下。

```
public class MainActivity extends AppCompatActivity {
  private NewLandECloud cloud;
  private Button btnFindButton;//获取经纬度
  private Button loginYunButton;//登录云平台
  private EditText edtJingdu;
  private EditText edtWeidu;

  private Context contest;
  // 云平台连接成功与否
  private boolean isConnsuccess = false;
  //是否停止采集的标志
  private boolean isStop;


  @Override
  protected void onCreate(Bundle savedInstanceState) {
      super.onCreate(savedInstanceState);
      setContentView(R.layout.activity_main);
      contest = this;
      loginYunButton = findViewById(R.id.loginYunButton);
      btnFindButton  = findViewById(R.id.btnFindButton);
      edtJingdu = findViewById(R.id.edtJingdu);
      edtWeidu = findViewById(R.id.edtWeidu);
      cloud = new NewLandECloud();
  }
  //消息对象
  Handler handler = new Handler() {
      @Override
      public void handleMessage(Message msg) {
          super.handleMessage(msg);
          if (msg.what == 2) {
              loginYunButton.setText("云平台连接成功");

          } else if (msg.what == 3) {
              loginYunButton.setText("登录云平台");
          } else if (msg.what == 1) {
              Map<String, String> map = (Map) msg.obj;
              if (map != null) {
                  edtJingdu.setText(map.get("longitude"));
                  edtWeidu.setText(map.get("latitude"));
              }
          }
      }
  };
  //登录云平台的按钮事件
  public void loginYun(View view) {
      //再一次点击时停止数据采集，显示登录云平台
      if(!(loginYunButton.getText().toString().trim().equals("登录云平台")))
      {
          loginYunButton.setText("登录云平台");
          isConnsuccess=false;
          isStop=true;
          return;
      }

      new Thread(new Runnable() {
          @Override
          public void run() {
              boolean isOk = false;
              try {
```

```
                    // 登录云平台
                    isOk = cloud.login(NewLandECloud.Account, NewLandECloud.Password);
                    if (isOk) {
                        //发送登录成功的消息
                        Message msg = new Message();
                        msg.what = 2;
                        handler.sendMessage(msg);
                        //设置登录成功的标志为真
                        isConnsuccess = true;
                    } else {
                        //发送登录失败的消息
                        Message msg = new Message();
                        msg.what = 3;
                        handler.sendMessage(msg);
                        //设置登录成功的标志为假
                        isConnsuccess = false;
                    }
                } catch (IOException e) {
                    e.printStackTrace();
                }
            }
        }).start();
    }

    //获取数据的按钮事件
    public void getData(View view) {
        if (!isConnsuccess) {
            Toast.makeText(this, "请先登录云平台", Toast.LENGTH_LONG).show();
            return;
        }
        if (btnFindButton.getText().toString().trim().equals("获取经纬度")) {
            isStop = false;
            new Thread(new Runnable() {
                @Override
                public void run() {

                    while (!isStop) {

                        try {
                            //获取多个设备中多个传感器的数据值
                            Map<String, String> map  = cloud.getDevicesData(NewLandECloud.devIds);

                            Message msg = new Message();
                            msg.what = 1;
                            msg.obj = map;
                            //把采集到的数据通过消息机制发出去
                            handler.sendMessage(msg);
                            Thread.sleep(2000);
                        } catch (Exception e) {
                            e.printStackTrace();
                        }
                    }
                }
            }).start();

            btnFindButton.setText("结束");
        } else {
            //结束采集
            isStop = true;
```

```
                btnFindButton.setText("获取经纬度");
            }
        }
}
```

运行程序，结果如图 10-31 所示。

图 10-31 登录并获取云平台上的经纬度数据

(3) 使用高德地图 SDK 配合经纬度数据实现定位。

获取到经纬度数据后，可以利用高德地图实现根据经纬度值实现定位(查询地图位置)的功能，需要按以下步骤进行。

① 创建高德地图新应用并获取 key。高德开放平台网址为：https://lbs.amap.com，按提示注册并登录后，进入控制台→应用管理→“新建应用”对话框，输入应用名称(自定义)、选择应用类型(自行选择)后，单击“新建”按钮，即可创建好应用，如图 10-32 和图 10-33 所示。

图 10-32 “新建应用”对话框

图 10-33 园区定位应用

单击图 10-33 中的“添加”按钮，可以为应用添加 Key，如图 10-34 所示。

图 10-34 中需要填写发布版和调试版安全码 SHA1，要生成 SHA1，得先获得 Key Store。Key Store 是软件的数字签名，当新版程序和旧版程序的数字证书相同时，Android 系统才会认为这两个程序是同一个程序的不同版本。如果新版程序和旧版程序的数字证书不相同，Android 系统则认为它们是不同的程序，并产生冲突，会要求新程序更改包名。

Android Studio 生成 apk 对应的 Key Store 的步骤是：单击 Build，选择 Build Bundle(s)/APK(s)→Build APK→Create new，自定义保存的路径和文件名并填写相关信息，记下创建 KeyStore 时的密码，在 Build Type 中选择签名模式为 release，完成后单击 Finish

按钮，如图 10-35 所示。

图 10-34　为应用添加 Key

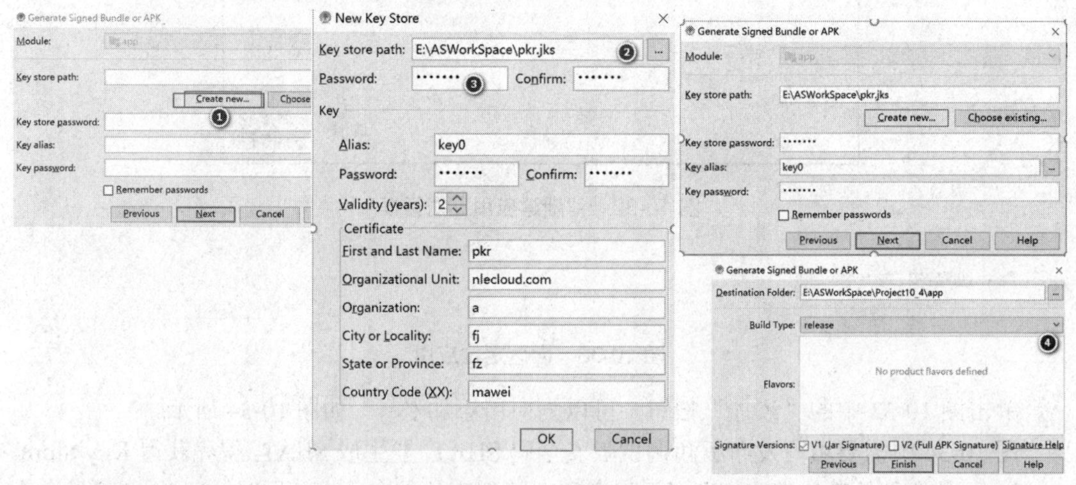

图 10-35　生成 Key Store

生成发布版本的安全码 SHA1 需要使用 apk 对应的 Key Store，先切换到.android 所在路径(一般在用户文件夹下面的.android 文件夹中)，再使用命令 keytool -list -v -keystore apk 的 Key Store，如图 10-36 所示。

```
C:\Users\pkr\.android>keytool -list -v -keystore E:\ASWorkSpace\pkr.jks
输入密钥库口令:
密钥库类型: JKS
密钥库提供方: SUN

您的密钥库包含 1 个条目

别名: key0
创建日期: 2021-4-10
条目类型: PrivateKeyEntry
证书链长度: 1
证书[1]:
所有者: CN=pkr, OU=nlecloud.com, O=a, L=fj, ST=fz, C=mawei
发布者: CN=pkr, OU=nlecloud.com, O=a, L=fj, ST=fz, C=mawei
序列号: 270498a5
有效期为 Sat Apr 10 14:47:09 CST 2021 至 Wed Apr 04 14:47:09 CST 2046
证书指纹:
         MD5:  1D:1D:72:E9:02:1D:B6:A6:2C:4C:41:DA:51:66:7D:40
         SHA1: EB:48:E8:4C:75:85:2F:95:FC:32:CC:B6:22:83:DC:E2:70:0D:A4:70
         SHA256: EE:14:1A:98:E9:92:C6:8E:8C:A0:00:17:23:E8:C8:6A:35:86:C0:9F:6F:19:77:32:82:DE:65:23:7B:B9:0B:F5
签名算法名称: SHA256withRSA
主体公共密钥算法: 2048 位 RSA 密钥
版本: 3

扩展:

#1: ObjectId: 2.5.29.14 Criticality=false
SubjectKeyIdentifier [
KeyIdentifier [
0000: C7 BF E9 7D 91 88 32 5E   F3 CF F5 FE 51 D7 9F D4  ......2^....Q...
0010: 16 69 9D 1B                                        .i..
```

创建Key Store时的密码

图 10-36 生成发布版的 SHA1

生成调试版本的安全码 SHA1 需要使用 debug.keystore(在用户文件夹下面的.android 文件夹中)，命令为：keytool -list -v -keystore debug.keystore，密钥库口令默认为 android，如图 10-37 所示。

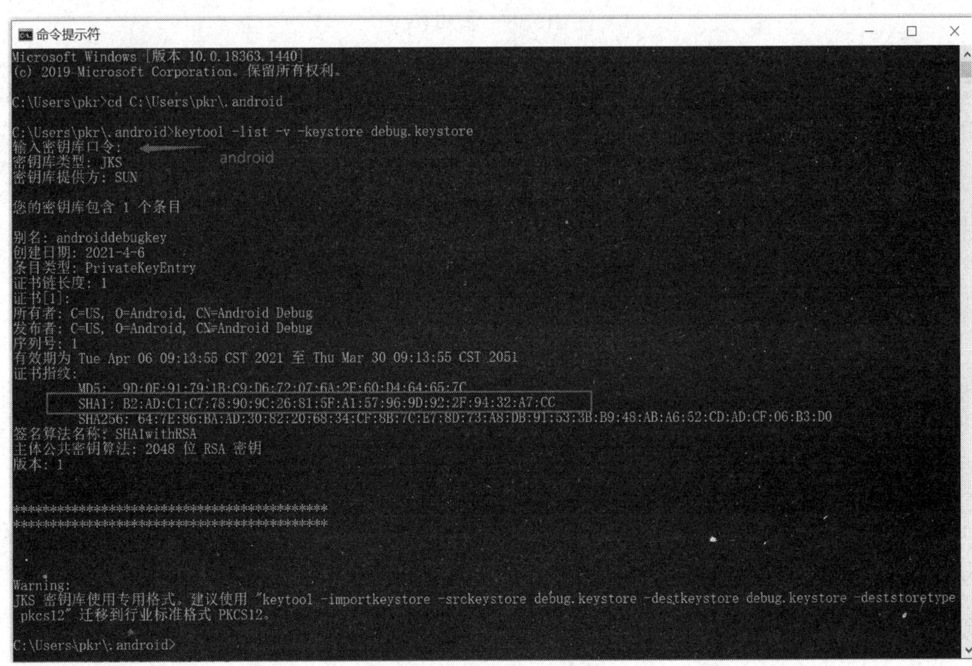

图 10-37 生成调试版本的 SHA1

将生成的 SHA1 和包名填写到图 10-38 中相应的位置后，单击“提交”按钮，完成给应用添加 Key 的操作，如图 10-38、图 10-39 所示。

图 10-38　添加 Key

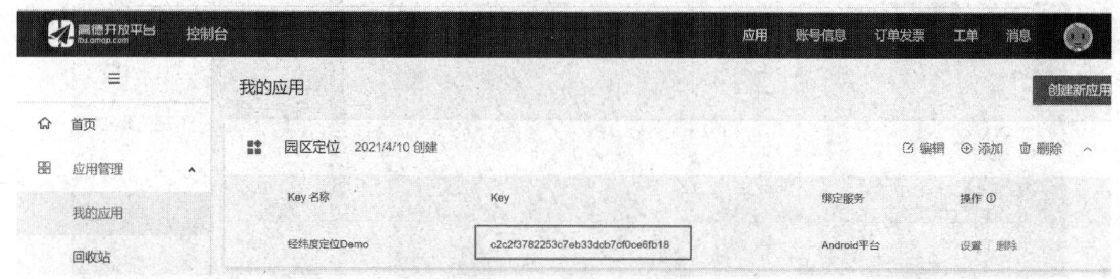

图 10-39　查看应用对应的 Key

② 在 Project10_4 项目里使用高德地图 SDK。在项目里集成 SDK 有以下两种方法。

方法一：将下载的定位 SDK jar 文件复制到工程。

方法二：通过 Gradle 集成 SDK，即添加相关的依赖。

这里采用的是方法二，在 app/build.gradle 中添加高德地图的依赖并同步项目，代码如下。

```
dependencies {
......
    implementation 'com.amap.api:3dmap:latest.integration'
    implementation 'com.amap.api:location:latest.integration'
}
```

③ 在 Project10_4 项目里配置 Key。在 AndroidManifest.xml 的 application 标签中配置 Key，value 为图 10-39 中的 Key，实际应用时要改成自己的 Key，代码如下。

```
<meta-data android:name="com.amap.api.v2.apikey"
    android:value="c2c2f3782253c7eb33dcb7cf0ce6fb18">
</meta-data>
```

④ 在 Project10_4 项目里声明地图定位服务。在 AndroidManifest.xml 的 application 标签中声明 service 组件，代码如下。

```
<service android:name="com.amap.api.location.APSService"></service>
```

⑤ 在清单文件中添加权限。在 AndroidManifest.xml 中添加权限，代码如下。

```
<!-- 用于进行网络定位 -->
<uses-permission android:name="android.permission.ACCESS_COARSE_LOCATION" />
<!-- 用于访问 GPS 定位 -->
<uses-permission android:name="android.permission.ACCESS_FINE_LOCATION" />
<!-- 获取运营商信息，用于支持提供运营商信息相关的接口 -->
<uses-permission android:name="android.permission.ACCESS_NETWORK_STATE" />
<!-- 用于访问 Wi-Fi 网络信息，Wi-Fi 信息会用于进行网络定位 -->
<uses-permission android:name="android.permission.ACCESS_WIFI_STATE" />
<!-- 这个权限用于获取 Wi-Fi 的获取权限，Wi-Fi 信息会用来进行网络定位 -->
<uses-permission android:name="android.permission.CHANGE_WIFI_STATE" />
<!-- 用于访问网络，网络定位需要上网 -->
<uses-permission android:name="android.permission.INTERNET" />
<!-- 用于读取手机当前的状态 -->
<uses-permission android:name="android.permission.READ_PHONE_STATE" />
<!-- 写入扩展存储，向扩展卡写入数据，用于写入缓存定位数据 -->
<uses-permission android:name="android.permission.WRITE_EXTERNAL_STORAGE" />
<!-- 用于申请调用 A-GPS 模块 -->
<uses-permission android:name="android.permission.ACCESS_LOCATION_EXTRA_COMMANDS" />
<!-- 用于申请获取蓝牙信息进行室内定位 -->
<uses-permission android:name="android.permission.BLUETOOTH" />
<uses-permission android:name="android.permission.BLUETOOTH_ADMIN" />
```

⑥ 处理运行时权限(在 Android 6.0 后需要)。如果 Project10_4 项目运行在工控平板上，不需要处理运行时的异常，因为工控平板是 Android 5.1 的，如果运行在 Android 6.0 以后的版本，则需要处理运行时的异常。

在 MainActivity 中添加处理运行时权限的方法，并在 onCreate()方法中调用，代码如下。

```
public class MainActivity extends AppCompatActivity {
     ......
    @Override
    protected void onCreate(Bundle savedInstanceState) {
      ......
        requestPermission(this);//处理运行时权限
    }

    //Android 6.0 之后需要动态申请权限
    public  boolean isPermissionRequested;
    public  boolean requestPermission(Activity activity) {
        if (Build.VERSION.SDK_INT >= 23 && !isPermissionRequested) {
            ArrayList<String> permissionsList = new ArrayList<>();
            String[] permissions = {
                    Manifest.permission.ACCESS_COARSE_LOCATION,
                    Manifest.permission.INTERNET,
                    Manifest.permission.ACCESS_FINE_LOCATION,
                    Manifest.permission.ACCESS_NETWORK_STATE,
                    Manifest.permission.ACCESS_WIFI_STATE,
                    Manifest.permission.CHANGE_WIFI_STATE,
```

```
                    Manifest.permission.READ_EXTERNAL_STORAGE,
                    Manifest.permission.READ_PHONE_STATE,
                    Manifest.permission.WRITE_EXTERNAL_STORAGE,
                    Manifest.permission.ACCESS_LOCATION_EXTRA_COMMANDS,
                    Manifest.permission.BLUETOOTH,
                    Manifest.permission.BLUETOOTH_ADMIN,
            };

            for (String perm : permissions) {
                if (PackageManager.PERMISSION_GRANTED != checkSelfPermission(perm)) {
                    permissionsList.add(perm);
                    // 进入到这里代表没有权限
                }
            }
            if (!permissionsList.isEmpty()) {
                Toast.makeText(activity,"请先开启权限方可使用! ",Toast.LENGTH_LONG).show();
                //  ToastUtils.toast(context,"请先开启权限方可使用! ");
                String[] strings = new String[permissionsList.size()];
                activity.requestPermissions(permissionsList.toArray(strings), 0);

                Log.e("pkr","in1");
                return  false;
            }else {
                Log.e("pkr","in2");
                return  true;
            }
        }else {
            Log.e("pkr","in3");
            return  true;
        }
    }
}
```

⑦ 通过经纬度在地图上显示定位位置。在 activity_main.xml 中原来预留地图显示的位置添加地图控件，代码如下。

```
<LinearLayout
    android:layout_width="match_parent"
    android:layout_height="0dp"
    android:layout_weight="1" >
    <com.amap.api.maps.MapView
        android:id="@+id/map"
        android:layout_width="match_parent"
        android:layout_height="match_parent"/>
</LinearLayout>
```

在 MainActivity 中使用高德地图对象 aMap，并用 Activity 的生命周期管理地图控件的生命周期，将从云平台获取的经纬度数据封装到 latLng 对象中，通过 aMap 对象提供的方法在地图中显示位置，代码如下。

```
public class MainActivity extends AppCompatActivity {
     ......
    private MapView mapView;
    private AMap aMap;
    private LatLng latLng;// 以纬度和经度表示的地理坐标点

    @Override
    protected void onCreate(Bundle savedInstanceState) {
      ......
        mapView = (MapView) findViewById(R.id.map);
```

```
        mapView.onCreate(savedInstanceState);//必须写
        // 初始化AMap对象
        if (aMap == null) {
            aMap = mapView.getMap();
        }
        requestPermission(this);//处理运行时权限
    }
    //消息对象
    Handler handler = new Handler() {
        @Override
        public void handleMessage(Message msg) {
        ......
               else if (msg.what == 1) {
                Map<String, String> map = (Map) msg.obj;
                if (map != null) {
                    edtJingdu.setText(map.get("longitude"));
                    edtWeidu.setText(map.get("latitude"));
                    latLng = new LatLng(Double.valueOf(map.get("longitude")), Double.valueOf(map.get("latitude")));//地点经纬度
                    final Marker marker = aMap.addMarker(new MarkerOptions().position(latLng).title("福州").snippet("DefaultMarker"));//添加标志
                    aMap.moveCamera(CameraUpdateFactory.newLatLngZoom(latLng, 18));//显示指定的地点
                }
            }
        }
    };
    //登录云平台的按钮事件
    public void loginYun(View view) {
      ......
    }
    //获取数据的按钮事件
    public void getData(View view) {
     ......
    }
    @Override
    protected void onResume() {
        super.onResume();
        mapView.onResume();
    }
    @Override
    protected void onPause() {
        super.onPause();
        mapView.onPause();
    }
    @Override
    protected void onSaveInstanceState(Bundle outState) {
        super.onSaveInstanceState(outState);
        mapView.onSaveInstanceState(outState);
    }
    @Override
    protected void onDestroy() {
        super.onDestroy();
        mapView.onDestroy();
    }
    //Android 6.0之后需要动态申请权限
    public  boolean isPermissionRequested;
    public  boolean requestPermission(Activity activity) {
      ......
```

```
        }
}
```

上述代码中的经纬度数据只实现了从云平台上对应项目中获取，读者理解后，也可以改写代码，实现获取手动输入的经纬度，在地图上显示定位位置。

3. 运行结果

运行程序，登录云平台成功后，先接受运行时权限许可，地图先显示北京市的定位信息，单击“获取经纬度”按钮后，可以看到从云平台上获取了经纬度数据并回填到控件上，同时定位信息也显示在地图控件上，如图 10-40 所示。

图 10-40　定位程序的运行结果

项目总结

本项目主要讲解了网络技术与定位技术的使用，包括用 WebView 浏览网页、用 HttpURLConnection 和 OkHttp 访问网络、用北斗定位模块获取经纬度数据并配合高德地图进行定位，为后续开发网络相关的功能打好基础。

思考与练习

10-1　简述 WebView 的使用过程。

10-2　简述使用 HttpURLConnection 连接网络并获取数据的过程。

10-3　简述使用 OkHttp 访问网络的过程。

10-4　简述使用北斗定位模块实现定位的过程。

项目11 园区环境实时监测系统

【项目描述】

本项目实现园区环境的实时监测。监测的数据来源有两种：第一种是采集到的传感器数据经 LoRa 网关上传到物联网云平台；第二种是采集到的传感器数据经 NB-IoT 上传到物联网云平台。

第一种数据流向：利用物联网工程实训系统 2.0 中的 PC 端智能环境云软件生成模拟的传感器数据，通过串口下发到 6 个模拟传感器 NewSensor(NS)的设备上。NS 通过 LoRa 通道或 LoRa 节点把数据传递给 LoRa 网关。LoRa 网关通过轮询方式采集 NS 传感器数据并将数据上传至物联网云服务平台，由云平台长期保存。Android 客户端程序从云平台取数据进行展示。

第二种数据流向： NB-IoT 节点上接传感器，经过采集后通过 NB-IoT 网络上传至物联网云平台，并由云平台长期保存。Android 客户端程序从云平台取数据进行展示。

本项目使用的设备如图 11-1 所示。

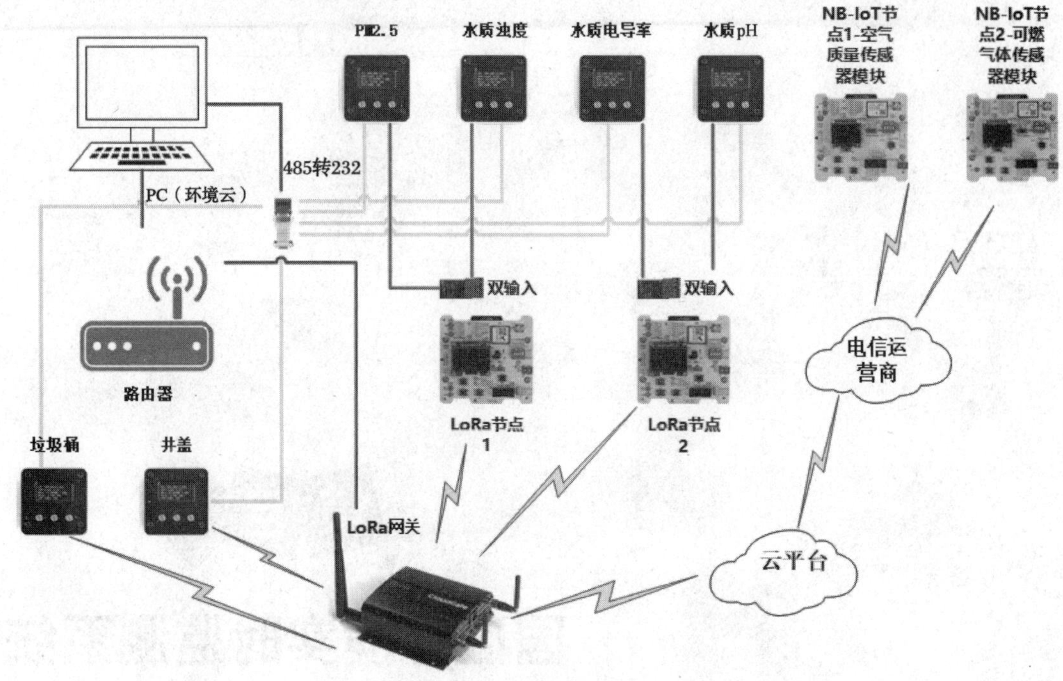

图 11-1　拓扑图

项目的实现过程如图 11-2 所示。

图 11-2　实现过程

项目运行结果如图 11-3 和图 11-4 所示。

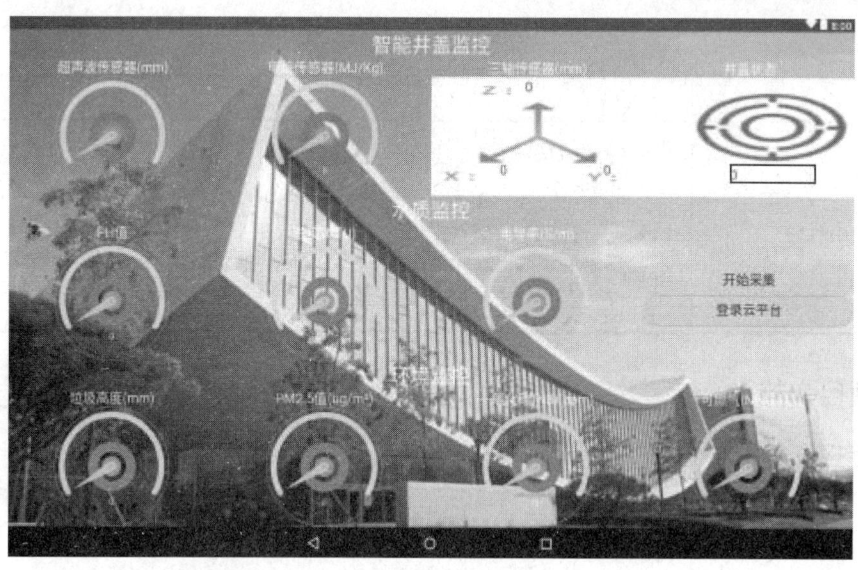

图 11-3　数据采集界面

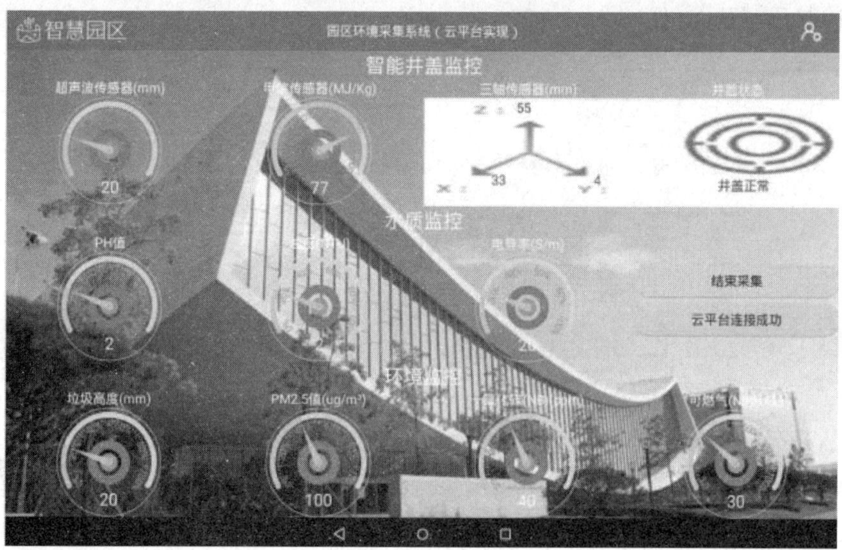

图 11-4　运行结果

【学习目标】

技能目标：

能综合设计制作 UI 界面。
能综合应用 Android 的网络通信技术。
能综合应用 Android 的多线程技术。
能编写与物联网云服务平台交互的 Android 程序。
具体任务清单如表 11-1 所示。

表 11-1　任务清单

任务课时	12 课时	任务组员数量	建议 3 人
任务组采用设备	PC 安装好智能环境云 NS 模块六个 LoRa 节点两个 LoRa 网关一个 NB-IoT 节点两个		

【项目实施】

1. 连接设备并进行调试

本项目用到的设备比较多，所有的设备连接、烧写、配置可参考物联网工程实训系统中相关的操作，为确保项目成功实施，需注意以下事项。

1)　LoRa 网关配置注意事项

LoRa 网关参数配置如图 11-5 所示。

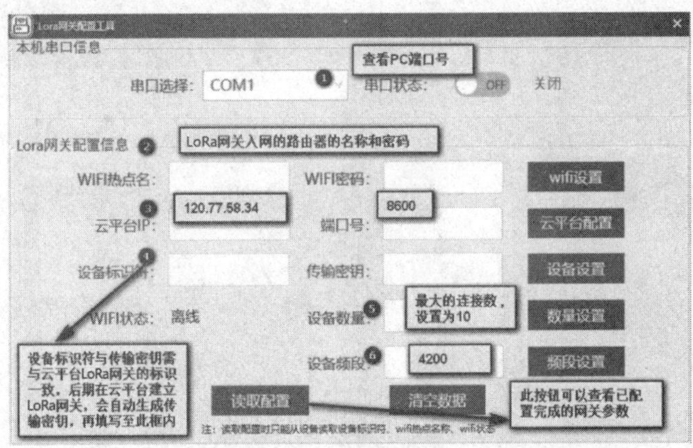

图 11-5　LoRa 网关参数配置

2)　NS 配置注意事项

NS 传感器的数据来源于环境云，可参考物联网工程实训系统中关于智能环境云的设置，用 NS 配置工具配置 NS 时具体的传感器名、标识名、地址如图 11-6 所示。

2. 搭建物联网云平台项目

1)　注册与登录物联网云平台

可参考物联网工程实训系统中的操作步骤，此处略。

2)　新增项目和添加网关设备

登录成功后，进入开发者中心的新增项目页面，如图 11-7 所示。

第一步：在开发者中心的项目列表页面上，单击左上角的“新增项目”按钮，在弹出的“添加项目”页面中输入“项目名称”和“行业类别”，“行业类别”可选择“智慧城市”，单击“下一步”按钮完成新增项目，如图 11-8 所示。

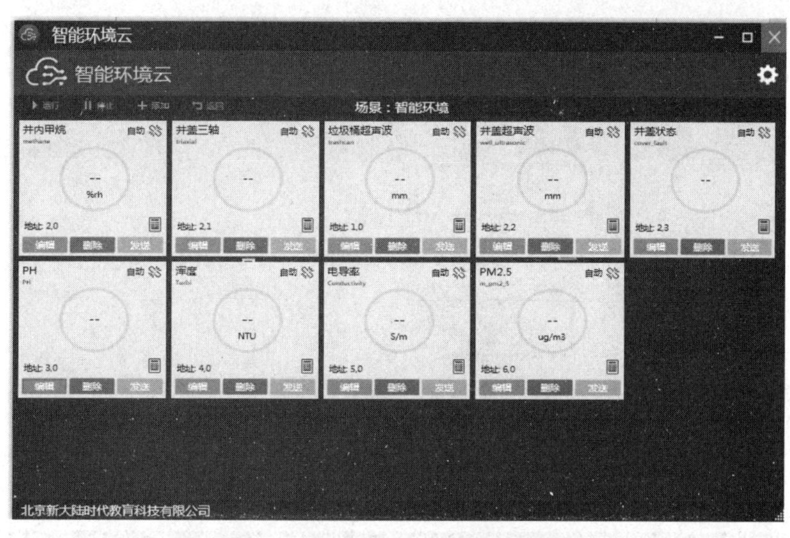

图 11-6　智能环境云配置参数界面

图 11-7　开发者中心的新增项目页面

第二步： 新项目添加完成后，进入“添加设备”页面，输入“设备名称”和“设备标识”(标识名必须唯一)，单击“确认添加设备”按钮完成设备添加，如图 11-9 所示。

图 11-8　新增项目

图 11-9　“添加设备”页面

第三步： 添加设备完成后，自动返回到开发中心的项目列表页面。在此页面上可以看到刚才新增的项目，此项目中有一个新增设备，如图 11-10 所示。

第四步： 在开发者中心的设备列表页面上，单击 1 个设备 0 个传感器，进入设备列表页面，可查看到刚添加的设备，如图 11-11 所示。

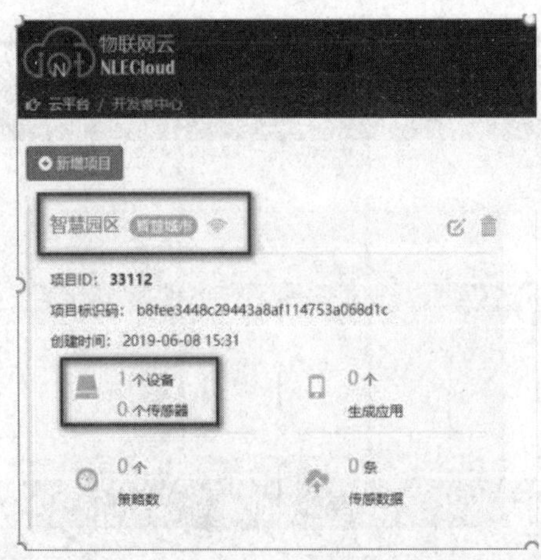

图 11-10　开发者中心的项目列表页面

图 11-11　项目的设备列表页面

第五步：在项目的设备列表页面上，单击“添加设备”按钮，再添加其他设备，如“一氧化碳(NB)”和“可燃气(NB)”，如图 11-12 和图 11-13 所示。

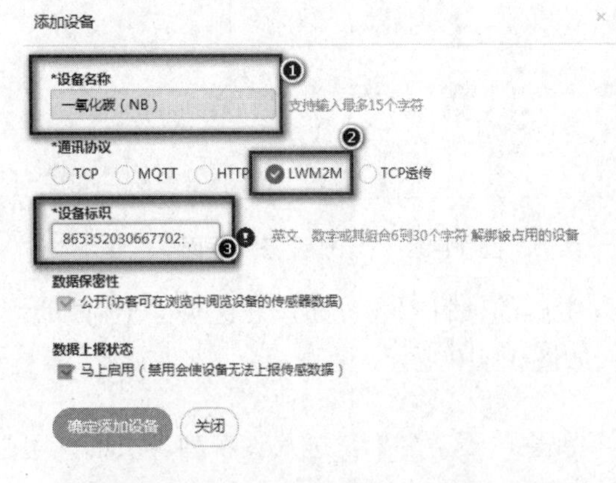

图 11-12　添加一氧化碳(NB)设备

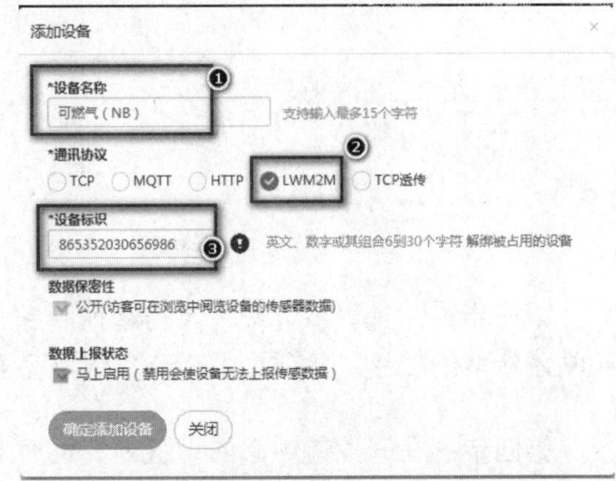

图 11-13　添加可燃气(NB)设备

第六步：设备添加完成后，再查看项目的设备列表页面，如图 11-14 所示。

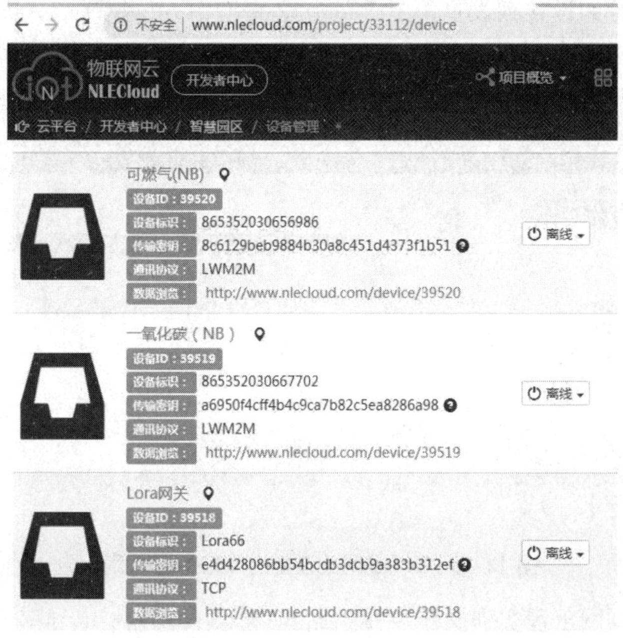

图 11-14　项目的设备列表页面

3)　给 LoRa 网关设备添加传感器

本项目使用 6 个 NS 设备用于模拟真实传感器，其中 NS1、NS2 为 LoRa 版，NS3～NS6 为 AO 版。需要注意的是，采用 LoRa 模式的 NS1 和 NS2(内置 LoRa 节点)中数据直接被 LoRa 网关采集，采用 AO 模式的 NS3～NS6 输出 4～20mA 电流(需外加 LoRa 节点)，才能被 LoRa 网关采集。

6 个 NS 设备，分别用于模拟表 11-2 所列传感器。

表 11-2　NS 设备说明

<table>
<tr><th>设备名称</th><th colspan="4">传感器名称</th><th>传输方式</th><th>监测场合</th></tr>
<tr><td>NS1</td><td colspan="4">超声波</td><td rowspan="2">LoRa 传输</td><td>垃圾桶</td></tr>
<tr><td>NS2</td><td>甲烷</td><td>三轴</td><td>超声波</td><td>井盖状态</td><td>井盖</td></tr>
<tr><td>NS3</td><td colspan="4">pH</td><td rowspan="4">AO 传输</td><td rowspan="3">水质环境</td></tr>
<tr><td>NS4</td><td colspan="4">浊度</td></tr>
<tr><td>NS5</td><td colspan="4">水质电导率</td></tr>
<tr><td>NS6</td><td colspan="4">PM2.5</td><td>空气环境</td></tr>
</table>

NS1 代表智能环境中垃圾桶，超声波标识测距垃圾桶是否溢满。

NS2 代表智能环境中的井盖，用一个 NS 设备同时模拟 4 种传感器设备，包括井盖内甲烷值，三轴即为井盖位置的坐标点，超声波为井盖内水位的高度，井盖状态为井盖当前信息。

NS3～NS6 分别代表监测水质和空气环境的传感器：pH、浊度、水质电导率、PM2.5。

下面详细介绍如何在云平台上添加上述 NS 传感器。

第一步：在项目的设备列表页面上，单击设备名“LoRa 网关”，进入传感器管理页面(见图 11-15)，再单击添加传感器图标 ，进入添加传感器管理页面。

图 11-15　LoRa 网关设备的传感器管理页面

第二步：按物联网工程实训操作步骤添加 pH 监测传感器、电导率监测传感器、浊度监测传感器、PM2.5 值监测传感器、井盖状态监测传感器、井盖超声波监测传感器、井盖三轴监测传感器、井内甲烷监测传感器、智能垃圾桶监测传感器等，添加结果如图 11-16 所示。

图 11-16　添加传感器

4)　给 NB-IoT 设备添加传感器(一氧化碳)

第一步：在项目的设备列表页面上，单击设备名“一氧化碳(NB)”，进入传感器管理页面，当新增设备时采用 LWM2M 通信协议设备，系统默认会添加以上两个传感器，如图 11-17

所示。

图 11-17　一氧化碳(NB)设备的传感器管理页面 1

第二步：本设备只监测一氧化碳值(标识名 CarbonMonoxidesStr)，所以可以删除标识名为 FlammableGas 的传感器，单击此传感器右边的 ⊗ 图标实现删除，删除后如图 11-18 所示。

图 11-18　一氧化碳(NB)设备的传感器管理页面 2

5)　给 NB-IoT 设备添加传感器(可燃气)

第一步：在项目的设备列表页面上，单击设备名“可燃气(NB)”，进入传感器管理页面，当新增设备时采用 LWM2M 通信协议，系统默认会添加以上两个传感器，如图 11-19 所示。

图 11-19　可燃气(NB)设备的传感器管理页面 1

第二步：本设备只监测可燃气值(标识名 FlammableGas)，所以可以删除标识名为 CarbonMonoxidesStr 的传感器，单击此传感器右边的 ⊖ 图标实现删除，删除后如图 11-20 所示。

图 11-20　可燃气(NB)设备的传感器管理页面 2

3．云平台 API 在线调试工具的使用

要登录并获取云平台数据，需要调用云平台提供的一套 API。接口概览网址为 http://api.nlecloud.com/doc/api，如图 11-21 所示。

API接口概览 - 新大陆物联网云平

不安全 | api.nlecloud.com/doc/api

物联网云 NLECloud 开发文档　平台概述　快速入门　设备接入　应用开发　资源下载　开发者中心

目录

1. 接口概览
2. 详细列表
   - 2.1.帐号API
   - 2.2.项目API
   - 2.3.设备API
   - 2.4.设备传感器API
   - 2.5.传感数据API
   - 2.6.命令API

接口概览

组别	HTTP	API地址	描述
帐号API	POST	users/login	用户登录（同时返回AccessToken）
项目API	GET	projects/{projectid}	查询单个项目
	GET	projects	模糊查询项目
	GET	projects/{projectid}/sensors	查询项目所有设备的传感器
设备API	GET	devices/datas	批量查询设备最新数据
	GET	devices/status	批量查询设备的在线状态
	GET	devices/{deviceid}	查询单个设备
	GET	devices	模糊查询设备
	POST	devices	添加个新设备
	PUT	devices/{deviceid}	更新某个设备
	DELETE	devices/{deviceid}	删除某个设备
设备传感器API	GET	devices/{deviceid}/sensors/{apitag}	查询单个传感器
	GET	devices/{deviceid}/sensors	模糊查询传感器
	POST	devices/{deviceid}/sensors	添加个新传感器
	PUT	devices/{deviceid}/sensors/{apitag}	更新某个传感器
	DELETE	devices/{deviceid}/sensors/{apitag}	删除某个传感器

图 11-21　云平台接口概览页面

单击左边“详细列表”菜单栏，在右边可以得到请求方式和地址及参数说明等，如单击“2.1.账号 API”后，结果如图 11-22 所示。

图 11-22　登录 API

Android 程序从云平台获取数据，可以参照相关格式进行数据请求并从响应中获取对应的请求结果。为了让读者明白请求和响应结果的用法，先利用云平台提供的 API 接口进行数据测试，以验证传感器数据确实上传到了云平台。接下来介绍 API 在线调试工具的使用。

1)　测试登录云平台是否成功

打开 API 在线调试工具，如图 11-23 所示。

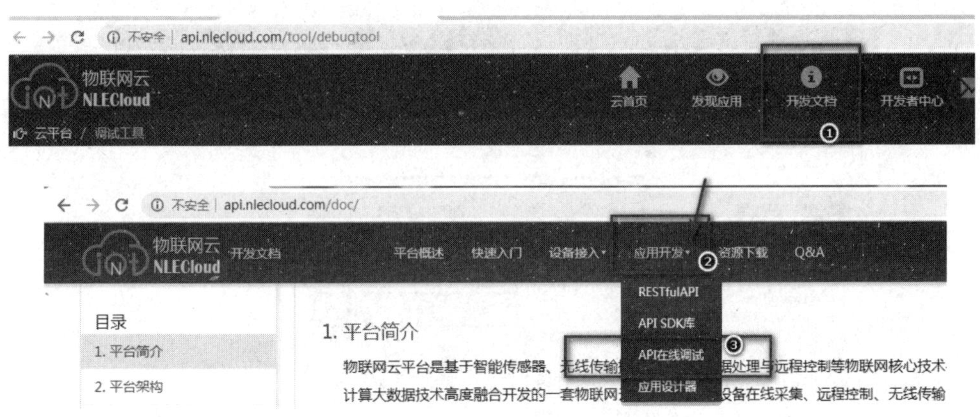

图 11-23　进入 API 在线调试界面

从“登录账号 API”可以看出，用户名和密码发送到云平台，采用的是 POST 方式，如果成功则返回 AccessToken 值，之后的每个请求都必须携带 AccessToken 值，所以需要保存这个 AccessToken 值，如图 11-24 所示。

图 11-24 中①选择 Post，②输入登录云平台的账号和密码(注意要输入自己的账号和密码，此处是模拟的账号和密码)，③单击“发送请求”按钮，④返回结果。如果登录成功，把返回结果中的 AccessToken 值复制并粘贴到③处。

2)　测试批量查询设备最新数据

有了 AccessToken 值后，如果要批量获取设备最新数据，可以选择“设备 API”，查阅请求方式和地址，如图 11-25 所示。

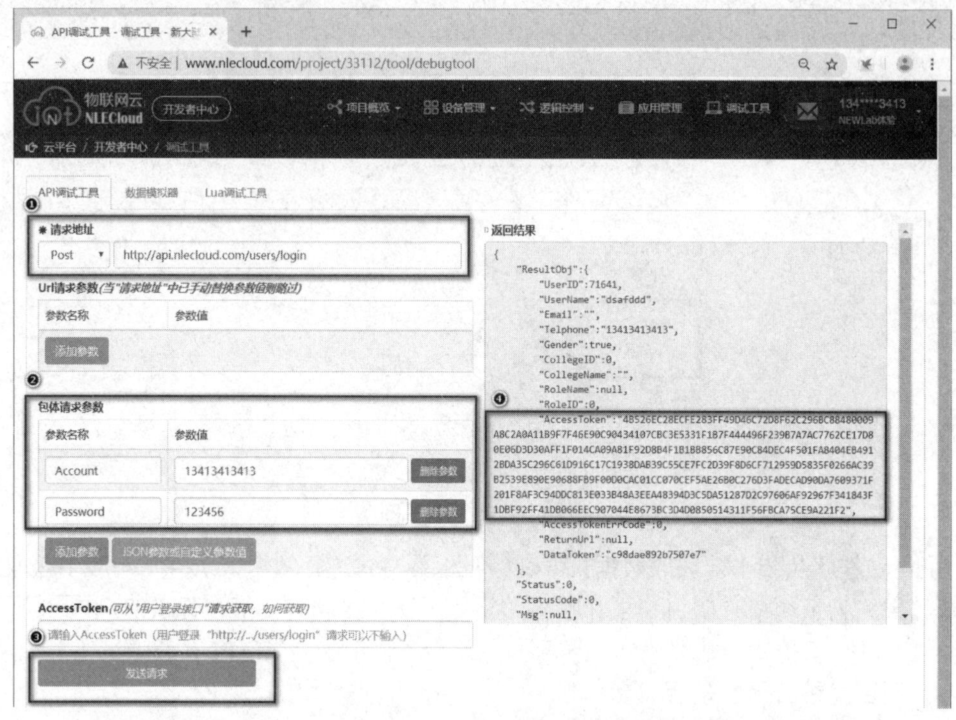

图 11-24　测试登录

图 11-25　设备 API 说明

接着回到 API 测试工具页面，按请求方式和地址及 URL 请求参数填写图 11-26，其中②的参数值分别从图 11-25 中获取，参数填完单击“发送请求”按钮，从右边可以看到返回结果列出了每个设备中包含的传感器的 ApiTag 和对应的值 Value，如果 Java 程序想获取这些值，只需要按格式解析即可，如图 11-26 所示。

其他 API 接口有兴趣的读者可继续测试，返回结果分析可参考 API 接口说明文档。

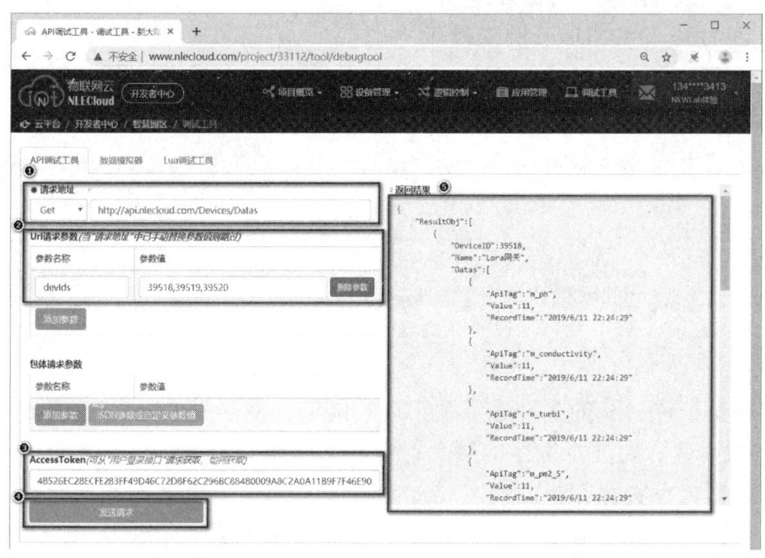

图 11-26　批量获取数据

4. 实时监测园区环境的功能实现

经过上述操作，初步验证了空气环境、水质环境、井盖状态、垃圾桶状态等传感器数据的实时监测数据已上传至物联网云服务平台，现需要编写 Android 端应用程序实时显示园区环境监测值。首先需要添加访问网络和解析 JSON 格式的依赖库，制作云平台参数设置和保存界面，编写保存和读取云平台参数的类，编写使用 OkHttp 的类进行网络访问，制作实时显示环境监测数据的界面，再通过启动线程定时从云平台获取实时数据，通过消息机制更新 UI 界面的显示值。

1)　园区环境实时监测功能的目录结构

本项目是在项目 9 的基础上实现园区环境的实时监测，涉及的目录结构如图 11-27 所示，后续功能对应的代码编写位置读者可参考本结构图。

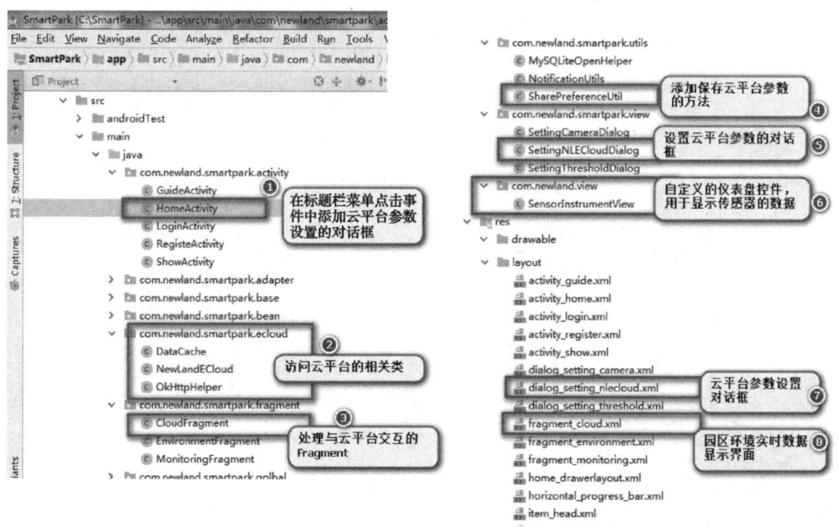

图 11-27　园区环境实时监测目录结构

2) 添加 OkHttp 和 FastJson 的依赖库

本项目访问物联网云平台通过 OkHttp 来完成，从云平台返回数据的格式是 JSON 格式，用 FastJson 来解析，所以需要在工程中添加这两个依赖库。

打开项目 9 的工程 SmartPark，在 app/build.gradle 文件的闭包 dependencies 中添加如下代码：

```
1.  dependencies {
2.     ......
3.      implementation 'com.squareup.okhttp3:okhttp:4.0.0-RC3'
4.      implementation 'com.alibaba:fastjson:1.2.58'
5.  }
```

因为本项目是在项目 9 的基础上继续实现功能的，项目 9 中用到的摄像头的 jar 包中已包含 OkHttp，所以读者要注意上述的第 3 行代码此时要去掉。

3) 制作云平台参数保存界面

在 layout 下新建 dialog_setting_nleclout.xml，用于设置和保存登录云平台的参数和要获取的设备 id，效果如图 11-28 所示，此处不展开介绍。

图 11-28　云平台参数保存界面

4) 编写缓存云平台参数的相关类

第一步，在 SharePreferenceUtil 类中添加以下云平台相关参数的 GET 和 SET 方法：

```
public class SharePreferenceUtil {
.....
  //云平台 NB 甲烷
    public void setNB2IDYun(String id) {
        editor.putString("flammableGasDeviceIdNB",id);
        editor.commit();
    }
    public String getNB2IDYun() {
        return sp.getString("flammableGasDeviceIdNB","39520");
    }

```

```
    //云平台NB 一氧化碳
    public void setNB1IDYun(String id) {
        editor.putString("carbonMonoxideDeviceIdNB",id);
        editor.commit();
    }
    public String getNB1IDYun() {
        return sp.getString("carbonMonoxideDeviceIdNB","39519");
    }

    //云平台LoRaid
    public void setLoRaIDYun(String id) {
        editor.putString("gatewayDeviceIdLoRa",id);
        editor.commit();
    }
    public String getLoRaIDYun() {
        return sp.getString("gatewayDeviceIdLoRa","39518");
    }

    //云平台端口
    public void setPortYun(String port) {
        editor.putString("port",port);
        editor.commit();
    }
    public String getPortYun() {
        return sp.getString("port","80");
    }

    //云平台域名
    public void setAddressYun(String address) {
        editor.putString("address",address);
        editor.commit();
    }
    public String getAddressYun() {
        return sp.getString("address","api.nlecloud.com");
    }

    //云平台密码
    public void setPasswordYun(String pass) {
        editor.putString("password", pass);
        editor.commit();
    }
    public String getPasswordYun() {
        return sp.getString("password","");
    }

    //云平台账号
    public void setUsernameYun(String name) {
        editor.putString("loginname", name);
        editor.commit();
    }
    public String getUsernameYun() {
        return sp.getString("loginname","");
    }
......
}
```

第二步，编写配置参数的缓存类。

新建包com.newland.smartpark.ecloud，在包下新建DataCache.java类实现配置参数的缓存，并添加以下方法：

```
package com.newland.smartpark.ecloud;
import android.content.Context;
import com.newland.smartpark.utils.SharePreferenceUtil;
import java.util.Map;
public class DataCache {
    private Context context;
    public Map<String, String> map;
    public SharePreferenceUtil sputil;
    public DataCache(Context context) {
        this.context = context;
        sputil = SharePreferenceUtil.getInstant(context);
        readECloudConnParam();
    }

    //项目标题
    public final String TITLE = "智慧园区";

    public static String address;//="api.nlecloud.com"物联网云服务平台网址
    public static String port;//="80"物联网云服务平台端口号

    public static String gatewayDeviceIdLoRa;//="39518"LoRa 网关 ID
    public static String carbonMonoxideDeviceIdNB;//="39519"一氧化碳设备 ID(NB-Iot 设备)
    public static String flammableGasDeviceIdNB;//="39520"可燃气设备 ID(NB-Iot 设备)

    public static String loginname;//="此处换成自己的账号" 云平台账号
    public static String password;//="此处换成自己的密码"云平台密码

    public Map<String, String> getMap() {
        return map;
    }

    /**
     * 保存物联网云服务平台地址和端口号
     *
     * @param address 物联网云服务平台地址
     * @param port    物联网云服务平台端口号
     * @return
     */
    public boolean saveECloudConnParam(String address, String port,
                                       String loginname, String password,
                                       String gatewayDeviceIdLoRa,
                                       String carbonMonoxideDeviceIdNB,
                                       String flammableGasDeviceIdNB) {
        this.address = address;
        this.port = port;
        this.loginname = loginname;
        this.password = password;
        this.gatewayDeviceIdLoRa = gatewayDeviceIdLoRa;
        this.carbonMonoxideDeviceIdNB = carbonMonoxideDeviceIdNB;
        this.flammableGasDeviceIdNB = flammableGasDeviceIdNB;
        sputil.setAddressYun(address);
        sputil.setPortYun(port);
        sputil.setUsernameYun(loginname);
        sputil.setPasswordYun(password);
        sputil.setLoRaIDYun(gatewayDeviceIdLoRa);
        sputil.setNB1IDYun(carbonMonoxideDeviceIdNB);
        sputil.setNB2IDYun(flammableGasDeviceIdNB);
        return true;
    }

```

```
61.     /**
62.      * 读取配置文件参数，并初始化类变量(缓存变量)
63.      *
64.      * @return 传感器标识名
65.      */
66.
67.     public void readECloudConnParam() {
68.         address = sputil.getAddressYun();
69.         port = sputil.getPortYun();
70.         loginname = sputil.getUsernameYun();
71.         password = sputil.getPasswordYun();
72.         gatewayDeviceIdLoRa = sputil.getLoRaIDYun();
73.         carbonMonoxideDeviceIdNB = sputil.getNB1IDYun();
74.         flammableGasDeviceIdNB = sputil.getNB2IDYun();
75.     }
76.
77.     //为其他类变量添加get方法
78.     public String getTitle() {
79.         return TITLE;
80.     }
81.
82.     public static String getAddress() {
83.         return address;
84.     }
85.
86.     public static String getPort() {
87.         return port;
88.     }
89.
90.     public static String getGatewayDeviceIdLoRa() {
91.         return gatewayDeviceIdLoRa;
92.     }
93.
94.     public static String getCarbonMonoxideDeviceIdNB() {
95.         return carbonMonoxideDeviceIdNB;
96.     }
97.
98.     public static String getFlammableGasDeviceIdNB() {
99.         return flammableGasDeviceIdNB;
100.    }
101.
102.    public static String getLoginname() {
103.        return loginname;
104.    }
105.
106.    public static String getPassword() {
107.        return password;
108.    }
109. }
```

上述代码中第 9～13 行在构造方法中初始化 SharePreferenceUtil 对象，并进行数据的读取，如果之前保存过就读取出来设置到控件上，如果没有保存过就读到默认值。saveECloudConnParam()和 readECloudConnParam()方法分别用于保存和读取相关参数。

5) 实现云平台参数的保存和读取

在包 com.newland.smartpark.view 中新建 SettingNLECloudDialog.java 类，加载的布局是 layout/dialog_setting_nlecloud.xml，用于设置和保存云平台的域名、端口号、登录账号和密码、要访问的设备 ID 参数。在构造方法中调用缓存类 DataCache 的方法进行数据读取并显示到控件上，在对话框的“确定”按钮被单击时调用缓存类 DataCache 的方法进行保存。代码如下：

```
public class SettingNLECloudDialog extends Dialog {
    //云平台参数设置界面的各项控件
    private EditText loginnameYun, passwordYun, addressYun, portYun;
    private EditText gatewayDeviceIdLoRa, carbonMonoxideDeviceIdNB, flammableGasDeviceIdNB;
    private Button btn_yun_cancel;//取消按钮
    private Button btn_yun_confirm;//保存按钮
    private Context context;
    //用于记录各项参数的输入值
    private  String loginname;
    private  String password;
    private String address;
    private String port;
    private String loraid;
    private String nbid1;
    private String nbid2;
    private  DataCache dataCache;//缓存类对象

    public SettingNLECloudDialog(@NonNull Context context) {
        super(context, R.style.Dialog);
        //关联布局文件
        this.setContentView(R.layout.dialog_setting_nlecloud);
        this.context = context;
        dataCache = new DataCache(context);

        initView();  //初始化组件
        initData();  //初始化数据
        addListener();  //添加事件侦听
    }

    //初始化组件
    private void initView() {
        loginnameYun = findViewById(R.id.loginnameYun);
        passwordYun = findViewById(R.id.passwordYun);
        addressYun = findViewById(R.id.addressYun);
        portYun = findViewById(R.id.portYun);
        gatewayDeviceIdLoRa = findViewById(R.id.gatewayDeviceIdLoRa);
        carbonMonoxideDeviceIdNB = findViewById(R.id.carbonMonoxideDeviceIdNB);
        flammableGasDeviceIdNB = findViewById(R.id.flammableGasDeviceIdNB);
        btn_yun_confirm = findViewById(R.id.btn_yun_confirm);
        btn_yun_cancel = findViewById(R.id.btn_yun_cancel);
    }

    //初始化数据
    private void initData() {
        loginnameYun.setText(dataCache.getLoginname());
        passwordYun.setText(dataCache.getPassword());
        addressYun.setText(dataCache.getAddress());
        portYun.setText(dataCache.getPort());
        gatewayDeviceIdLoRa.setText(dataCache.getGatewayDeviceIdLoRa());
        carbonMonoxideDeviceIdNB.setText(dataCache.getCarbonMonoxideDeviceIdNB());
        flammableGasDeviceIdNB.setText(dataCache.getFlammableGasDeviceIdNB());
    }

    //判断各项输入是否为空
    private boolean isInputNull() {
        loginname = loginnameYun.getText().toString().trim();
```

```
        password = passwordYun.getText().toString().trim();
        address = addressYun.getText().toString().trim();
        port = portYun.getText().toString().trim();
        loraid = gatewayDeviceIdLoRa.getText().toString().trim();
        nbid1 = carbonMonoxideDeviceIdNB.getText().toString().trim();
        nbid2 = flammableGasDeviceIdNB.getText().toString().trim();

        if (loginname.isEmpty() || loginname.length() == 0) {
            Toast.makeText(context, "用户名不能为空",
                    Toast.LENGTH_SHORT).show();
            return false;
        }
        if (password.isEmpty() || password.length() == 0) {
            Toast.makeText(context, "密码不能为空",
                    Toast.LENGTH_SHORT).show();
            return false;
        }
        if (address.isEmpty() || address.length() == 0) {
            Toast.makeText(context, "云平台的域名不能为空",
                    Toast.LENGTH_SHORT).show();
            return false;
        }
        if (port.isEmpty() || port.length() == 0) {
            Toast.makeText(context, "云平台的端口不能为空",
                    Toast.LENGTH_SHORT).show();
            return false;
        }
        if (loraid.isEmpty() || loraid.length() == 0) {
            Toast.makeText(context, "云平台的端口不能为空",
                    Toast.LENGTH_SHORT).show();
            return false;
        }
        if (nbid1.isEmpty() || nbid1.length() == 0) {
            Toast.makeText(context, "一氧化碳(NB)ID不能为空",
                    Toast.LENGTH_SHORT).show();
            return false;
        }
        if (nbid2.isEmpty() || nbid2.length() == 0) {
            Toast.makeText(context, "甲烷(NB)ID不能为空",
                    Toast.LENGTH_SHORT).show();
            return false;
        }
        return true;
    }
      //添加事件侦听
    private void addListener() {
        //设置确定点击事件
        btn_yun_confirm.setOnClickListener(new View.OnClickListener() {
            @Override
            public void onClick(View v) {
                if (!isInputNull()) {//如果输入为空
                    return;
                }

                //保存云平台地址和端口号
                dataCache.saveECloudConnParam(address, port, loginname, password, loraid, nbid1, nbid2);
                Toast.makeText(context, "保存成功", Toast.LENGTH_LONG).show();

                //对话框消失
```

```
                SettingNLECloudDialog.this.dismiss();
            }
        });
        //设置取消点击事件
        btn_yun_cancel.setOnClickListener(new View.OnClickListener() {
            @Override
            public void onClick(View v) {
                //对话框消失
                SettingNLECloudDialog.this.dismiss();
            }
        });
    }
}
```

在 HomeActivity 的标题栏菜单点击事件中，添加保存云平台配置参数的对话框的处理代码，当单击时让对话框显示出来：

```
//标题栏菜单点击事件
@Override
public boolean onOptionsItemSelected(MenuItem item) {
    switch (item.getItemId()) {
        ......
        case R.id.action_set_cloud://云平台参数
            //实例化对话框
        SettingNLECloudDialog dialogYun = new SettingNLECloudDialog(this);
            //设置区域外不可取消
            dialogYun.setCanceledOnTouchOutside(false);
            dialogYun.show();
            break;
    }
    return true;
}
```

6) 编写显示园区实时环境数据的界面类

编辑 layout/fragment_cloud.xml，数据显示界面如图 11-29 所示。因为界面中用到了较多的控件，同时还用到了自定义的仪表盘控件，代码较多且仪表盘的算法过于复杂，这里不便展开，该控件源代码会在课程资源包中提供，读者可直接复制到工程中使用。

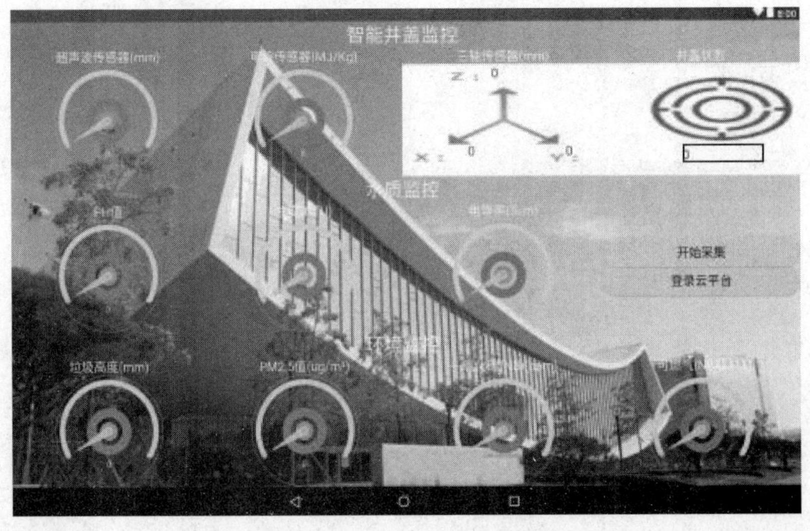

图 11-29 数据显示界面

仪表盘控件定义放在包 com.newland.view 中，类名为 SensorInstrumentView，在 layout/fragment_cloud.xml 中使用时可参照以下用法：

```
<com.newland.view.SensorInstrumentView
            android:layout_width="0dp"
            android:layout_weight="1"
            android:layout_height="match_parent"
            android:id="@+id/well_ultrasonicLab" />
```

在 CloudFragment.java 中设置最大值和实时更新值的用法如下：

```
 well_ultrasonicLab = rootView.findViewById(R.id.well_ultrasonicLab);
//设置井盖水位(超声波)最大值
well_ultrasonicLab.setMaxLange(最大值);
//更新井盖水位(超声波)
 well_ultrasonicLab.setProgress(值);
```

7) 封装 Http 请求工具类 OkHttpHelper

在包 com.newland.smarpark.ecloud 中新建 OkHttpHelper 类，用于封装 OkHttp 的相关操作，此处与项目 10 任务 3 相同，不再展开解释。代码如下：

```
package com.newland.smartpark.ecloud;
import java.io.IOException;
import okhttp3.MediaType;
import okhttp3.OkHttpClient;
import okhttp3.Request;
import okhttp3.RequestBody;
import okhttp3.Response;

public class OkHttpHelper {
    //声明媒体类型为 JSON
    private static final MediaType JSON = MediaType.get("application/json; charset=utf-8");
    private static final String TOKEN_KEY = "AccessToken";
    //使用 okhttp
    public static OkHttpClient client = new OkHttpClient();
    //保存 token 值(登录一次后，获取的 token 值可以重复多次使用)
    public static String accessToken;

    public static String get(String url) throws IOException {
        Request.Builder builder = new Request.Builder();
        if (accessToken != null)
            builder.addHeader(TOKEN_KEY, accessToken);

        Request request = builder.url(url).build();
        Response response = client.newCall(request).execute();
        return response.body().string();
    }

    public static String post(String url, String json) throws IOException {
        RequestBody requestBody = RequestBody.create(JSON, json);
        Request.Builder builder = new Request.Builder();
        if (accessToken != null)
            builder.addHeader(TOKEN_KEY, accessToken);
        Request request = builder.url(url).post(requestBody).build();
        Response response = client.newCall(request).execute();
        return response.body().string();
    }
}
```

8) 封装物联网云服务平台请求操作类

在包 com.newland.smarpark.ecloud 中新建 NewLandECloud.java 实现云平台的相关操作，所使用的 API 地址来源于物联网云平台 api.nlecloud.com，云平台中 API 的使用在项目 10 任务 3 中已做过介绍，此处不再展开说明。代码如下：

```
package com.newland.smartpark.ecloud;

import java.io.IOException;
import java.util.HashMap;
import java.util.Map;
import java.util.logging.Logger;
import com.alibaba.fastjson.JSONArray;
import com.alibaba.fastjson.JSONObject;

public class NewLandECloud {
    private static Logger logger = Logger.getLogger("NewLandECloud");

    private final String baseURL;
    private static final String PROTOCOL = "http://";
    private static final String LOGIN = "/Users/Login";
    private static final String DEVICES = "/Devices";
    private static final String DATAS = "/Datas";

    public NewLandECloud(){
        this(null, null);
    }
    public NewLandECloud(String ip, String port){
        if(ip != null && port != null) {
            baseURL = PROTOCOL + ip + ":" + port;
        }else {
            baseURL = "http://api.nlecloud.com";
        }
    }

    /**
     * 登录物联网云服务平台
     * @param loginname
     * @param password
     * @return
     * @throws RuntimeException
     * @throws IOException
     */
    public boolean login(String loginname, String password) throws RuntimeException, IOException {
        String url = baseURL + LOGIN;
        String json = "{'Account':'"+loginname+"','Password':'"+password+"','IsRememberMe':'true'}";

        String response = OkHttpHelper.post(url, json);
        logger.info(response);

        JSONObject jsonObj = JSONObject.parseObject(response);
        Integer status = jsonObj.getInteger("Status");
        if(status.equals(0)) {
            OkHttpHelper.accessToken = jsonObj.getJSONObject("ResultObj").getString("AccessToken");
            logger.info("accessToken = " + OkHttpHelper.accessToken);
            return true;
        }else {
            throw new RuntimeException("连接云平台失败！");
        }
    }
```

```java

    /**
     * 批量查询设备最新数据
     * @param deviceId 多个设备编号
     * @return 多个传感器值
     * @throws IOException
     */
    public Map<String, String> getDevicesData(String ... deviceId) throws IOException {
        if(deviceId == null || deviceId.length == 0) {
            return null;
        }

        StringBuilder url = new StringBuilder(baseURL).append(DEVICES).append(DATAS);
        url.append("?devIds=").append(deviceId[0]);
        for (int i = 1; i < deviceId.length; i++) {
            url.append(",").append(deviceId[i]);
        }

        String response = OkHttpHelper.get(url.toString());
        logger.info(response);
        System.out.println("传感器数据："+response);
        JSONObject jsonObj = JSONObject.parseObject(response);
        Integer status = jsonObj.getInteger("Status");
        if(!status.equals(0)) {
            throw new RuntimeException("您的请求操作失败，请检查您传入的参数");
        }
        //保存采集到的数据
        Map<String, String> data = new HashMap<>();
        //解析云平台返回的 JSON 数据
        JSONArray deviceAry = jsonObj.getJSONArray("ResultObj");
        int deviceSize = deviceAry.size();
        //遍历多个设备：LoRa 网关、NB-Iot(一氧化碳)、NB-Iot(可燃气)
        for (int i = 0; i < deviceSize; i++) {
            JSONObject deviceObj = deviceAry.getJSONObject(i);
            JSONArray sensorAry = deviceObj.getJSONArray("Datas");
            int sensorSize = sensorAry.size();
            //遍历获取每个设备中多个传感器的值，并保存到 Map 类型的对象中
            for (int j = 0; j < sensorSize; j++) {
                JSONObject sensorObj = sensorAry.getJSONObject(j);

                String apiTag = sensorObj.getString("ApiTag");
                String value = sensorObj.getString("Value");
                data.put(apiTag, value);
            }
        }

        return data;
    }
}
```

本项目只用到两个功能：一个是登录云平台，另一个是凭设备 ID 获取传感器数据。更多的 API 使用可查阅物联网云平台的相关解释说明。

9) 登录云平台

第一步，编辑类 CloudFragment，加载的布局文件为 fragment_cloud.xml，在类中添加布局文件 fragment_cloud.xml 中各控件对应的变量和缓存类的操作对象等，并声明一个消息对象，具体如下：

```java
public class CloudFragment extends BaseFragment {
    private View rootView;
    //井盖超声波
```

```
    private SensorInstrumentView well_ultrasonicLab;
    //井内甲烷
    private SensorInstrumentView methaneLab;
    //井盖三轴中 z 轴
    private TextView triaxialLab_z;
    //井盖三轴中 x 轴
    private TextView triaxialLab_x;
    //井盖三轴中 y 轴
    private TextView triaxialLab_y;
    //井盖状态
    private TextView cover_faultLab;
    //pH
    private SensorInstrumentView m_phLab;
    //浊度
    private SensorInstrumentView m_turbiLab;
    //电导率
    private SensorInstrumentView m_conductivityLab;
    //PM2.5
    private SensorInstrumentView m_pm2_5Lab;
    //垃圾桶超声波
    private SensorInstrumentView trashcan_ultrasonicLab;

    //一氧化碳(NB-Iot 设备)
    private SensorInstrumentView carbonMonoxideLab;
    //可燃气(NB-Iot 设备)
    private SensorInstrumentView flammableGasLab;

    // 开启自动采集云平台数据按钮
    private Button getDataFromYun;
    // 云平台连接成功与否
    private boolean isConnsuccess = false;
    //是否停止采集的标志
    private boolean isStop;
    //登录云平台按钮
    private Button loginYun;
    //数据缓存类对象
    private DataCache dataCache;

    //注意：设置各传感器的最大值，此处是模拟数据，读者应该按真实数据进行设置
    int well_ultrasonicLab_max = 100;
    int methaneLab_max = 100;
    int m_phLab_max = 14;
    int m_turbiLab_max = 10;
    int m_conductivityLab_max = 100;
    int m_pm2_5Lab_max = 250;
    int trashcan_ultrasonicLab_max = 100;
    int carbonMonoxideLab_max = 100;
    int flammableGasLab_max = 100;



    @Override
    public View onCreateView(LayoutInflater inflater, ViewGroup container, Bundle savedInstanceState) {
        //fragment 只加载一次
        if (null != rootView) {
            ViewGroup parent = (ViewGroup) rootView.getParent();
            if (null != parent) {
                parent.removeView(rootView);
            }
```

```
63.         } else {
64.             rootView = inflater.inflate(R.layout.fragment_cloud, null);
65.         }
66.         return rootView;
67.     }
68.
69.     @Override
70.     public void onActivityCreated(@Nullable Bundle savedInstanceState) {
71.         initView();//初始化控件
72.         dataCache = new DataCache(getContext());
73.         loginYunListener();//登录云平台的按钮事件
74.         getDataListener();//采集数据的按钮事件
75.     }
76.
77.  //消息对象
78.     Handler handler = new Handler() {
79.         @Override
80.         public void handleMessage(Message msg) {
81.             super.handleMessage(msg);
82.           }
83.     };
84.  }
```

第二步，上述代码中的第 71 行，initView()方法用于初始化控件。示例代码如下：

```
private void initView() {
    well_ultrasonicLab = rootView.findViewById(R.id.well_ultrasonicLab);
    well_ultrasonicLab.setMaxLange(well_ultrasonicLab_max);

    methaneLab = rootView.findViewById(R.id.methaneLab);
    methaneLab.setMaxLange(methaneLab_max);

    triaxialLab_z = rootView.findViewById(R.id.triaxialLab_z);
    triaxialLab_x = rootView.findViewById(R.id.triaxialLab_x);
    triaxialLab_y = rootView.findViewById(R.id.triaxialLab_y);
    cover_faultLab = rootView.findViewById(R.id.cover_faultLab);

    m_phLab = rootView.findViewById(R.id.m_phLab);
    m_phLab.setMaxLange(m_phLab_max);

    m_turbiLab = rootView.findViewById(R.id.m_turbiLab);
    m_turbiLab.setMaxLange(m_turbiLab_max);

    m_conductivityLab = rootView.findViewById(R.id.m_conductivityLab);
    m_conductivityLab.setMaxLange(m_conductivityLab_max);

    m_pm2_5Lab = rootView.findViewById(R.id.m_pm2_5Lab);
    m_pm2_5Lab.setMaxLange(m_pm2_5Lab_max);

    trashcan_ultrasonicLab = rootView.findViewById(R.id.trashcan_ultrasonicLab);
    trashcan_ultrasonicLab.setMaxLange(trashcan_ultrasonicLab_max);

    carbonMonoxideLab = rootView.findViewById(R.id.carbonMonoxideLab);
    carbonMonoxideLab.setMaxLange(carbonMonoxideLab_max);

    flammableGasLab = rootView.findViewById(R.id.flammableGasLab);
    flammableGasLab.setMaxLange(flammableGasLab_max);

    getDataFromYun = rootView.findViewById(R.id.getDataFromYun);
    loginYun = rootView.findViewById(R.id.loginYun);
}
```

第三步，登录云平台。

在登录按钮中启用一个线程，调用云服务平台请求操作类中的登录方法，传入从缓存类中获取到的用户名和密码进行登录，如果成功发送消息类型为 2 的消息，如果失败则发送消息类型为 3 的消息，同时设置登录标志的状态。其代码如下：

```
private void loginYunListener()
{
    //登录云平台
    loginYun.setOnClickListener(new View.OnClickListener() {
        @Override
        public void onClick(View v) {
            //以下完成云平台登录
            NewLandECloud nlecloud = new NewLandECloud(DataCache.getAddress(), DataCache.getPort());
            String tname = DataCache.getLoginname();
            String tpass = DataCache.getPassword();
            if (tname == null || tname.equals("0") || tpass == null || tpass.equals("0"))  //DataCache.getLoginname(), DataCache.getPassword()
            {
                Toast.makeText(getContext(), "请先设置相关参数", Toast.LENGTH_LONG).show();
                return;
            }
            //再一次单击时停止数据采集，显示登录云平台
            if(!(loginYun.getText().toString().trim().equals("登录云平台")))
            {
                loginYun.setText("登录云平台");
                isConnsuccess=false;
                isStop=true;
                getDataFromYun.setText("开始采集");
                return;
            }

            new Thread(new Runnable() {
                @Override
                public void run() {
                    boolean isOk = false;
                    try {
                        //登录云平台
                        isOk = nlecloud.login(DataCache.getLoginname(), DataCache.getPassword());
                        if (isOk) {
                            //发送登录成功的消息
                            Message msg = new Message();
                            msg.what = 2;
                            handler.sendMessage(msg);
                            //设置登录成功的标志为真
                            isConnsuccess = true;
                        } else {
                            //发送登录失败的消息
                            Message msg = new Message();
                            msg.what = 3;
                            handler.sendMessage(msg);
                            //设置登录成功的标志为假
                            isConnsuccess = false;
                        }
                    } catch (IOException e) {
                        e.printStackTrace();
                    }
```

```
                    }
                }).start();
            }
        });

}
```

10) 使用线程实时采集传感器数据并通过消息机制显示在界面上

第一步， 采集数据。

在采集数据按钮中启用一个线程，调用云服务平台请求操作类中采集数据的方法，传入从缓存类中获取到的所要采集的设备 id，采集到的数据通过发送消息类型为 1 的消息传递到主 UI，同时设置采集标志 isStop 的值。具体代码如下：

```
private void getDataListener()
{
    //采集数据
    getDataFromYun.setOnClickListener(new View.OnClickListener() {
        @Override
        public void onClick(View v) {
            if (!isConnsuccess) {
                Toast.makeText(getContext(), "请先登录云平台", Toast.LENGTH_LONG).show();
                return;
            }
            if (getDataFromYun.getText().toString().trim().equals("开始采集")) {
                isStop = false;
                new Thread(new Runnable() {
                    @Override
                    public void run() {
                        DataCache dataCache = new DataCache(getContext());
                        //获取设备id
                        NewLandECloud ecloud = new NewLandECloud(null, null);
                        String LoRaGateId = dataCache.getGatewayDeviceIdLoRa(); //LoRaId
                        String carbonMonoxideIdNB = dataCache.getCarbonMonoxideDeviceIdNB();  //一氧化碳(NB)Id
                        String flammableGasIdNB = dataCache.getFlammableGasDeviceIdNB();    //可燃气(NB)Id

                        while (!isStop) {

                            try {
                                //获取多个设备中多个传感器的数据值
                                Map<String, String> map  = ecloud.getDevicesData(LoRaGateId, carbonMonoxideIdNB, flammableGasIdNB);
                                Message msg = new Message();
                                msg.what = 1;
                                msg.obj = map;
                                //把采集到的数据通过消息机制发送出去
                                handler.sendMessage(msg);
                                Thread.sleep(1000);
                            } catch (Exception e) {
                                e.printStackTrace();
                            }
                        }
                    }
                }).start();

                getDataFromYun.setText("结束采集");
```

```
                } else {
                    //结束采集
                    isStop = true;
                    getDataFromYun.setText("开始采集");
                }
            }
        });
    }
```

第二步，显示数据。

在消息对象的 handleMessage()方法中判断，如果消息类型是 1，取出数据，更新界面的值。具体代码如下：

```
//消息对象
Handler handler = new Handler() {
    @Override
    public void handleMessage(Message msg) {
        super.handleMessage(msg);
        if (msg.what == 2) {
            loginYun.setText("云平台连接成功");

        } else if (msg.what == 3) {
            loginYun.setText("登录云平台");
        } else if (msg.what == 1) {
            Map<String, String> map = (Map) msg.obj;
            if (map != null) {

                //更新UI界面

                //更新井盖水位(超声波)
                double val1 = Double.valueOf(map.get("well_ultrasonic"));
                int tval1 = (int) (val1 / well_ultrasonicLab_max * 100);
                well_ultrasonicLab.setProgress(tval1);

                //更新甲烷值
                double val2 = Double.valueOf(map.get("methane"));
                int tval2 = (int) (val2 / methaneLab_max * 100);
                methaneLab.setProgress(tval2);

                //更新井盖三轴值
                String triaxial = (map.get("triaxial"));
                if (triaxial != null && !triaxial.isEmpty()) {
                    String[] triaxial_xyz = triaxial.split("\\|");
                    if (triaxial_xyz.length == 3) {
                        triaxialLab_x.setText(triaxial_xyz[0]);//+ "mm"
                        triaxialLab_y.setText(triaxial_xyz[1]);
                        triaxialLab_z.setText(triaxial_xyz[2]);
                    }
                }

                //更新井盖状态
                boolean coverIsOk = Boolean.parseBoolean(map.get("cover_fault"));
                if (coverIsOk) {
                    cover_faultLab.setText("井盖正常");
                } else {
                    cover_faultLab.setText("井盖故障");
                }

                //更新pH数据
                double val3 = Double.valueOf(map.get("PH"));
```

```
                int tval3 = (int) (val3 / m_phLab_max * 100);
                m_phLab.setProgress(tval3);

                //更新浊度的数据
                double val4 = Double.valueOf(map.get("Turbi"));
                int tval4 = (int) (val4 / m_turbiLab_max * 100);
                m_turbiLab.setProgress(tval4);

                // 更新电导率的数据
                double val5 = Double.valueOf(map.get("Conductivity"));
                int tval5 = (int) (val5 / m_conductivityLab_max * 100);
                m_conductivityLab.setProgress(tval5);

                //更新 PM2.5 的数据
                double val6 = Double.valueOf(map.get("m_pm2_5"));
                int tval6 = (int) (val6 / m_pm2_5Lab_max * 100);
                m_pm2_5Lab.setProgress(tval6);

                //更新垃圾桶超声波监测数据
                double val7 = Double.valueOf(map.get("trashcan"));
              int tval7 = (int) (val7 / trashcan_ultrasonicLab_max * 100);
                trashcan_ultrasonicLab.setProgress(tval7);

                //更新 NB-IoT 设备监测数据(一氧化碳)
               double val8 = Double.valueOf(map.get("CarbonMonoxideStr"));
                int tval8 = (int) (val8 / carbonMonoxideLab_max * 100);
                carbonMonoxideLab.setProgress(tval8);

                // 更新 NB-IoT 设备监测数据(可燃气)
                double val9 = Double.valueOf(map.get("FlammableGas"));
                int tval9 = (int) (val9 / flammableGasLab_max * 100);
                flammableGasLab.setProgress(tval9);
            }
        }
    }
};
```

5. 运行结果

检查设备连接无误后，发布程序到工控平板，先单击“智慧园区”，出现侧滑菜单，如图 11-30 所示。

图 11-30 “智慧园区”侧滑菜单

选中侧滑菜单中的“园区环境”，如图 11-31 所示。

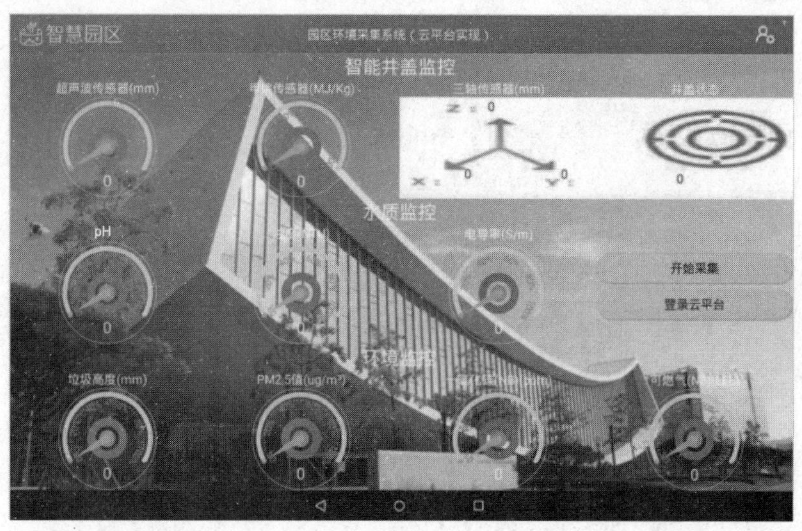

图 11-31　园区环境

单击右上角的图标，弹出对话框，设置云平台的参数，设置好后单击“保存”按钮，如图 11-32 所示。

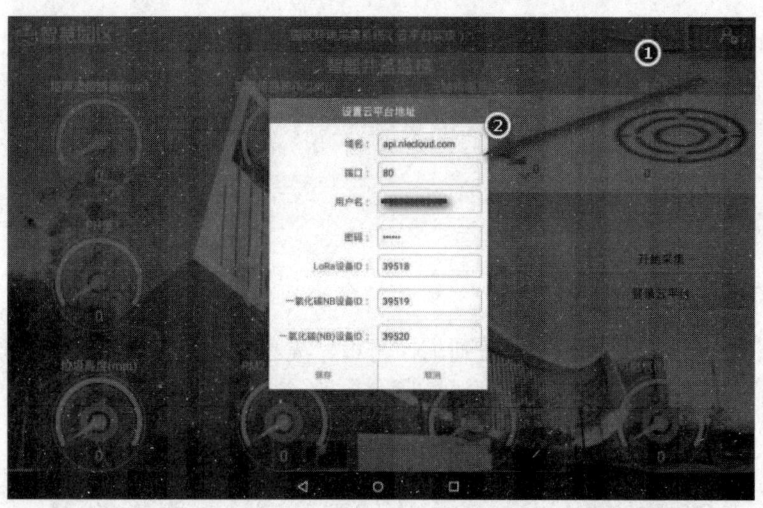

图 11-32　云平台参数设置

回到园区监控界面，单击“登录云平台”，登录成功后，可以看到 LogCat 的日志输出，已经返回了 AccessToken，如图 11-33 所示。

图 11-33　日志输出 AccessToken

单击“开始采集”，可以看到数据采集成功并显示在界面上，如图 11-34 所示。

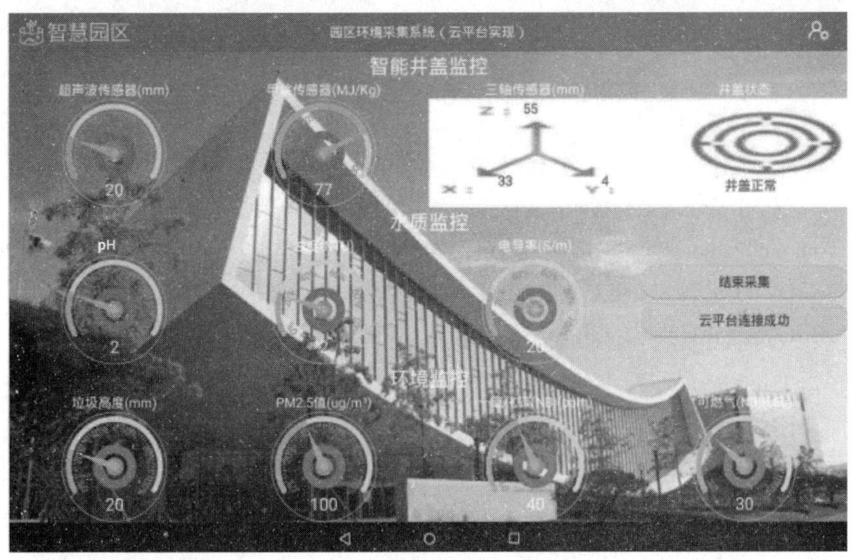

图 11-34　数据采集并显示

查看 LogCat 日志输出，也可以采集返回的 JSON 格式数据，如图 11-35 所示。

图 11-35　LotCat 日志输出 JSON 格式的传感器数据

至此，园区环境实时监测功能已实现，完整的项目源代码可参阅随书资料中的 SmartPark 工程源代码。

参 考 文 献

[1] 焦战. Android 物联网应用程序开发[M]. 北京：机械工业出版社，2016.
[2] 郭霖. 第一行代码-Android[M]. 2 版. 北京：人民邮电出版社，2016.
[3] 欧阳燊. Android Studio 开发实战从零基础到 App 上线[M]. 2 版. 北京：清华大学出版社，2018.
[4] 明日科技. Android 开发详解(全彩版)[M]. 长春：吉林大学出版社，2018.
[5] 廖建尚. 企业级物联网开发与应用[M]. 北京：电子工业出版社，2018.